The Weidenfeld and Nicolson Natural History
THE LIFE OF BIRDS
VOLUME II

The Weidenfeld and Nicolson Natural History

Editors:
Dr L. Harrison Matthews FRS
Professor J. Z. Young FRS

The Life of Birds

Volume II

JEAN DORST

*Professor at the National Museum
of Natural History, Paris*

Translated by
I. C. J. Galbraith

WEIDENFELD AND NICOLSON
LONDON

Weidenfeld and Nicolson
11 St. John's Hill London SW11

ISBN 0 297 17041 4

Printed by Cox & Wyman Ltd,
London, Fakenham and Reading

Contents

List of Plates

The author and publisher would like to thank the following for providing photographs for this volume: Jacana, Paris; pl. 18, Brun; pl. 19, Brosselin; pl. 20, Suinot; pl. 21, Prevost; pl. 23, 25, Dubois; pl. 24, Kalifa; pl. 27, Sundance; pl. 28, Visage; pl. 29, Milwaukee; pl. 31, Brosset. A. Fatras, Paris; pl. 22, L. Ricciarini, Milan; pl. 26, E. Robba; pl. 30, M. Pasotti; pl. 32, G. Tomsich.

Chapter 19

The Sea

Marine birds

The sea as a whole is an extremely complex environment with very high productivity at all trophic levels. Solar energy, fixed by microscopic algae and diatoms, is transferred to a great many populations made up of innumerable individual animals. The composition of these communities is ultimately determined by the physical conditions of the environment. These conditions are far from being uniformly fulfilled, which explains the considerable differences in the richness of marine animal communities. Certain zones of the sea constitute highly productive environments, of which birds take advantage through adaptations which have been easily acquired since their mode of locomotion was preadapted to this way of life.

Although vertebrates had a distant marine origin, birds have all evolved on dry land, but some of them very soon turned to exploiting the resources of the sea. The majority of the oldest-known birds are marine species – though this is partly explained by their greater chances of fossilization – and it is among marine birds that the most primitive types such as penguins, pelecaniforms and larids are found. We have already seen that from some points of view adaptation to the marine environment is less complete among birds than among mammals. All birds must lay their eggs on land, to which they are thus tied for at least a part of the year. As Murphy has said, a bird feeds where it wishes, but nests where it can. Furthermore – apart from penguins which are incapable of flight and which pass part of their annual and daily cycles in the water – even the most pelagic birds are primarily aerial animals, which spend the greater part of their time in flight although they can swim and dive.

Even discounting those groups (such as grebes, divers, some ducks and geese, and waders) which frequent the sea only occasionally or while wintering, no fewer than 267 species of birds are primarily marine (though this is no more than 3 per cent of all avian species, whereas the sea covers 70 per cent of the surface of the earth). The marine species belong to the

351

following systematic groups, those families marked with an asterisk being exclusively marine:

SPHENISCIFORMES	*Spheniscidae (penguins)
PROCELLARIIFORMES	*Diomedeidae (albatrosses)
	*Procellariidae (petrels, shearwaters)
	*Hydrobatidae (storm petrels)
	Pelecanoididae (diving petrels)
PELECANIFORMES	*Phaethontidae (tropic birds)
	Pelecanidae (pelicans)
	*Sulidae (gannets, boobies)
	Phalacrocoracidae (cormorants)
	*Fregatidae (frigate birds)
LARIFORMES	Stercorariidae (skuas)
	Laridae (gulls, terns)
	Rhynchopidae (skimmers)
	*Alcidae (auks, guillemots, puffins)

Even among those of these groups whose distribution includes fresh waters, a high proportion of the species is adapted to marine life. This is true of the Laridae among others. Apart from a few species such as the Black-headed Gull, the majority of gulls (*Larus*) are marine, and so are most terns, among which only a few tropical species and the marsh terns (*Chlidonias*) are birds of freshwater. Most cormorants frequent the sea-shore. On the other hand all the penguins and all the albatrosses and petrels, frigate birds, tropic birds, gannets and auks are strictly restricted to salt water. These birds form the essential part of marine bird communities, of which they are the most characteristic representatives.

To these entirely or mainly marine types must be added the isolated representatives of predominantly terrestrial or freshwater groups, which have secondarily become marine. Eiders and scoters among the Anatidae, and a few egrets among the Ardeidae, take their food from marine resources throughout the year, and many of them live at sea outside the breeding season. Grebes, divers, ducks, a few geese and phalaropes, keep to the sea only while wintering. While they do not play as important a part as the groups mentioned above, these birds undoubtedly belong to marine biocenoses for part of the year, temporarily exploiting them for food.

Distribution

Many seabirds seem at first sight to be evenly distributed across the seas, but in fact the densities of populations and the specific composition of seabird faunas vary enormously. The distinguishable life zones are defined by the physical properties of the surface waters – the temperature (not that of the air), and the content of dissolved mineral salts and gases. These easily measurable factors act upon the marine biocenoses as wholes, determining their productivity. The richness and diversity of the phytoplankton and zooplankton are directly dependent upon them, and as a result the populations of secondary consumers equally dependent. They are thus characteristic of the various marine zones, and can usefully serve as environmental indicators which reveal the quality of the water. Some antarctic petrels such as the Snow Petrel are linked to the coldest waters, whereas the tropic birds belong to waters which are warm, saline and clear. Where oceanographic conditions change, along boundaries which are as clearly defined as those on the continents, the avifauna also usually changes in composition and density.

The physical and biotic conditions of the oceans alter greatly in passing from one pole to the other. The antarctic waters – poor in most mineral salts, but rich in nitrogen compounds (0·5 parts per million of nitrogen as against 0·1 parts in tropical waters), and rich in dissolved gases also – are of very high productivity. In passing towards the equator the temperature and total salinity of the water increase. These seas are much less favourable to life, and their biological communities are impoverished. From the equator northwards, more and more favourable conditions are regained, and as a result the animal communities of the arctic are comparable in richness with those of the antarctic. Thus cold waters are distinctly more productive than warm ones and the biomass reaches much higher values than elsewhere. If one plots the quantity of organisms per unit volume of sea-water against latitude, one obtains a very characteristic curve, clearly showing the impoverishment of the equatorial and tropical zones. Birds, although at the ends of complex food-chains, faithfully reflect these variations in the total biomass of the water. The diversity of bird species also is much greater in the cold parts of the oceans than in the tropics. The South Pacific is occupied by 128 species (51 per cent of marine species), and the North Pacific by 107 (40 per cent), whereas the Indian Ocean has only seventy-three species (27 per cent) and the Mediterranean twenty-four (9 per cent). These differences are accompanied by a parallel variation in the density of the populations, which are very dense over the cold seas and much sparser over warm ones.

One can thus distinguish a series of concentric zones around the world, with boundaries which should parallel the equator, but are disturbed in some areas by currents which modify the oceanographic conditions. The cold currents are especially important, since they allow flourishing communities to become established at latitudes where the 'normal' conditions would be much less favourable. Similarly the meeting of several currents results in zones of high productivity. Oceanographic conditions are also modified locally by the upwelling of bottom waters. In stirring up, the waters carry to the surface the nutrients which result in high productivity. Birds are good indicators of these marine zones, and consequently the distribution of marine birds allows the oceans to be divided into a number of regions, according to the species which live there and their numerical densities. Nowhere among terrestrial environments is the nature of the habitat defined by such precise and measurable characters. Thus the ecology of seabirds can be studied more objectively than those of terrestrial birds.

AUSTRAL ZONE

The austral zone surrounds the antarctic continent and extends over the cold seas of the southern hemisphere. It is divided into two sub-zones by the *antarctic convergence*, a very distinct oceanographic boundary. The waters south of this line are cold (from $-1°$ to $+3.5°C$ in summer), and low in salinity owing to the melting of ice and considerable precipitation, which exceed the evaporation. However, though the total salinity of these waters is only of the order of 33 per ml, their content of nitrogen and phosphate is high, and they are almost saturated with oxygen. Naturally, the water temperature increases progressively on passing from the Antarctic northwards, but it rises abruptly at the line of the antarctic convergence, which marks the limit of icebergs though they often stop short of it. Beyond the convergence is the subantarctic zone, whose waters are warmer (from $+5.5°$ to 14.5 °C in summer), more saline (34 to 34.5 per ml) and still richer in phosphate and nitrogen. It is also characterized by violent and constant westerly winds (the 'roaring forties' of sailors) which drive the waters towards the east and north-east.

Although the antarctic avifauna is rather homogenous, there are marked differences between marine communities. Among those invertebrates which are of great importance as food for birds, the euphausids – the 'krill' of whalers – disappear north of the antarctic convergence and are replaced by other forms. The avifauna also changes across this line: although there are elements in common, the same general types of birds are mostly represented by different species and genera on either side. Thus to the south one meets the Emperor Penguin *Aptenodytes forsteri*, the Adelie Penguin *Pygoscelis adeliae*, the Light-mantled Sooty Albatross *Phoebetria palpebrata*, the Silver-grey Petrel *Fulmarus glacialoides*, and a series of petrels such as the Pintado Petrel *Daption capense*, the Snow Petrel *Pagodroma nivea* and the

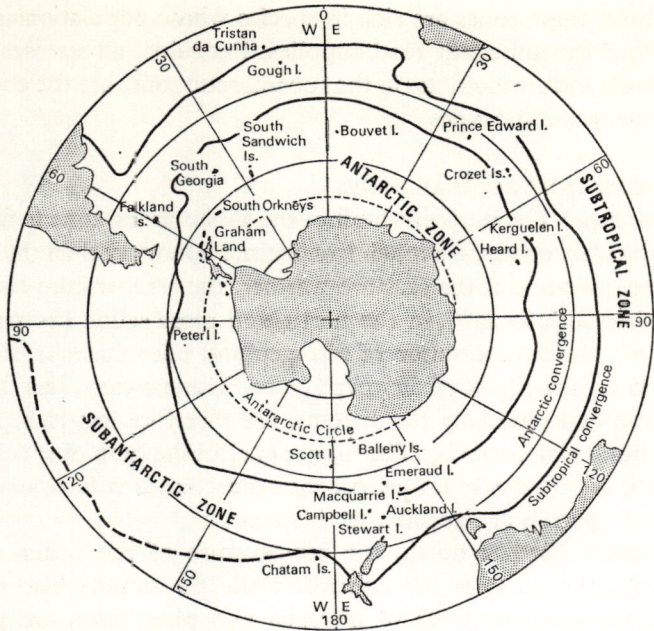

Figure 73. The southern zone. The two heavy lines encircling the antarctic continent mark the limits of the antarctic and subtropical convergences.

Dove Prion *Pachyptila desolata*. To the north, these are replaced by the King Penguin *Aptenodytes patagonica*, the Magellan Penguin *Spheniscus magellanicus*, the Gentoo Penguin *Pygoscelis papua*, the Rockhopper Penguin *Eudyptes magellanicus*, several albatrosses such as the Black-browed Albatross *Diomedea melanophrys*, the Royal Albatross *D. epomophora* and the Sooty Albatross *Phoebetria fusca*, petrels such as the Broad-billed Prion *Pachyptila vittata*, and various shearwaters such as the Greater, Sooty and Pink-footed Shearwaters *Puffinus gravis*, *P. griseus* and *P. creatopus*. Cormorants make their appearance in the subantarctic zone, whereas they are absent farther south. Thus the antarctic convergence forms a distinct biogeographic boundary, roughly following a parallel of latitude. South Georgia and Macquarie Island are inhabited by the same diving petrel *Pelecanoides georgicus* since, though at opposite sides of the pole, both are within the convergence. In contrast Aukland Island, not far from Macquarie Island, together with a series of islands leading from South Georgia to the Falkland Islands, all of which are north of the convergence, are populated by a related but distinct species, *P. urinatrix*. The same type of distribution is found in other groups.

355

As a whole, these zones are rich in species whose populations are large, as a result of the abundant food supplies. Penguins, albatrosses, petrels, diving petrels and cormorants (in the subantarctic zone) are the commonest and most characteristic types.

SUBTROPICAL AND TROPICAL ZONES

Subtropical zone: The northern limit of the subantarctic zone is formed by the *subtropical convergence*, a much less distinct boundary than the antarctic convergence. Beyond it the seas are warmer, the temperature rising from 15·5° in the south to 23°C in the north, and more saline (35 per mil on average), but the concentration of nitrogen and phosphates is diminished together with the content of dissolved gases. These waters, less favourable for plankton and especially for diatoms, are therefore less rich, and their communities less flourishing. The numbers of birds, both of species and of individuals, are markedly lower. All the antarctic birds have disappeared while certain petrels are peculiar to this zone.

Tropical zone: Farther north, the temperature of the water rises still further, reaching 29°C at the equator, while the salinity also increases. However, the concentrations of nitrogen and phosphates are very low. These seas are therefore poor in plankton, and can support only relatively simple and highly specialized biocenoses. Birds are rare, except in special circumstances, and mostly gather on the continental shelves, avoiding the open oceans. They form a series of species peculiar to the intertropical zone – several terns, especially noddies (*Anous*) and the White Tern (*Gygis*), frigate birds, tropic birds, boobies and several albatrosses confined to the warm Pacific seas.

TEMPERATE, BOREAL AND ARCTIC ZONES

Continuing towards the north, the temperature of the sea water drops again, while the salinity also decreases and the content of dissolved gases rises. A much greater standing crop biomass supports a richer avifauna, and the birds of the cold arctic seas are abundant and highly diversified. The most characteristic groups are gulls, skuas, certain terns such as the Arctic Tern *Sterna paradisea*, cormorants, and especially auks – puffins, razorbills, and guillemots – as well as several petrels and storm petrels.

The arctic avifauna shows scarcely any relationship to that of the antarctic, except that a few species have a bipolar distribution, on either side of the intertropical zone from which they are absent. This is true of the Northern and Southern Fulmars *Fulmarus glacialis* and *F. glacialoides*, and of the Northern Gannet *Sula bassana*, represented by the Cape Gannet and

Australian Gannet *S. capensis* and *S. serrator*. Otherwise arctic and antarctic birds are clearly different, belonging to systematic groups which are not closely related.

However, these two faunistic assemblages do show certain resemblances, due to convergence. The arctic gulls (although a few species of *Larus* do occupy the antarctic they are much less diversified there) are replaced by certain petrels – species which use their wings only for flying. Little Auks, puffins and other small auks are the equivalent of the antarctic diving petrels, the first group having evolved from gulls or their ancestors, and the second from petrels. They show considerable resemblances in way of life and in the shape of their wings, which serve both for flight and for propulsion under water. Large auks and guillemots have no parallels in the southern hemisphere, whose many species of penguin similarly lack equivalents in the northern seas – except for the Great Auk, whose wings functionally approached the flippers of penguins though they were of different form. Despite certain differences, the Lariformes and Procellariiformes have thus evolved convergently, although not all the evolutionary stages attained by one stock are represented in the other.

Certain birds of the arctic zone have a circumboreal distribution, because the same ecological conditions are found at the same latitudes all around the pole. This is true of the Arctic Tern *Sterna paradiseae*, which nests in America as well as in Eurasia. However, the ranges of most northern seabirds are markedly narrower than those in the Antarctic, in accordance with the distribution of land and sea in the northern hemisphere. In the Antarctic the circumpolar oceans are not interrupted by land, so that the birds are distributed in concentric zones. Some, such as albatrosses, even circumnavigate the globe during their annual migrations. In the Arctic on the other hand, the distribution of continental masses breaks up the oceans into separate basins, so that most circumboreal species have evolved local races – evidence of isolation due to geographical circumstances. For example the Common Kittiwake *Larus tridactylus* is represented in the Atlantic by the nominate race, and in the Pacific by the race *pollicaris*. Sometimes this isolation has led to the formation of distinct geographically representative species, such as the Common Gull *Larus canus* in the Old World and the Ringbilled Gull *L. delawarensis* in the New. In some cases whole faunas have been able to develop in one of the boreal oceans. Thus the Alcidae, which are relatively poorly developed in the Atlantic, are remarkably diversified in the North Pacific, attaining generic status (*Brachyrhamphus, Endomychura, Synthliboramphus, Ptychoramphus, Cyclorrhynchus, Aethia* and *Cerorhinca*) as a result of isolation in this basin. Fragmentation of the arctic seas has thus encouraged the development of local faunas whereas the uniformity and continuity of the antarctic oceans have produced marine faunas which are homogeneous, with distribution zones which are purely latitudinal.

ZONES TRAVERSED BY CURRENTS

Stretches of sea may thus be divided into a certain number of zones which are more or less concentric and parallel to the equator, like the climatic zones. However, this simple arrangement is considerably modified by marine currents and upwellings of deep waters. Cold currents are especially important, notably those which carry water from the Antarctic sometimes as far as the equator, resulting in real climatic and oceanographic reversals.

In particular the Pacific coasts of South America are bathed by the Humboldt Current, which flows from the Antarctic like a powerful river along the coasts of Chile and Peru. At the latitude of the frontier with Ecuador it flows obliquely westwards to the islands of the Galapagos, beyond which it disappears beneath warmer surface waters. The cold waters of this current are incredibly rich, and populated by marine animals at such a density that observers have spoken of 'thick plankton soup' and 'black clouds of marine organisms'. From these microscopic animals spring flourishing food chains, with very numerous birds representing some of the last links. As well as some widely distributed species, they include a high proportion of endemic birds, such as the Humboldt Penguin *Spheniscus humboldti*, the Guanay Cormorant *Phalacrocorax bougainvillei*, the Peruvian Booby *Sula variegata*, the Inca Tern *Larosterna inca*, several petrels, and the Peruvian Diving Petrel *Pelecanoides garnoti*. The Guanay Cormorant, the Peruvian Booby and the local race of the Brown Pelican (*Pelecanus occidentalis thagus*) are especially abundant. Nesting on desert islands in colonies numbering millions of individuals – the Chinchas Islands off Peru alone support four to five million cormorants – these three birds are the producers of guano, a fertilizer famous for its nitrogen content. The average annual production from Peru is 250,000 tons.

The creation by the Humboldt Current of an abnormal oceanographic situation at low latitudes explains certain paradoxical distributions, and the northwards extension of characteristically antarctic species. Apart from several petrels, the most striking example is that of the penguins, whose specialized representatives are established in Peru (the Humboldt Penguin *Spheniscus humboldti*) and even as far north as the Galapagos (the Galapagos Penguin *S. mendiculus*) – that is, as far as the equator and even a little beyond.

A comparable situation is found in any region of the world where currents and upwellings of cold water modify the simple model. Such results are produced by the Benguella Current along the Atlantic coasts of southern Africa, the cold current which bathes Morocco and Mauretania, and the cold upwellings off the Cape Verde Islands – a zone rich in living things and therefore the wintering territory for many pelagic birds. These upwellings of deep waters are often very narrowly localized, so that oceanographers know well-marked areas of cold water within warm tropical seas. Because of their productivity, clearly greater than that of the nearby zones, it is not surprising that seabirds concentrate there to fish, and establish their nesting colonies in the neighbourhood. The study of the local distribution of birds in Polynesia shows many correlations of this kind.

The relative ease with which the relations between physical conditions of the aquatic environment and the density of birds may be established, makes this ecological correlation remarkably obvious in the seas.

Adaptations of birds to the marine environment

In order to live in the marine environment, birds need to show a certain number of adaptations, morphological as well as ecological. The sea presents them with very special atmospheric conditions, with only one surface on which to alight and that disturbed by waves, with a lack of shelter, and with a highly specialized salt-rich diet.

We have seen how birds are adapted to swimming and diving, and it is enough to recall that all seabirds have feet which are webbed for aquatic locomotion. The wings of puffins and diving petrels serve for propulsion during dives, which prevents these birds from flying economically and easily. There is a clear conflict between the properties of an aquatic locomotor organ, and of a wing capable of sustaining the bird in the air. Many other birds use only their feet in swimming, keeping their wings folded against their bodies. We may also recall that the plumage of seabirds, like that of waterbirds generally, is particularly impervious to wetting.

Life above the wastes of ocean demands a particular form of wing, capable of sustaining the animal in flight for long periods with economy of effort. Many seabirds are excellent gliders, knowing how to use variations in wind velocity to travel with a minimum of effort. A series of adaptations to atmospheric conditions may thus be detected in the morphology of the wing and in its modes of use (See Chapter 2).

In addition various adaptations to the marine diet are found in the bill, the alimentary canal, and methods of grasping food. All truly marine birds are

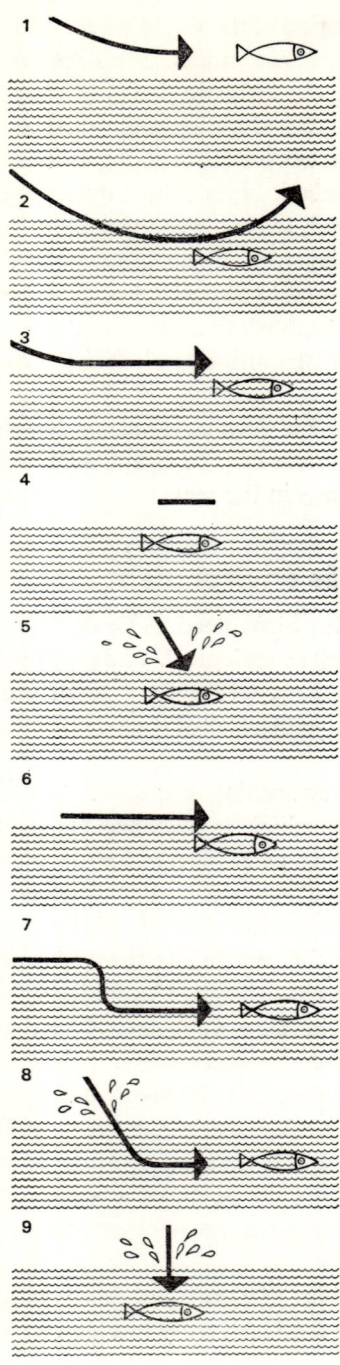

Figure 74. Some fishing methods used by seabirds. 1 & 2 diving in mid-flight. 3. glancing flight. 4. hovering. 5. diving vertically. 6. fishing on the surface. 7. swimming and then chasing underwater. 8. diving and chasing. 9. diving directly.

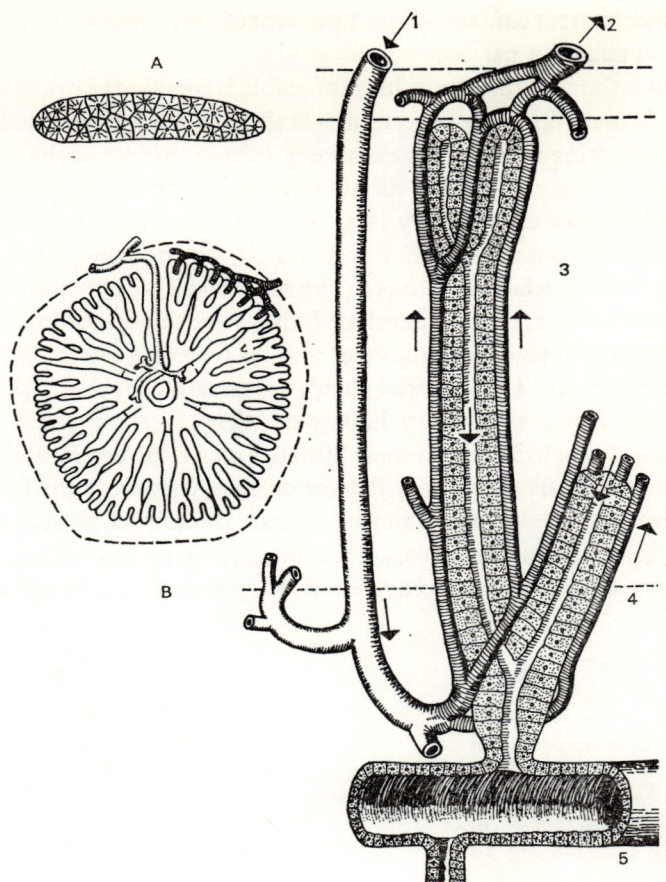

Figure 75. Microscopical structure of a gull's nasal gland. 1. arteriole. 2. veinule.
3. secretory tubule. 4. connective tissue. 5. main canal.   A. the whole gland.
B. section of a lobe.

carnivores. Seabirds have long intrigued biologists by their behaviour in
relation to salt water. Mammals, including man, cannot habitually drink
seawater, since the capacity of the kidneys for saline excretion is decidedly
limited. The intake of saline solution, at the concentration found in seawater, involves the loss from the body of an additional quantity of water,
resulting in dehydration of the tissues. Not only does seawater not relieve
thirst, it aggravates it. Seabirds however, like marine reptiles such as
turtles and iguanas, feed exclusively on plants or animals rich in salt, and can
drink nothing but salt water without suffering, so that they must be able to
balance their water losses, and eliminate the harmful excess of salts. Their

361

internal environment contains about 1 per cent of salt, whereas their food and drink contain about 3 per cent.

It was once thought that the kidneys of seabirds functioned in a specialized way. This is not true, since measurements show that a gull needs to secrete twice as much urine as it has drunk seawater, in order to excrete the contained salt. In this birds are inferior to certain mammals such as whales and desert rodents, which can excrete very large amounts of salt: some rodents can survive with only seawater to drink. Seabirds, on the other hand, secrete excess salt from specialized glands in the nasal region. Long known in seabirds, in which they are highly developed, they were thought to lubricate and protect the nasal fossae from seawater and its corrosive effects on delicate mucous membranes. These paired glands are situated at the anterior angles of the orbits, above which they hollow out deep furrows which in some species entirely encircle the cavities. Situated above the eye in most forms, and between eye and nostril in the Pelecaniformes, they are found in all birds which frequent the sea or sea-shore: waders, ducks and geese, penguins, gulls and other Laridae, pelicans, cormorants, gannets, frigate birds, petrels, shearwaters, albatrosses and some herons. In groups with only a few marine

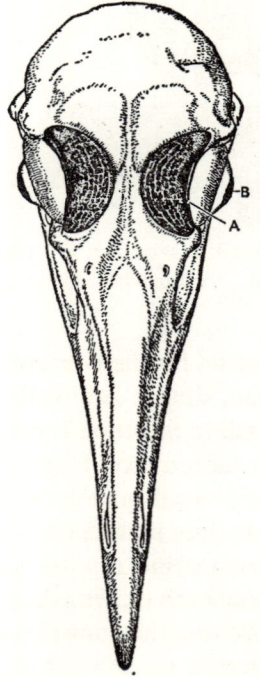

Figure 76. The structure of the nasal glands is almost the same in all seabirds. In gulls, the gland (A) is situated above the eye (B).

species, these are the only ones to show these structures – for example the scoters among ducks, and the exclusively marine species among the waders. The development of these glands is more or less proportional to the degree of adaptation of the birds in question to marine life.

Their development varies somewhat from species to species. They are more highly developed, and function more effectively, in the truly pelagic species than in littoral ones with a mixed diet such as the gulls. Their development depends quite closely on the salt content of their food and drinking water, since one can alter their size in experimental subjects by adjusting the amount of salt in the diet.

Each of these glands, which opens by two ducts into the nostril, shows the microscopic structure of a kidney. The blood flow to these glands in a gull is from 1.9 to 4.7 cm³/min/g, which is of the same order as that to a man's kidney (McFarland & Warner). Between the meshes of the capillaries are tubular glands closed at one end. Here the salt-charged blood is actively filtered by the cells which form the walls of the tubules. The central canals of the tubules open into collecting ducts, and these into the excretory duct. The secretions flow into the nasal fossae, from which they are expelled in the form of fine colourless transparent droplets which run out along the bill. In cormorants and gannets, whose nostrils are closed, the secretions flow through the interior nares into the mouth, from which they are expelled. Equivalent organs have been found in the marine iguanas of the Galapagos and in turtles, whose ducts lead not into the nasal fossae but into the internal angles of the eye.

The efficiency of the nasal glands has been shown experimentally. A gull forced to take 134cm³ of seawater – that is 10 per cent of its total weight, corresponding to about eight litres for a man – eliminated the equivalent amount of salt in less than three hours. During this time the nasal gland secreted only two-thirds of the volume simultaneously secreted by the kidneys, yet excreted 90 per cent of the salt. Thus they are capable of eliminating salt by concentrating it and expelling it in a minimal quantity of liquid – their secretion contains 5 per cent of salt compared to 3 per cent in seawater. Thus they function so that the body gains in water from drinking seawater, whereas the kidneys lose water in working to get rid of saline solution.

Thus these organs provide a kind of supplementary kidney, specialized for the excretion of salt. They are simpler than kidneys, which can secrete a greater variety of substances, depending on the physiological state of the body. In contrast the nasal glands can excrete nothing but chlorine, sodium (with traces of potassium) and water. In contrast to the kidneys the nasal glands work only intermittently, depending on the concentration of salt in the blood. The injection of saline solutions into the circulatory system stimulates them to excrete, no doubt by way of a centre in the central nervous system. It is quite clear that without this adaptation the marine environment would never have been conquered (Schmidt-Nielsen 1959, 1960).

Ecological divisions of seabirds

Birds which derive their diets from marine resources, although they may nest relatively far from the sea (up to 30km air-line distance for petrels which nest in the interior of certain mountainous islands), may be divided into three groups according to their feeding places and the ways in which they disperse.

The first group includes those species which never leave the intertidal zone (except during migration, when like terrestrial birds they may fly over the high seas). They seek their food on the beaches, sandbanks or rocks, in the zone uncovered by the tide or in very shallow water. The waders (Charadriiformes), most marine ducks, and some passerines and egrets (Ardeidae) do this.

The second group includes those birds which as a general rule frequent only the littoral waters. They do not transgress the limits of the continental shelves, even during their migrations which most often have the character of wide-scale dispersions. Among these are the Laridae – gulls, terns and skuas – the Alcidae – auks and their relatives – the penguins, and to some extent the gannets and frigate birds.

The third group comprises the pelagic birds, which are by far the best adapted to the seas. Although tied to the land for reproduction, these regularly frequent the high seas, not only during large-scale migrations but also on feeding flights during the breeding season. This is true of the Procellariiformes – albatrosses, shearwaters and petrels – as well as some frigate birds, gannets and auks.

This ecological division corresponds quite closely to the systematic one. The Procellariiformes are the most pelagic of all birds, being almost the only ones met with on the high seas. Their dispersions across the globe are enormous, although their nesting areas are mostly extremely local. Thus the Short-tailed Shearwater *Puffinus tenuirostris* frequents almost the whole Pacific, but nests only on a few islets off Tasmania. Similarly the Great Shearwater *Puffinus gravis* is distributed at different times of year across the whole Atlantic, but nests only on Tristan da Cunha. Nevertheless, a few exceptions to this agreement between systematic position and type of dispersion need to be noted. Thus among the Laridae, a distinctly littoral family, the Kittiwake *Larus tridactyla* is pelagic outside the breeding season. Among the terns, of predominantly littoral distribution, the Arctic Tern *Sterna paradisea* undertakes pelagic migrations which make it the most widely-travelled bird.

Of course, these categories are not rigidly defined. Intermediate states exist, of which the gannets provide a good example. Furthermore, the populations of many birds, which are tied to coastal districts while nesting, build up greatly in areas remote from the coasts outside the breeding season.

Adaptations of littoral birds

In general these birds do not pass beyond the limits of the continental shelves, their radius of action is limited to relatively short distances around the nest, and their feeding flights never involve the long voyages of pelagic species. They form well-balanced communities in which diversification of ecological niche is shown partly in diet and choice of fishing grounds, and partly in choice of nesting site (Fisher & Lockley 1954).

These species are biologically differentiated in both diet and choice of fishing grounds, owing to the ecological preferences of their prey. Thus in the Crimea four species of tern nest in mixed colonies: the Sandwich Tern *Sterna sandvicensis*, the Common Tern *S. hirundo*, the Little Tern *S. albifrons* and the Gull-billed Tern *Geochelidon nilotica*. Although they defend the nest sites jointly against predators, each species acts independently of the others. The Sandwich Tern fishes offshore, the Common Tern in littoral waters, and the Little Tern in much shallower waters quite close to the coast. As for the Gull-billed Tern, it hunts mainly on land, seeking insects and lizards on the plains, and thus does not actually belong to the marine trophic ecosystem.

Comparison between closely related species is equally interesting among European cormorants, the Shag *Phalacrocorax aristotelis* and Common Cormorant *P. carbo*. Although both species nest in cliffs overhanging the sea, and hunt fish below the surface of the water, they occupy distinct ecological niches. The Shag prefers cavities and cracks between boulders, whereas the Cormorant nests on fairly wide ledges and even in open spaces. The Shag fishes in the high seas and the Cormorant in shallower waters and estuaries, and as a result they feed on different species of fish, especially since the Shag takes mainly free-swimming fishes and the Cormorant bottom-living species. A similar ecological specialization prevents competition between sympatric species of guillemot (Lack 1945).

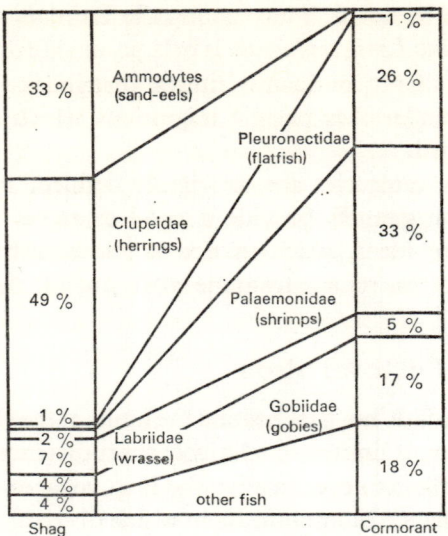

Figure 77. The diets of the Shag, *Phalacrocorax aristotelis*, and Cormorant, *P. carbo*. Though these two birds occupy the same biotopes during the breeding season, they concentrate on different types of fish and so do not compete.

Dietary competition between birds of the same community is often merely apparent. When food is superabundant because of the proliferation of a particular species of fish or crustacean, all the members of the community turn to this prey and thus seem to share the same diet, but in such a case there cannot be competition which occurs only during relative scarcity. Under conditions of competition each species shows clearly marked preferences for particular prey which it specially seeks, although it continues to feed on other prey at lower frequencies. Ecological specialization must be looked for in such preferences, rather than in the qualitative range of prey taken. Thus in Spitzbergen the Kittiwake, Fulmar, Ivory Gull, Iceland Gull, Eider, Brünnich's Guillemot, Black Guillemot and Puffin all feed on the crustacean *Thysanoessa inermis*, whose populations provide an enormous biomass. However, each seeks in addition a particular preferred prey, ranging from various crustacea to molluscs and fishes. Thus all true competition is avoided, while the whole available biomass is used. Similarly the fishing zones of closely related species which nest in the same area are often distinct. Thus in the Galapagos the Blue-footed Booby *Sula nebouxi* fishes in very shallow littoral waters, whereas the Blue-faced Booby *S. dactylatra* fishes farther offshore, and the Red-footed Booby *S. sula* decidedly on the high seas. Thus

366

their fishing zones, and as a result the prey on which they feed, are distinct (Nelson 1968).

Specialization by different species is also shown in nesting sites. To return to the example of the Galapagos boobies, the Blue-footed Booby nests predominantly on flat ground such as the bottoms of craters, the Blue-faced Booby especially on sloping ground, and the Red-footed Booby in bushes. In Europe banks and grassy ledges are occupied both by Puffins, which dig burrows under them, and by gannets which establish their nests open to the sky. Vertical cliffs interrupted by narrow ledges are the nesting sites of Kittiwakes and Guillemots, and Cormorants which seek sheltered niches. Cracks in the rocks hide the eggs of storm petrels. Fulmars occupy the edges of precipitous cliffs. Terns nest on flat ground and gulls especially on more or less accessible ground, showing the same flexibility in choice of nesting site as in their feeding habits. Each of these birds thus occupies a particular kind of site, thus reducing competition and allowing the best use of the breeding areas.

Littoral birds show a number of common tendencies in breeding biology, which distinguish them from pelagic species (Lack 1967). A very few have nidifugous young, like the eiders (*Somateria*), which nest on ground accessible to predators and for which an accelerated juvenile development is an indisputable advantage. More are semi-nudifugous, among these are the gulls and terns, which may also nest on ground accessible to predators, but are protected by their colonial breeding and collective defence. All the others, having nidicolous young nest only in places inaccessible to terrestrial predators – especially on narrow ledges little frequented even by avian predators, or in true burrows like puffins. They nest in colonies which are often large, though not so huge as those of pelagic birds. This dispersal of smaller populations, with the multiplication of 'bases' from which the birds go fishing, allows a better use of the resources of the sea and an economy in energy during the gathering of food – which, unlike pelagic species, littoral birds do not go very far to seek.

Littoral birds usually have a distinctly higher rate of reproduction than pelagic ones. The average number of eggs in a clutch is twice as high as in the latter, which lay single eggs. This agrees with their opportunity to rear more young, offered by their markedly nearer food-gathering sites – a correlation which is confirmed between littoral birds according to the distances of their fishing grounds. Cormorants, which fish in the immediate vicinity of their nests, lay three to four eggs, whereas boobies which fish farther off lay only one or two, spending more time in feeding flights. This greater accessibility of the fishing grounds is equally reflected in the frequency with which the young are fed. Whereas pelagic species come to the nest for this purpose only once a day, the young of littoral species are fed much more frequently – up to once an hour among gulls – while boobies are once more intermediate between the two categories. This more frequent feeding results in more rapid development and correspondingly shorter incubation times. The young Shag *Phalacrocorax aristotelis* leaves the nest around fifty-five days, while the young Manx Shearwater *Puffinus puffinus* does not do so until about seventy days, although by that time it is about four times as large. Littoral birds reach sexual maturity and their first breeding season sooner. Gulls, terns and cormorants nest from their third year, whereas petrels wait until their fifth or sixth years and albatrosses still longer.

Thus the population dynamics of the two groups are different. The population growth rate is high among littoral birds because of their larger clutches, which compensates for a higher mortality among the young as well as among the adults. The young are subject to considerable predation, especially from nest pilferers such as certain gulls. (As an unexpected result, the protection of seabirds in Europe has greatly benefited the Herring Gull whose populations have grown remarkably, while their severer predation on other birds has been especially serious for auks, some colonies of which have consequently been greatly reduced.) The adult mortality rate has been estimated at 15 per cent for the Shag, compared with only 5 per cent for the Manx Shearwater. Thus the growth rate is adjusted to the mortality rate, one compensating the other in relation to the nature of the habitat.

Comparison of various data on reproduction is especially interesting when littoral and pelagic species of the same family are compared. The Kittiwake and Sooty Tern *Sterna fuscata*, both pelagic though belonging to the predominantly littoral groups Larinae and Sterninae respectively, may be compared with related species, to yield data summarized in the following table (simplified from Lack):

	GULLS		TERNS	
	Black-headed Gull	Kittiwake	Common Tern	Sooty Tern
	L. ridibundus	*L. tridactyla*	*S. hirundo*	*S. fuscata*
Habitat	Littoral	Pelagic	Littoral	Pelagic
Usual clutch size	3	2	2(3)	1
Days to incubation	23	27	21	29.5
Days to fledging	30	43	30	60
Feeding frequency to older young	1 per hour	1 in 4 hours	3–4 per hour	1 per day
Years to first breeding	2	3–4	2–4	6
Annual mortality of adults	18%	12%	25%	16–20%

Thus the various reproductive characteristics are not inherent in given systematic groups, but may be interpreted as clearly ecological adaptations to particular habitats.

Pelagic birds

Pelagic birds, especially albatrosses and petrels, show a series of adaptations which are still more specialized than those of littoral birds. These are reflected in their external morphology, especially in the long narrow wings. In contrast to littoral birds, breeding and feeding sites are often very far apart, corresponding to the exploitation of much wider marine resources. Thanks to their locomotor apparatus, these birds have thus been able to penetrate into a world which would otherwise be closed to them, and to colonize minute islands lost in the wastes of ocean. This is an enormous advantage in protecting them from their natural enemies, since most of such oceanic islets are free from mammalian carnivores. The ease with which the Procellariiformes travel over the seas also results in the group being universally distributed around the world, though with a marked antarctic predominance. Most of the species have very wide distributions although their breeding areas are circumscribed often on minute islets. Nevertheless, the possession of long wings is a serious handicap on the ground, and especially during take-off. These birds are clumsy on land, where they move about with the more difficulty since their legs are often short. Albatrosses need real runways, and many petrels use boulders, small heights and sometimes the edges of cliffs, from which they launch themselves into space.

The ecological specialization of pelagic birds also involves their diet and methods of taking food. They all feed at sea – except the Giant Petrel which, as a facultative scavenger, feeds on detritus carried by the waves and on the eggs and young of other marine species which it kills before dismembering them. The food of all the other species is composed almost exclusively of macroplankton: cephalopods, fishes, pelagic crustacea and ctenophores, as well as detritus and corpses carried by the tides. The plankton is known to make daily vertical migrations ascending to the surface at twilight and during the night. This explains the circadian rhythms of the Procellariiformes, which are largely nocturnal and thus able to profit from the abundance of food available during the night. The Swallow-tailed Gull *Creagrus furcatus* of the Galapagos is similarly nocturnal, in accordance with its diet which is based on cephalopods, and is thus able to protect its nest by day from the attacks of frigate birds (Hailman).

Methods of taking food vary from group to group. The storm petrels

369

(Hydrobatidae) flutter over the water, seeming at times to walk on it as they make use of their webbed feet. Held by the wind, they pick up prey which they sight on the surface. In contrast, other Procellariiformes settle on the water to feed. Thus an albatross, after locating an area rich in prey by sight and perhaps by smell, settles and swims about, fishing by picking up prey in its hooked bill powered by a very mobile and greatly extensible neck, and not fearing to swallow cephalopods half a metre in length. Some Procellariiformes, especially the genus *Pterodroma*, have an intermediate method of catching food by seeking their prey in flight, settling for an instant to seize it and immediately taking off again. Prions of the genus *Pachyptila* have a very unusual fishing method, corresponding to the specialization of their bills, the bases of which are provided with systems of horny plates, rather like those of certain ducks. Prions swim in the position of a surf-rider on the surface, propelling themselves vigorously with their feet, submerging their heads and skimming the surface layer, their bills functioning as filters mainly taking pelagic crustaceans. Many petrels can dive to chase their prey. While shearwaters (*Puffinus*) readily plunge into the sea and 'fly' under the water, this is the habitual fishing method of the diving petrels (*Pelecanoides*). Notably different from the other Procellariiformes – in their short wings (which prevent them from gliding) and especially firm thoracic cages which effectively protect them against the repeated shocks of violent entry into the water – these birds dive from a great height and travel underwater by rapidly beating their wings. Uniquely they fish underwater, mainly on small fishes (anchovies and Atherinidae) and pelagic crustacea (notably *Munida*).

Various Procellariiformes often assemble to fish in mixed flocks. In the Antarctic fourteen species, nearly all of different genera, have been seen to fish in the same place, while in favourable areas their combined numbers are incredible. The existence of mixed flocks agrees with their dietary specialization, which is still poorly understood in detail. Albatrosses feed on large prey – especially cephalopods whose eye-lenses and horny beaks are found in their stomachs, and also fishes (some of which are up to 45cm long), crustacea, and refuse of all sorts. Surprisingly, they readily feed on siphonophores, whose powerful stings are well known. Shearwaters also very often feed on cephalopods, and some of them on fish – such as the Manx Shearwater which takes sardines, young herrings and others. Certain Procellariiformes are carrion feeders and gather eagerly round the sites where whale carcases are being dismembered, hungrily gobbling the fragments. This is true of the Giant Petrel, whose diet varies with latitude since it is a carrion feeder in the north of its range but clearly a predator in the south, attacking the eggs and young of penguins and petrels. Small petrels, especially the storm petrels, naturally take smaller prey, and small fishes, crustacea and their larvae (especially Euphausiidae) form the most obvious part of their diet.

When one or a group of prey species is abundant, Procellariiformes (like littoral birds) all make up a single community harvesting this common asset. However, when the available biomass is less abundant, the dietary specialization of each species prevents undue competition for food, and allows the rational exploitation of the whole available supply.

Adaptation of pelagic birds to their environment is shown equally in their modes of reproduction. Except for certain rare species, such as the sooty albatrosses *Phoebetria* which nest as solitary pairs, they are all gregarious and gather in larger or smaller colonies. Most show undeniable social behaviour, such as the collective displays of albatrosses and the community flights in which shearwaters and petrels gather. However, the colonies of others are not communities properly speaking, since there are no true social links between the individuals. They are mere assemblages of pairs, which the scarcity of nesting sites has forced to gather in a restricted area.

Albatrosses and fulmars nest in the open, the first seeking grassy areas on slopes from which their take-off is easier, and the second rocks and precipitous cliffs. Their strength and size protect their broods from potential natural enemies. All other species nest in cavities. Some, such as the Pintado Petrel and some storm petrels, nest in cracks in the rocks. Others use subterranean tunnels worn by erosion, such as Audubon's Shearwater *Puffinus lherminieri* which near Martinique occupies the interior of Hardy Islet, pierced like a gruyere cheese. Many others dig real burrows which open at the surface of the soil and penetrate to several metres in depth. The Snow Petrel *Pagodroma nivea* digs a chamber in hard snow. This subterranean nesting protects the birds from their enemies and also from cold and wind. Real underground towns are populated every breeding cycle, each pair finding their own former burrow which they merely have to repair.

Such specialization in modes of nesting avoids any troublesome competition between members of the same community, and allows different species to occupy a restricted area. Thus the Wandering Albatross establishes itself in grassy areas where it builds its nests on the surface, while the soil beneath it is tunnelled by prions *Pachyptila*. Thus the limited space is simultaneously occupied by a diurnal bird nesting in the open, and a nocturnal bird nesting underground. Thus too, each of the five species of the Procellariiformes which nest on Whero Island off New Zealand digs its nest in a different layer of the soil (Richdale).

All the Procellariiformes lay a single egg, which is always white without any cryptic coloration. They could not raise larger broods, because they have to seek food for their young so far away. The eggs are laid on the same dates with astonishing regularity from year to year, since pelagic birds lay at a fixed time. Thus Short-tailed Shearwaters *Puffinus tenuirostris*, although they make wide migrations across the Pacific, return every year at the end of September to their breeding grounds in southern Australia and Tasmania. After repairing their nests they go to sea again, presumably to mate. They return about November 20th, and all the eggs of a population numbering millions are laid within the following twelve days, with a maximal frequency between the twenty-fourth and twenty-sixth (Marshall & Serventy).

The incubation period of Procellariiformes is always long, especially in comparison with those of other birds of the same size. Thus it is thirty-five days in the Storm Petrel *Hydrobates pelagicus* (but only 20-22 days in the Little Tern *Sterna albifrons* of similar size), and reaches eighty days in the albatrosses. This long incubation corresponds with a general slowing of the development of the young, which continues after they hatch. Possibly this peculiarity gives the embryo greater resistance to cold: this is often intense despite the protection of the burrow, while the distances which the parent birds must cover in feeding flights are often such that the egg lies unincubated for a long time.

The young, which are always nidicolous, live alone in the nest for a very long time. Since the parents go to fish at great distances from the nest, it is important that the amount of food brought back from each trip should be large. Thus the Sooty Tern carries fish and cephalopods amounting to a fifth of its own weight. The length of the journeys presents the problem of digestion of the prey by the adult: since seabirds digest fish quickly, there is a danger of the load being digested too much, but it seems that mucus secreted by the digestive canal covers the prey and protects it from the action of the enzymes. In addition the proventriculus of some seabirds, especially petrels, secretes an abundant oil which the bird vomits when molested (Matthews). In albatrosses it has been shown that this oil is regurgitated by adults as a food for the young, and this is also true no doubt of petrels (Rice & Kenyon). Such a secretion allows great economy in weight, the food components of the prey being stored in the form of fats, and then secreted by special glands as oil with a nutritive value five to ten times as great, weight for weight, as that of the fish which served as the raw materials for its manufacture. This mechanism provides the young with a foodstuff whose energy value is very high because of its fat content. Proteins must be provided by the residue of partly digested food in the alimentary canal of the adult, but are nevertheless not abundant in the diet of the nestlings. This explains the slow growth of Procellariiformes, whose young are fed only at long intervals (Ashmole & Ashmole 1967).

Young storm petrels fly between their fifty-seventh and sixty-sixth days, petrels between the forty-fifth and 131st, and Wandering Albatrosses – the slowest-developing of all birds – not before the 220th day. Some tropical species, such as the Great-winged Petrel *Pterodroma macroptera* and Laysan Albatross *Diomedea immutabilis*, develop particularly slowly for their size, no doubt because of the relative poverty of warm waters prevents them from ensuring abundant food for their young. In contrast polar species have much shorter development times, such as the Fulmar in the Arctic (forty-nine days) and the Snow Petrel in the Antarctic (forty-five days). This may be regarded as an adaptation to polar and subpolar environments, where food supplies are richer. Since the favourable period is short at these high latitudes, only birds of rapid development can nest there.

All Procellariiformes accumulate large amounts of fats towards the end of their development, before they fly. One result of this is that many of these young birds are gathered by man for the sake of their fat ('mutton-birds'). Most of them finish distinctly heavier than the adults – by up to 50 per cent in many species, 70 per cent in the Pintado Petrel and Leach's Petrel, and even 80 per cent in the Short-tailed Shearwater. A similar increase in weight is shown also by tropic birds and some boobies, whose distinctly pelagic tendencies are well known. This deposition of fatty reserves may be considered as a response to the conditions of the marine environment. The risks of sea fishing, especially as a result of the weather, are such that the nestlings must be in a condition to resist prolonged periods of fasting. The presence of fatty reserves, which also allow better defence against cold, is an unquestionable advantage to them. This advantage no doubt continues when the time comes for them to fly. Many Procellariiformes abandon their young after rearing them, when their accumulated fats allow them to face the expenditure of energy involved in the growth of plumage and in their first flights. After taking wing the young are independent and are no longer fed by their parents, which have often already left on migration, and their reserves serve them in good stead until they begin to succeed in fishing.

The fecundity of pelagic birds as a whole is low. Procellariiformes lay only one egg, and do not produce replacement clutches. Further, while the small species have annual reproductive cycles, this is not true of the large ones. Among the albatrosses especially, the antarctic species nest only every two years, and some apparently at still longer intervals, so that their populations can increase only slowly. Curiously however, certain pelagic species have a reproductive cycle of less than one year – notably the Sooty Tern *Sterna fuscata* which regularly nests every nine months on Ascension Island, and the Madeiran Storm-Petrel which does so every six months in the Galapagos. Such cycles are possible only where climatic conditions remain constant throughout the year.

Corresponding to this low fecundity is a mortality rate distinctly lower than those of terrestrial or littoral birds. The Manx Shearwater has an annual mortality rate of 5 per cent, compared to 15 per cent for the Shag. Comparison between the pelagic Kittiwake (12 per cent) and other gulls such as the Black-headed Gull (18 per cent) is equally eloquent. After the juvenile mortality, which as always is high, the annual mortality rate of pelagic birds settles to a much lower level. Thus they renew their populations distinctly more slowly than littoral and terrestrial birds, especially since they are potentially very long-lived.

Population regulation among seabirds

Seabirds are often incredibly numerous. There are 2 to 2.5million individuals of the Great Shearwater *Puffinus gravis* in the breeding season on Tristan da Cunha, where the whole species concentrates in order to nest. Hundreds of millions of Short-tailed Shearwaters nest on several islands of the Bass Strait between Australia and Tasmania. The colonies of the birds which produce guano on the Peruvian coast, notably the Guanay Cormorant, number on the average sixteen million individuals. These high densities are all the more striking since these birds congregate to breed in well-defined colonies which are occupied year after year. Certain colonies of Fulmars in Baffin Land, Canada, contain more than a hundred thousand nests. Terns, auks, shearwaters, gannets and certain cormorants all form enormous colonies, for which penguins are equally famous.

This strong tendency to gregariousness, shown by 93 per cent of seabirds as against 13 per cent as a whole, has led to a series of strongly marked social behaviour patterns. The territory is considerably reduced in area, to a few square feet among those birds which defend only the nest itself, while territorial behaviour is more intense at this level. Independently of the territorial behaviour of each pair, the whole colony sometimes shows communal territorial behaviour, all the individuals defending its boundaries. This collective defence, especially marked among terns, is in contrast not shown by other equally gregarious birds, such as penguins. These differences no doubt result partly from the varying effects of predation and competition. Colonial birds have also developed a series of signals, which serve especially to prevent all unnecessary aggression. These partly auditory and partly visual signals are remarkably well developed among seabirds.

The advantages of gregariousness, to some seabirds at least, are indisputable: terns for example are thereby assured of better defence against their enemies. For others there is no disadvantage in nesting communally and conspicuously with no attempt at concealment, since their colonies are established on offshore islets or oceanic islands, which carnivorous mammals have not reached. Among other influences, it is the small area of such places which constrains the birds to congregate. Some species can modify their behaviour according to circumstances – such as the Eider *Somateria mollissima*, which nests in scattered pairs in places accessible to the Arctic Fox *Alopex lagopus*, and colonially on islands inaccessible to this predator. It is noteworthy that the coloration of seabirds, and of their eggs and young, is not cryptic, in contrast to shorebirds which nest where predation on the broods is heavy.

The concentration of seabirds in limited areas is in accordance with the richness of the sea in foodstuffs. Colonies are situated near the most productive areas, where oceanographic conditions result in abundance of foods. The standing crop biomass available within a restricted area makes it sufficient for great numbers of seabirds, despite the large quantities needed. In terrestrial communities biomass is rarely concentrated in a restricted area but is more uniformly distributed, so that terrestrial birds need to disperse much more widely.

It must be stressed that colonial seabirds exploit their food resources over considerable areas, although liable to concentrate at a given moment in a restricted area because of the abundance of prey, as for example over a shoal of fish. They are in general very good flyers, able to cover great distances. The fishing grounds of Guanay Cormorants may be several tens of kilometres from their nests, while the Procellariiformes go to fish at still greater distances. Thus Manx Shearwaters *Puffinus puffinus* nesting on Skokholm, off South Wales, regularly prey in the Bay of Biscay on shoals of sardines and other fish, which they bring back to their young. The distance of these fishing grounds reduces the competition between individuals, and the depletion of the stocks on which they prey.

It is difficult, lacking precise figures, to estimate the effect of seabirds on marine communities. Most of them are gross feeders, so that their takings must be considerable. Thus each Guanay Cormorant daily swallows up to 430g of fish, principally of the Anchovy *Engraulis ringens*. The total takings of the birds which produce guano are estimated as 2.5 million tons or more a year. Kittiwakes alone fish six million of the crustacean *Thysanoessa inermis*

a

375

daily at a single point on the coast of Spitzbergen, where many other birds join them in the hunt. It is obvious that such hauls can reduce stocks of the populations on which the birds feed, at least locally. This leads to the question of the limitation of seabirds populations and their regulation by natural factors. At first sight it seems that, as in terrestrial habitats, the birds must be limited by the amount of food available for themselves and their young. The action of this factor has been long suspected and is suggested by many examples two of which are classical: one negative and the other positive. The first of these again relates to the birds which produce guano on the Peruvian coast. It is known that the normal oceanographic conditions there are periodically upset when the cold waters of the Humboldt Current are masked over huge areas by warm waters coming from the north, carried by the Niño Current. This entirely alters the salinity and temperature, and by disorganizing the food chains results in a reduction in the biomass available to birds. The cormorants, boobies and pelicans which make up the enormous colonies then suffer severe losses, resulting in massive reduction of their populations, which are thus directly linked to their food supply. Their stocks only build up again when food is once more abundant after the restoration of normal oceanographic conditions.

The positive example is provided by the Fulmar, whose populations have considerably increased during recent decades while its breeding range has extended, notably along the coasts of the British Isles. This has been attributed to the results of intensive human exploitation of the sea. First whaling made available considerable quantities of offal which was spread about during flensing. When whaling ceased because of the reduction in whale populations, trawling replaced it as a source of waste produced during the preparation of the fish. This new food supply, accessible even when sea conditions are too bad for the Fulmars to fish for themselves, has allowed the populations to grow, by increasing breeding success and the survival rate of the still-inexperienced young. It is characteristically the young birds which colonize new areas (Fisher 1952). However, other authors believe that the first cause of the numerical and geographical expansion of the Fulmar is to be sought in the appearance of a new genotype, advantageous to the species and bringing about a new state of equilibrium with its environment (Salomonsen 1965). The increased food supply would not alone explain the increase of only the most southerly populations of Fulmars, since those of the Arctic have remained stationary despite the development of fisheries in the northern oceans. It is not impossible to imagine that the two factors have acted conjointly, the appearance of a new genetic combination favourable to the species allowing it to make use of a supplementary source of supply in both the original area and that which remained to be colonized. This example shows how the genetics of populations must be taken into consideration at the same time as their ecology and variations in their diet. A readjustment of the genotype, even if necessary, only becomes effective if it is favourable in the light of environmental conditions.

The part played by the food supply as a factor limiting seabird populations has been studied, notably by Lack, Wynne-Edwards and Ashmole (1963). After examining the arguments put forward by them all, it seems as though in this case the amount of food must on the whole be the most important factor, as is clear from observations on populations of various species – notably the Sooty Tern *Sterna fuscata* and Lesser Noddies *Anous tenuirostris* of Ascension Island (Ashmole). Food supply acts exclusively upon the annual increase in population, by influencing the success of broods and renewal of the reproducing individuals. During the breeding season seabirds, tied to their nesting site, can exploit only the very limited stock of prey in the area round the site which they can economically search. Although this area may be large, it is limited, since overlong feeding journeys would make the search too demanding of energy. Therefore, if the population of a colony increases, competition for food heightens to the point at which the adults become incapable of finding sufficient food for themselves and their young. Brooding success will decrease until it compensates for the mortality of adults and young, thus halting the increase in population and finally reversing it when food gathering becomes still more difficult. Density-dependent variation in brooding success may thus be a means of controlling the population, the limiting factor being the amount of food which the birds of a colony can economically take. If this is true populations must be not far short of the limiting capacity, and in fact considerable losses have been observed among the young of many such populations, caused by famine and its consequences.

The low fecundity of pelagic birds is explicable in terms of a pair's difficulty in collecting enough food for itself and even a single nestling. This also explains why they begin to breed at relatively advanced ages, since only very experienced adults would succeed in raising their young one. Among the Procellariiformes, shearwaters do not nest before five to six, and albatrosses not before eight to eleven years. If there were considerable, unexpected and accidental losses throughout the population, these unemployed individuals would probably begin to nest and thus take the place of older and more experienced birds. Thus populations will regenerate from a stock of 'reserve breeders'. Such individuals do occasionally attempt to nest under normal, conditions, in competition with experienced adults; but since they are pushed out by the latters' dominance to marginal nest sites their breeding success is usually much lower, and they scarcely affect the growth of populations.

Such competition for food acts only around nesting colonies, as a result of the increased needs of the adults – for displays, the physiological demands of reproduction, and feeding flights – and their young. No doubt the depletion of stocks of prey, within the radius of action of the birds around their nesting sites, explains their departure on migration. Except when breeding, seabirds can distribute themselves over much wider areas. Being no longer tied to these sites, they disperse on a greater or lesser scale depending on the availability of their prey. At such times food is probably not a limiting factor since the birds are widely dispersed and can move about according to its availability, though no presently available estimate allows this to be verified.

Apart from the amount of specific foods available at the nesting places, the following are also limiting factors though of unequal importance. Predation plays only a minor part in the limitation of pelagic birds, which are for most of the time remarkably well protected against mammalian predators. While the young of coastal species pay fairly heavy toll to avian predators – especially to gulls (such as the Herring Gull, Great Blackbacked Gull and their relatives) which plunder eggs and nests, and some raptors – the adults are practically immune from predators. Their lack of natural adaptations against predators explains the harm which can be done by carnivores, especially rats introduced by man to certain oceanic islands. These have taken heavy toll of petrels on sub-antarctic islands by plundering their nests. Pacific albatrosses have been seen to let themselves be eaten alive by rats while they were on their nests. The number of nesting sites may be a limiting factor in certain cases. Each species searches for well-defined conditions for the preparation of their nests, which the same pair often occupies year after year. A great increase in population sometimes takes place as soon as they have access to new territories. Thus in Peru the guano-producing birds, especially the Guanay Cormorant, have begun to nest on the mainland since man has artificially walled off certain rocky promontories, thus preventing the intrusion of terrestrial carnivores (foxes) which until then had prevented the establishment of such colonies.

However, expansion of the colonies and consequential growth of populations are only possible if there is an unused surplus of food. Many colonies do not occupy all the possible nesting sites. Sometimes a species could very well colonize new territories, such as a neighbouring island which remains uninhabited although it offers suitable biotopes. This seems to show that here, in the last resort, only the food supply within the area frequented during the breeding season regulates the populations of seabirds, so that their equilibrium must be achieved by a very precise mechanism.

Seabird migrations

Conditions at a particular point at sea are far from uniform throughout the year. The courses of currents and upwellings of deep waters vary with the seasons, altering oceanographic conditions and resulting in considerable fluctuations in the biocenoses and standing crop biomass. Seabirds are obliged to respond to the resulting environmental fluctuations by seasonal movements. Such exploitation of alternate habitats also avoids depleting the stocks of particular prey at a given place. Thus greater or less dispersal across the seas leads to better use of the resources of the whole marine environment, not restricted to areas around the sites to which the birds are tied during the breeding season.

Seabirds can be divided into two categories according to their modes of seasonal movements. Littoral birds disperse more or less widely as they extend over the high seas, but especially follow the boundaries of the littoral waters to which they are limited. In contrast pelagic birds indulge in much greater migrations, some of which are of oceanic scale, such as those of the Great Shearwater in the Atlantic and the Short-tailed Shearwater in the Pacific, while the large albatrosses circumnavigate the globe across the antarctic oceans.

The place of seabirds in marine biocenoses

In the absence of accurate quantitative information, it is difficult to estimate the feeding impact of birds on marine biocenoses. Around their colonies while they are breeding they deplete the stocks of those animals on which they feed, so that they do contribute to the local control of certain marine populations. However, outside these breeding areas and seasons, it is unlikely that birds have any real influence as predators. The productivity of the marine waters is such that their impact has only minor effect in the long term. The structure and dynamics of fish populations is such that renewal is rapid.

It is not surprising that birds are especially numerous and highly diversified as predators at their level in the food chains, the majority of seabirds being adapted to capturing animals. They often show food preferences, which allow species of the same community to avoid unduly fierce competition. However, some seabirds, notably certain microphagous petrels, are at a lower trophic level, living on plankton. Others are polyphagous, notably gulls and skuas which feed on rubbish and carcases of all kinds.

Since most are at the highest level of the food chains, seabirds have few natural enemies and predation mostly plays a negligible part in their mortality, especially of pelagic species. Gulls and some raptors feed freely on the eggs and young, and indeed on the adults, of coastal birds such as gannets and fulmars. Skuas specialize in plunder, and Giant Petrels are unrestrained carnivores in the Antarctic. Frigate birds and skuas live parasitically on gannets, terns and other species, forcing them to give up their prey. Sometimes seabirds fall victim to carnivorous fish and no doubt to Killer Whales, but the effect on seabird populations is presumably minor, except on penguins which are pursued by Killer Whales during their pelagic life. Thus seabirds play only a minor part in the transfer of energy through marine ecosystems.

Chapter 20

The Polar Environment

THE polar regions provide a particularly severe environment, which because of the rigour of its ecological conditions has been colonized by few species of animals. Temperatures are low, winds violent, and sunlight non-existent for part of the year, so that terrestrial habitats are poor, and often reduced to expanses of ice. The polar seas are much richer in contrast, and animals which find their food there can approach much closer to the poles than any terrestrial species. Birds have succeeded in establishing themselves in this environment. The Great Skua *Catharacta skua* freely crosses the Antarctic continent, and has been seen at the South Pole itself, as has Wilson's Petrel.

The profound differences between the geographical conditions of the arctic and antarctic regions are important to their bird faunas. The southern polar region is a true continent of high relief, surrounded by a complete girdle of oceans. This land is hidden by a thick ice-cap, containing more than 90 per cent of the world's ice. In complete contrast, the arctic polar region is an ocean surrounded by almost continuous continental masses, icebound and transformed into a vast flat extent. Only Greenland forms a terrestrial platform covered with a mantle of ice, comparable to the Antarctic. At least in winter the continental ice extends farther as ice-floes, which considerably increase the area of solid substrate. This contrast in geographical conditions has led to profound differences in the environments. The Antarctic consists almost entirely of ice almost devoid of animal life, in the midst of which emerge only a few rocky outcrops, so that the animal communities must be turned entirely towards the sea. In contrast the wide continental stretches of the Arctic extend to very high latitudes bearing a terrestrial environment of tundras with highly specialized conditions, such as is virtually absent from the Antarctic for lack of exposed soil. Thus there are two fundamentally different polar environments. One is marine, land or ice serving only as a substrate on which birds nest, and is mainly antarctic though found in the Arctic also. The other is terrestrial, the animals not relying on marine resources; and this environment is virtually absent from the Antarctic.

The limits of the polar regions are defined by climatic conditions. In the northern hemisphere the Arctic Circle pretty nearly delimits the northern polar regions. In the southern hemisphere the Antarctic Convergence forms the limits of the southern polar regions.

The Antarctic polar environment

Because of the special character of the Antarctic, the marine polar environment can best be studied there. At the same latitudes, temperatures are much lower (by an average of about 10°C) in the Antarctic than the Arctic. The 'pole of cold' is situated at Vostok at a latitude of 78°S, where extreme temperatures of − 88°C have been recorded, whereas coastal stations enjoy a less severe climate.

The most interesting habitat is provided by the 32,000km coastline of the Antarctic, beyond which the birds seldom stray since the sea provides their sole source of provisions. The huge ice-cap ends at the sea in vertical cliffs and the snouts of glaciers, which break off to form icebergs – sometimes gigantic – which are carried away by marine drift, or trapped by sea ice during the winter. On the continent, except in Graham Land, rocky outcrops occupy only very small areas. They rarely show as ridges, but often have relatively gentle slopes covered with pebbles.

Sea ice forms during the winter, and sometimes in summer at high latitudes, and reaches 1 to 1·5m in thickness. It forms enormous ice-fields, imprisoning the icebergs originating on the continent, and doubling the effective surface of the Antarctic continent during the winter. Even in the depths of winter, they enclose fissures and limited stretches of open water. During the summer the sea ice begins to melt, and a gigantic break-up follows during a general drift towards the west and sometimes the north. Between the icebergs and pieces of shattered ice-floe, the coastal waters are now open.

Thus the sea ice passes through an annual cycle. The relative lengths of these two phases vary between the different sectors of the antarctic. These facts are of prime importance in the biology of penguins, since the ice controls their access to the water in which they find their living.

The climate of this environment is extremely hard, characterized especially by very low temperatures. At the coast daily maxima sometimes exceed 3°C in summer, but despite this the temperatures are almost constantly below freezing – in Adelie Land the mean for the warmest month (December) varies between − 0·5°C and − 2·5°C – are in contrast to conditions during the true summer of the Arctic. In winter the thermometer frequently falls to − 35°C and − 40°C, and means for the coldest months are − 18° C or even lower. As a result, annual means are around − 10°C. The climate is still colder in the Ross and Weddell Seas, where the annual means are about −20°C to −25°C, and warmer in Graham Land.

Nevertheless, these coastal climates are generally much milder than those of the continental plateaux, avoided by most though not all birds. Naturally penguins, obliged to move on foot, scarcely leave the coasts, though exceptionally they have been reported tens of kilometres or even farther inland. In contrast petrels freely nest far from the coast, in accordance with a general tendency among birds of this group to nest in mountain ranges far from the sea. At various places in the Antarctic, notably at 72°S in Queen Maud's Land (Lovenskiold), Snow Petrels *Pagodroma nivea* have been reported nesting 300km from the coast at altitudes of between 1,500 and 2,000m, while Great Skuas *Catharacta skua* also nest in the neighbourhood.

The effect of low temperatures is considerably aggravated by the violent winds, which blow continuously in the antarctic region and are indisputably the dominant characteristic of its climate. Winds reaching 180 km/hr have been reported, while they often average between 60 and 80 km/hr. More than this, they carry huge loads of rounded ice-particles, and are known as *blizzards*. With a wind of 126 km/hr, 10 tons of ice an hour have been measured as carried across a metre length on the ground, and up to 2.6 tons an hour as passing through an area of one square metre, perpendicular to the wind. The cooling power of such an environment is enormous. While in temperate regions the temperature is the most important climatic factor, under polar conditions it is the wind. At a temperature of − 20°C and a windspeed of 130 km/hr, the blizzard doubles the cooling power of the wind, and this ice-laden wind is thermally equivalent to still air at − 180°C. Since at certain places on the Antarctic coast blizzards blow for 300 days a year, these polar environments present birds with the worst thermal conditions imaginable (Sapin-Jaloustre).

As a result of these circumstances, the terrestrial habitats of the Antarctic are exceedingly impoverished. The interior, forming the gigantic antarctic ice-cap, is a desert almost devoid of animals, and even the few rocky outcrops are near-deserts. Their flora is almost entirely limited to a few algae and mosses, and numerous lichens. Only two phanerogams, of which one is a grass, are found in the Antarctic, and these only in the Graham Land peninsula. Some lower insects, notably collembola, are known from certain places. Thus these habitats, like similar ones in the Arctic as in the interior of northern Greenland, could never support a community of terrestrial predators.

In contrast the polar seas are singularly rich. Antarctic waters have among the highest productivities in the world, because of their low temperatures and low total salinity, and develop complex food chains from their abundant plankton. The euphausid crustacea are especially important among the zooplankton, forming the keystone of the marine antarctic environment. The 'krill' (various species of *Euphausia*) forms the essential diet of polar whales and seals as well as some birds. Fish are abundant, with no fewer than sixty species adapted to cold waters. Thus the standing crop biomass is enormous, with great productivity limited not by the supply of nutrients but by light, which is insufficient for a large part of the year so that photosynthesis slows down. The same conditions hold in arctic waters bordering the ice at high northern latitudes. As a result the birds, like the mammals, which populate these environments are exclusively marine and occupy the ends of oceanic food chains, using the land and ice for nesting. Besides the specialization for marine life common to all seabirds, they have to be strictly adapted to the ambient climate of very high cooling power, and to an annual rhythm marked essentially by the extension and regression of the pack ice surrounding the Antarctic.

Antarctic polar avifaunas

Only about forty species nest south of the Antarctic Convergence, and many fewer on the continent itself. Only a few species such as the Fulmar have bipolar distributions, whereas most antarctic species are distributed all around the continent. This avifauna contains a very high proportion of endemic elements, as do most other antarctic faunas. This is in accordance with the special ecological conditions of the environment, and also with the long-standing isolation of this part of the world.

The most characteristic elements of the avifauna are the penguins. Among the seventeen species, all of southern distribution, only four can be considered as truly polar. Two even of these are not rigorously confined to the antarctic environment: the Chinstrap Penguin *Pygoscelis antarctica* (restricted mainly to the American sector, but now tending to become circumpolar) and the Gentoo Penguin *P. papua* actually extend their distributions across the Antarctic convergence, the last nesting especially on Crozet and Macquarie Islands. The Adelie Penguin *P. adeliae* is the most widely distributed and abundant all round the Antarctic. The Emperor Penguin *Aptenodytes forsteri* is the most polar of all and unrivalled as the best adapted of all birds to this environment. Its populations were formerly under-estimated, and there are at least 350,000 individuals according to recent censuses of the known colonies, a figure which will no doubt be increased still further when the whole Antarctic has been prospected. The Procellari-iformes are another dominant group in the Antarctic: sixteen species are known from the region, but only six of these nest on the continent. The most characteristic are the Giant Petrel *Macronectes giganteus*, the Snow Petrel *Pagodroma nivea*, the Antarctic Petrel *Thalassoica antarctica*, the Pintado Petrel *Daption capenses*, the Silver-grey Petrel *Fulmarus glacialoides* and Wilson's Petrel *Oceanites oceanicus*. All form more or less numerous colonies along the coasts, leaving extensive sectors empty. In addition other species, especially Sooty Albatrosses, regularly frequent the pack-ice. Of the Lari-formes, the Great Skua *Catharacta skua* is characteristic of the Antarctic, while wintering Arctic Terns *Sterna paradisea* are found along the edge of the sea ice. These truly polar species are for the most part replaced by others across the Antarctic convergence, as the Emperor Penguin *Aptenodytes forsteri* is replaced by the King Penguin *A. patagonica*.

All Antarctic birds are migratory. The alternation of a very unfavourable with a more clement season punctuates their annual cycles, which include large scale north–south displacements. Even the Emperor Penguin takes part in such movements, although its timing is reversed.

Defence against cold: the search for optimal environments

Antarctic birds, especially penguins, live where the air temperature is almost always below freezing and in nearby waters at 0° C or even less, so that their environments are always of very high cooling power. The most important factor which polar birds have to combat is cold. The methods employed are search for more favourable environments, reduced and well controlled heat-loss, and intense generation of heat demanding abundant food.

Local climates along the Antarctic coast are far from uniform, and a detailed study of the microclimates show that some are very different from the meteorologists' general climates. The heat-loss of a recumbent penguin, contained by the layer of air within 25cm of the ground, is less by a half than that which it suffers at the wind-speed recorded by normal meteorological techniques, while in a site sheltered by a wind-deflector it may not exceed a third or a quarter (Sapin-Jaloustre).

Thus birds can find microclimates which are much less unfavourable than meteorological readings would suggest – they only have to seek sites where the speed of the wind and hence its cooling power is reduced. This search for the most favourable conditions can be seen in the location of penguin colonies, which is largely in accordance with the physical conditions of the environment, and especially the microclimate. This behaviour is very clear in the Emperor Penguin, for which it seems to be an absolute necessity since this species breeds during the antarctic winter and must therefore avoid extreme conditions. Its colonies are established at sites where conditions are clearly favourable, sheltered as they are by glaciers or land relief. The Adelie Penguin is less committed to the search for the best-protected sites, since it nests during the summer when conditions are decidedly less rigorous. It establishes itself in situations well exposed to the wind, and thus avoids being covered with snow, and lies there sheltered by a lump of rock or a shallow hollow in the ground. It is known that certain antarctic petrels also choose the least exposed and best-sheltered sites for nesting. Their sub-terranean or semi-subterranean mode of nesting is especially favourable for thermoregulation, of the young as much as the adults.

The choice of sites for nests and colonies shows how antarctic birds take advantage of the least unfavourable ecological conditions, and no doubt to some extent explains their uneven dispersal and concentration at certain points on the continent. However, this is by no means the sole determining factor. These sites must be accessible, especially for penguins which travel only on foot, while Emperor Penguins – which nest when the sea ice blocks almost the entire coast – must also have open water not far away.

Defence against cold: thermoregulation

Despite the search for the most favourable surroundings, antarctic birds have to protect themselves from intense cold. Penguins especially fish submerged, and do not leave the water during the pelagic period of their annual cycle. Within the zones to which the truly antarctic penguins are confined, sea temperatures throughout the year are between $+1°C$ and $-2°C$. Cold water acts as a bottomless heat-sink: in slowly flowing water at $-1.9C$, heat is lost at the same rate as in a blizzard of 110 km/hr at $-20°C$. This cooling power is further increased by the fast swimming of the penguins themselves.

Thus polar birds find themselves under conditions which are markedly unfavourable to homoiotherms, since they have to reduce their heat losses so much, while penguins have to withstand long periods of fast during which they cannot be supplied with enough fuel. Since little chemical energy can be devoted to the generation of heat, physical control of heat loss is especially important. In fact penguins live wrapped in an insulating envelope which effectively reduces thermal interchange. At first sight, their thick layer of fat might be taken for thermal insulation, but in fact its protective role is negligible. The skin temperature is high, within 1 or 2°C of the central temperature, so that the body itself apparently has no cold 'shell' around the warm 'core'. In fact it is solely the plumage which, thanks to its quite remarkable insulating power, forms the penguin's shell. Its texture is very specialized, the feathers being massed against each other in a very dense layer, and evenly distributed over the body instead of being arranged in tracts separated by apteria as in other birds. Thus heat losses are reduced to a minimum, which can be made good by generation of heat with a limited expenditure of the chemical energy derived from food. When the bird is submerged, its body is in direct contact with the water only at the naked bill and feet, and at the flippers which are protected by a thin layer of feathers closely applied to the skin. Even more than those of other aquatic birds, the remarkably impervious feathers imprison a considerable layer of air. Thus

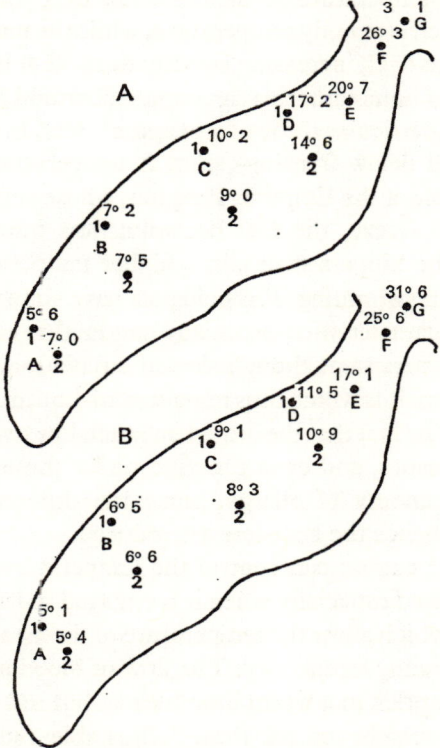

Figure 78. Diagrams showing differences in temperature (in °C) at various levels on the flippers of Adelie Penguins. A. from 17 birds within the colony. B. from 16 birds leaving the water.

387

thermal exchanges with the cold water must cross an air cushion, which considerably reduces heat losses by conduction, so that a penguin loses appreciable heat only through its naked parts and ill-protected flippers. These organs exposed to the cold must also reduce their heat loss, by working at low internal temperatures when necessary. Also the penguin must be able to lose the excess heat which is generated when it is active, or built up when environmental conditions are very favourable. Being permanently imprisoned in plumage whose insulating properties cannot be regulated, and able neither to sweat nor to pant, it can do this only through these exposed parts, which thus act as 'radiators' through which heat loss can be controlled. This regulation brings into play both physiological mechanisms and individual and collective behaviour.

In general, the limbs are maintained at distinctly lower temperatures than the body. In the Adelie Penguin, whose internal temperature is about $38.5°$C, the subcutaneous temperature of the toes is less by $5°$C, increasing steadily up to the tibia which is at body temperature, while the temperature of the tip of the flipper is 5 to $6°$C, increasing steadily to $26°$C at its base. This shows that some tissues of homoiotherms can remain alive and function at very low temperatures – in particular those of a penguin's feet, in contact with a very cold substrate well below freezing-point. A temperature of $0°$C has been measured in the feet of the Emperor Penguin, whose superficial layers were even beginning to freeze, the feet becoming less mobile but remaining perfectly alive. The flippers maintain a higher temperature, because they have to be active in swimming. Physiologists have shown that these tissues have specialized characteristics, especially conduction of nerve impulses at low temperatures, persistent though slowed circulation, and fatty reserves with lower melting points which thus remain semi-liquid at low temperatures (Irving & Krog). The fact that the limbs can attain low temperatures without damage to their tissues, and even function under these conditions, allows penguins great economy of energy, since the difference from ambient temperatures and hence the heat-loss are reduced.

Furthermore, a penguin can control the temperature of its extremities, increasing this at need especially when it is engaged in hard muscular work. The mechanisms which allow the temperature of these radiators to be raised or lowered are certainly circulatory. The flow of blood at the flipper varies considerably: a pinprick in a warm limb bleeds, but in a cold one does not. The mechanisms which control these 'adjustable radiators' are poorly known. As in the feet of some other aquatic birds, there are probably *rete mirabile*, or devices in which the veins and arteries run side by side, acting as heat-exchangers between blood flowing in opposite directions. It is also possible that there are artero-venous connections providing actual short-circuits. These arrangements must be under the control of a regulating nervous mechanism, allowing heat to be conserved or dissipated according to the bird's thermoregulatory requirements.

The organs which act as radiators are of reduced area in truly polar birds, in comparison with their close relatives living in less cold climates. Thus the bill of the Emperor Penguin is much shorter, and its flippers shorter and of smaller area, than those of the King Penguin which belongs to subantarctic zones where the climate is less severe. These differences conform to Allen's rule, of which they provide one of the few examples among birds, because of the specialized avian structure. Penguins also obey Bergman's rule (see Chapter 5): the Emperor Penguin is somewhat taller than the King Penguin and much fatter, its weight sometimes exceeding 40kg whereas its relative only reaches 20kg. This increased weight is certainly favourable to defence against the cold. Penguins also show different metabolic regimes corresponding to different conditions. The mean rectal temperature of the Emperor Penguin rises from 35° C to $36\cdot8^{\circ}$ C during pairing and incubation, and to more than 37° C at the end of the breeding cycle, and also varies according to the clustering of individuals. Thus when necessary this species can adopt an economical metabolic regime, extremely advantageous in the polar environment. Hypothermia, whether local or general, is abnormal in man and other animals which are not adapted to cold, and rapidly leads to irreversible damage. In the penguins on the other hand, it is a normal physiological state which can be indefinitely prolonged, and certainly provides another means of defence against the cold.

These physiological mechanisms are accompanied by behaviour patterns which reinforce their effect. During severe cold and when a penguin is inactive it seems to 'huddle', sitting on its feet which it withdraws into its ventral plumage, and plastering its flippers against its body. In contrast, in warm weather it widely exposes its feet and holds out its flippers in a crucified attitude. As well as these individual behaviour patterns, there is a collective defence against cold in accordance with the extreme gregariousness of these birds. When meteorological conditions are bad and especially during a blizzard, Emperor Penguins and young Adelie Penguins pack themselves so close together that they actually overlap, those on the outside turning their backs outwards, so that they form a compact mass known as a 'huddle'. The young gather in their own groups, or huddle together with the adults and thus gain the maximum benefit from this protection while their own thermoregulation is still imperfect. In this way the individual convection surfaces are replaced by a single collective surface, since a much larger mass than the individual bird is now involved, having a surface/volume ratio more favourable to reduced heat loss. Individuals massed in 'huddles' can thus live at a lower metabolic level, and economize their energy reserves. The following mean temperatures have been recorded for fasting male Emperor Penguins: in huddles 35·7°C; in groups 36·9°C; isolated 37·9°C. Thus the internal temperature can be varied by 2°C, being lowest in the members of a tortoise. Gregarious behaviour, by reducing individual heat losses, allows each bird to lower its energy metabolism, whereas isolated birds have to increase their heat production in order to maintain their temperature at a constant level and to defend themselves from the cold. The economy in energy resulting from this social thermoregulatory mechanism is great. The daily weight-loss of solitary birds is distinctly higher than that of gregarious ones (up to twice as great), and is lower the denser the groups in which the birds live. This highly developed gregarious behaviour is thus remarkably effective and may be considered as one of the birds' responses to the polar environment, partly explaining the survival of penguins in the Antarctic.

The annual cycle of the Adelie Penguin

The cycle of Antarctic birds is very closely linked to the annual climatic rhythm. The polar environment is marked by well-defined seasons, with large fluctuations in photoperiod. Although there is no true summer in the Antarctic like that of the Arctic tundras, the mean and extreme temperatures are milder than during the winter, so that the physical factors of the environment are less adverse for that part of the year, whereas they are almost unrelentingly severe during the other seasons. The cycle of the sea ice is even more important, since the extent of the ice can block or ease access to the marine food resources. The annual cycles of the birds must above all be closely adapted to that of the pack-ice, allowing them to feed under the best conditions when their needs are greatest. Each species of Antarctic bird has solved these problems in a different way. In particular, there is a fundamental difference between the Adelie and Emperor Penguins, which is related to their very different requirements.

The Adelie Penguin is a bird of lively and quarrelsome temperament and of moderate size, reaching a weight of from 4 to 6·5kg and a height of about 65cm. Its cycle may be clearly divided into six phases (Sapin-Jaloustre 1960), five of which concern breeding and take place on the coast between October and March, during the summer. The last phase is passed at sea during the winter, at the limit of floating ice. A breeding season on land with family life and territorial behaviour is contrasted with a period of sexual dormancy at sea without territorial behaviour, when the birds live alone or gregariously.

From the beginning of October, Adelie Penguins migrate inwards, from the northern limits of the ice towards the continent. Open water is sometimes found only at more than a thousand km from the coast, winds and blizzards are violent, and the temperature may fall to $-40°$ C. Nevertheless, the birds cover these distances on foot, unable to feed, and probably orient themselves astronomically by observing the sun. Exchanges between colonies are extremely rare, as much ringing has shown. The birds travel in groups of up to several hundred individuals, walking in indian file or tobogganing over the ice on their bellies. Having passed beyond the sea ice they reach the coast, and establish themselves on rocky surfaces which are free from snow during the summer. Since mates, which probably recognize one another by individual visual and acoustic characters, are remarkably faithful, the pairs re-form as in the previous year. They occupy the same sites, with a precision which is shown by many marking experiments, making use of an astonishing topographical memory. Thus the colony consists of a mosaic of territories, territorial behaviour being an essential element in the social life of this penguin. Each pair occupies a minute territory of 60-80cm diameter within the colony, defended with rare fury and playing an essential part in mutual display and copulation, which never take place outside this area. By ritualistic behaviour the pair build a nest of stones in the shape of a rough cup and mate from the time of their arrival, with a maximum frequency at the beginning of November. They do not feed, even if there is open water in the neighbour-hood. Laying begins towards 10 November and lasts for a fortnight, each female laying two eggs.

Meanwhile meteorological conditions have changed, temperatures becoming milder, winds less violent and daylight continuous. At the same time the sea ice has begun to melt, producing openings and channels of open water which announce the break-up. The females quit the colony after laying their second eggs, and go to feed after a month of fasting, leaving their clutches to the males which begin incubation. They return at the beginning of December to relieve the males which in their turn leave to feed, after two months of fasting which has resulted in weight losses of 40 per cent. Open water is now closer to the coast and the males are away for a shorter time, and the mates now brood alternately in a rapid rhythm.

When the chicks hatch after an incubation of thirty to thirty-seven days, it is full antarctic summer. Temperatures are relatively mild, winds less frequent, and the water open along almost the whole coast. Now the colony is the scene of intense activity. The adults go fishing to feed their chicks, and protect them by brooding so as to keep them warm and defended. The chicks develop, and towards their fifteenth day their thermal regulation is already efficient. From 15 January they gather in small groups, which join together to form crèches of ten, twenty and then thirty to fifty young. At the same time the adults abandon their territories and no longer show any defensive behaviour. The crèches appear to be under the care of a few adults, while the others come periodically to feed their young; but some at least of these supposed guardians seem to be merely unoccupied adults which happen to have established themselves near the crèche. Each parent comes to feed its own chick, which it recognizes by visible and audible peculiarities. From 20 January the moult begins, during which the down of the chicks is replaced by a plumage of true feathers, though this differs from that of the adults. Having completed their moult, the young leave the crèche and mix with the older birds.

Meanwhile, meteorological conditions have changed again, with the antarctic winter coming in between 15 February and the end of March. The immature birds, already seasoned by their first dives, disappear towards the north. The adults moult in their turn, gathering together on snowdrifts or rocks. For about a fortnight they are apathetic, and incapable of fishing on account of the state of their plumage, and this fast leads to a wasting which may amount to 40 per cent of their initial weight. When the moult is ended, they leave in their turn for the north.

From March to October Adelie Penguins winter at the edge of the sea ice, in conditions less severe than those on the ice-gripped continent. This partly liquid and partly solid environment allows them both to fish in open water and to rest on the floating ice, leaping out of the water onto the ice at one bound. No doubt some of the social bonds, which united them in the breeding colonies, are maintained in these flocks.

Thus the cycle of Adelie Penguins is closely linked to that of the seasons and of the pack-ice. They breed on land in a rocky environment during the summer, so as to take advantage of the open water near the coasts, and winter far from the continent near open water.

The annual cycle of the Emperor Penguin

The cycle of the Emperor Penguin, which frequents the same areas but not the same biotopes as the Adelie Penguin, is entirely different (Prévost 1961). This bird, the largest of all penguins, is unquestionably the best adapted to the hard conditions of the polar environment. It is for one thing the only one which need never leave the aquatic environment, since it most often nests on the sea ice and rarely ventures on to solid ground. Compared with that of the Adelie Penguin, its cycle is actually inverted. It remains at the coast for no less than ten months, from March to December. Its cycle may be divided into six periods, whose timing varies somewhat between colonies and from year to year in accordance with meteorological conditions, which markedly affect the freezing and breaking up of the sea ice.

On Adelie Land, the first of these penguins arrive during March, when the future site of the colony is covered with sea ice. They arrive at first as single birds, then in larger and larger groups until up to 500 individuals advance in procession, the influx spread over a whole month. At this stage the penguins are remarkably fat, some males having more than 10kg of fatty reserves. They now begin to search for mates, and pairing lasts until May. Pairing displays involve ritual behaviour patterns – such as bows, attitudes of the head and calls – which allow individual recognition. Mates remain together until laying, which begins during the first days of May and lasts until the beginning of June. The female gathers the single egg on to her feet and covers it with a kind of incubation pouch formed by a fold in the skin of her belly. No nest at all is built, since no suitable material is available. The pair now indulges in a display during which the female passes the egg to the male, who in turn receives it in his incubation pouch. The female at once leaves for the sea, so that no replacement can be laid should the egg be lost. At this stage the birds have fasted for forty-five to fifty days, and the females could not be expected to produce a second egg of such size (measuring 117-132×80-90mm, with an average weight of 447g and a maximum of 550g). Since their arrival at the colony they have lost 5 to 10kg, which is 17 to 30 per cent of their original weight. The males alone accomplish incubation of the eggs, which they carefully balance on their feet to keep them from the frozen ground, and cover with their ventral brooding folds. They can move about only with difficulty because of the burdens on their feet, but gather into 'huddles' as the weather deteriorates. The males take no further food for the whole duration of incubation – sixty-two to sixty-six days – meeting their water needs by eating snow. At this time the colony is peaceful and the birds very passive, their long fast committing them to a slow tempo. Though an individual cannot recognize its own egg, under normal circumstances eggs are not transferred.

The females return, about 20 June, sixty to seventy days after they left to feed in open water. They walk across the colony searching for their mates, which they probably recognize by ear since they produce a 'song' to which the males respond. Having found one another again, the pair indulges in mutual displays, during which the female takes back on to her own feet the

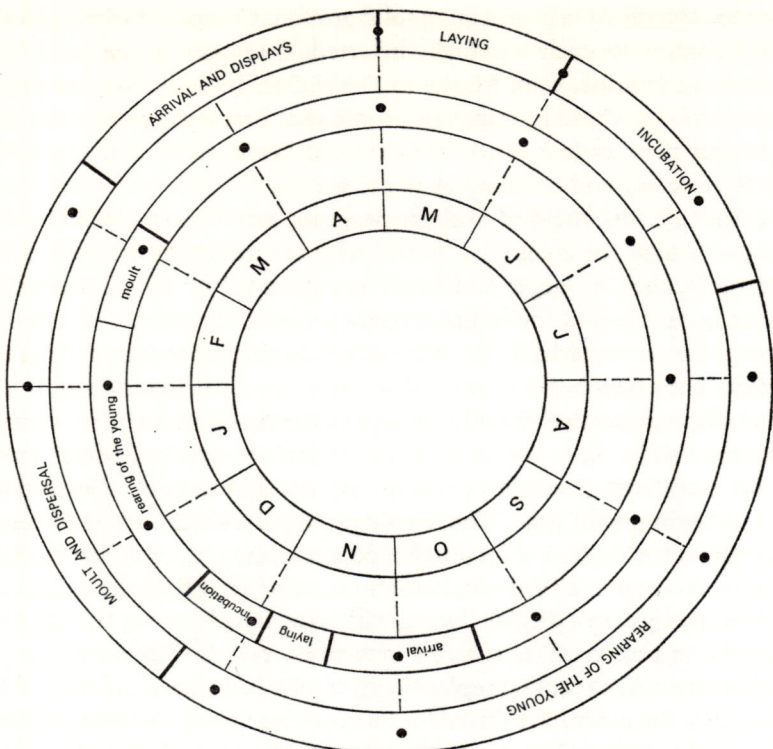

Figure 79. Annual cycles of the Emperor and Adelie Penguins (outer and inner labelled circles respectively; months shown on the innermost circle). Note how the phases of reproduction in these two species are desynchronized.

egg or hatched chick. The males are by now exhausted, by a fast which has lasted at least three months and often much more. They have lost 12 to 15kg, which may be up to 50 per cent of their original weight, and now leave for the sea in order to feed.

The chicks hatch at the beginning of July, and are reared in two very distinct phases. During the first phase the chick remains sheltered in the incubation pouch of its parent, especially the female. Very clumsy for the first few days, it gets steadily stronger, but remains in the pouch until towards its fortieth or fiftieth day. Finally it becomes too big to remain there, while the protection of the pouch has been made less and less necessary by its improving thermoregulation. On hatching it was defenceless against the cold, and only from the twentieth day did it begin to regulate its own temperature, and did so more and more easily as it grew older. Its speed of

394

development varies greatly with the quantity of food brought by the parents. The adult recognizes its chick, by voice as well as by sight, and attends only to its own young. Adults which are unoccupied, because they had not begun to breed or have lost their young, are desperate to rear the chicks of other birds, which they try to secure, and the young are often the unhappy victims of the resulting scuffles. The young are fed by regurgitation, but a broody male, despite its prolonged fast, can secrete from its oesophagus a substance exceptionally rich in proteins and fats. This secretion, rather like 'pigeon milk', provides a first food for the chick so that it can survive for several days, and even double its hatching weight, if the female is late in returning. Later it is fed with fish and crustacea, regurgitated by each parent as it returns from the sea.

The chicks begin to be independent during September, abruptly leaving their parents and going to live in crèches with others of their own age. Both parents now leave at the same time, in order to feed a young one whose demands have increased, and return to the colony only to feed the chick. They feed only their own, 'singing' in front of the crèche until it replies. Thus, though rearing in crèches appears to be communal, family life in fact persists. Rearing lasts for about five months, after which the chick begins to moult into immature plumage, and the juvenile down disappears in about another month.

Now the ice breaks up, and towards the middle of December the colony disperses. The adults moult, sometimes at the site of the colony and sometimes on the floating ice when it breaks up prematurely. Before moulting they feed heavily, so as to provide reserves for the thirty to forty days during which they will be immobilized and unable to feed, since they cannot dive while moulting. During the moult they therefore lose much weight. Once plumaged, the adults go to sea to feed and get fat again, but unlike Adelie Penguins they remain there very briefly, returning to the colony one or two months later. Thus the greater part of the Emperor Penguin's cycle is devoted to breeding and the rearing of its young.

The annual cycle of the Emperor Penguin is remarkably well adapted to the polar environment, and in particular to the rhythm of the sea ice on which all its reproductive phases take place. Its most notable feature is the reversal of the breeding cycle, laying and incubation taking place in winter during the full polar night – which necessitates hormonal control quite contrary to all that is known about the effects of photoperiod. This timing is necessary because of the slow development of the chick, which must finish its growth and become capable of independent life at sea by the time the sea ice breaks up. This means that the adults must have great resistance to cold, and be capable of protecting the egg and the chick from this adverse factor. They must also be able to withstand long periods of fasting, and to feed their chick in a special way while the ice extends for vast distances and obliges them to make long voyages for food. By the time the chick has passed its early stages of growth and needs more food, spring has already come and open water is nearer, so that feeding is easier.

The cycle of the Emperor Penguin is thus necessarily annual, being synchronized with the formation and breaking up of the sea ice. A close relative, the King Penguin *Aptenodytes patagonica*, occupies the subantarctic zone, where the climate is severe but much less so than on Antarctica. Comparison between these two species allows us to assess the Emperor Penguin's adaptations to the polar environment. We have already noted the difference in weight, the greater size of the Emperor being an undoubted advantage in defence against cold. Moreover its bill, flippers and feet are smaller, favouring more effective temperature control in very cold surroundings. The King is slimmer and has larger feet so that it can move easily among rocks as its relative cannot, and the latter does not need these adaptations since it seldom leaves the sea ice. These morphological differences are accompanied by still more important biological and physiological ones. The ability to store up considerable fatty reserves allows the Emperor to starve for prolonged periods. It can thus establish itself far from open water, to which it resorts only at very precise and widely-spaced periods – at least during the first part of the breeding season. In contrast the King lacks these abilities, which would in any case be useless to it since its nesting colonies are close to the open sea.

Further, defence against cold has forced the Emperor to abandon a series of individual behaviour patterns and to develop communal ones. The King establishes a territory which it defends against other members of the species and where it lays its egg, whereas the Emperor has no actual territory and is not tied to a particular point within the colony. While moving about it defends a virtual territory around itself, and carries its nest with it in the form of the incubation pouch. It has become supremely gregarious, as is shown especially by the formation of 'huddles' and the rearing of the young in crèches, while it has lost many agonistic behaviour patterns. However, all this does not mean that breeding and the rearing of young have become communal. The pairs actually remain united for the whole breeding season, and the young are still under the care of their parents even when in the crèche. Auditory and visual means of individual recognition enable the family to maintain its unity.

Finally, the two species of *Aptenodytes* show very different timing in the phases of their annual cycles. The King breeds during the southern summer, while its polar relative does so in midwinter. The Emperor's incubation lasts for a fortnight longer, while the chick grows more slowly and takes longer to rear. The King chick reaches the adult weight in three months, whereas in five months the Emperor chick only reaches the maximum juvenile weight, which is much less than the adult weight. However, young Emperors of only 10-15kg can live independently. The Emperor's long breeding season – ten months – leaves little time for purely vegetative activities. In contrast the King's cycle is much longer, despite a shorter breeding season. The whole cycle lasts from fourteen to sixteen months, which involves no major difficulty since even the most unfavourable season does not present the dramatic conditions of the antarctic winter. This penguin begins to nest at the beginning of summer (at the end of November in South Georgia). Rearing of the young is completed by the autumn, by which time they have laid down considerable reserves of fat which allow them to exist through the winter, feeding only very occasionally. They grow only slightly taller, and lose up to 40 per cent of their weight. The more frequent feeds available from the following spring allow them to complete their growth and to go to the sea from December onwards. Thus they take from ten to thirteen months to rear.

The adults moult when the young are independent, and are ready to breed once more in February. If they begin at once, their chick will be fit to stand the rigours of the following winter; but since it will be less developed at the same time than the previous year's, it will become independent later in the following summer. The parents will not begin to nest again during that season. Thus this penguin breeds only twice every three years, whereas the Emperor has a regular annual rhythm (Stonehouse).

The Emperor shows equally great differences from the Adelie Penguin, whose adaptations to the polar environment seem from some points of view to be less perfect. The latter's breeding cycle occurs during the summer, and must therefore be shortened so as to coincide with the least unfavourable period.[1] We have seen that it incubates and rears its young relatively quickly, so that adults and young can desert the colonies as soon as the sea begins to freeze, to migrate northwards and adopt the pelagic life which occupies a good part of their annual cycle. Thus the Adelie has solved the difficult problems of adapting to the polar environment in quite a different way from the Emperor. The reversal of the cycles of the two most strictly antarctic

[1] In the Yellow-eyed Penguin *Megadyptes antipodes*, similar to the Adelie in size but nesting in the milder climate of New Zealand, incubation lasts for 43 days as against a mean of 33 in the Adelie, and rearing of the young 106 days against a mean of 51.

penguins is also very important in the demands they make on the food resources of the ecological community.

The paradoxical cycle of the Emperor Penguin is its most remarkable adaptation to the antarctic biotope. Morphologically and ecologically, it is incomparably the terrestrial animal best adapted to the polar environment.

The annual cycles of Antarctic Petrels

Antarctic Procellariiformes as a whole have cycles similar to that of the Adelie Penguin (Prévost 1964, Mougin 1968). Many, such as the Fulmar, Wilson's Petrel and Pintado Petrel, completely desert the coasts of Antarctica during the winter. In contrast the Snow Petrel makes regular winter visits depending on the state of the sea ice, more and more frequently as spring approaches. So does the Giant Petrel, which begins to reoccupy its colonies during July. Most of these petrels, and especially the Pintado Petrel, Silver-grey Petrel *Fulmarus glacialoides* and Snow Petrel, have summer breeding cycles. They return to the coasts of Adelie Land in October, except for Wilson's Petrel which arrives later because of its greater sensitivity to cold. Generally the pairs re-form as in the previous year, and reoccupy their old nest sites. Laying takes place at the end of November and beginning of December, so that the eggs are incubated and the young reared during the most favourable period. Incubation – whose duration is somewhat variable, depending probably on weather conditions – begins in December when there is least snow and the temperature is relatively mild. The chicks can then be fed at a time when the consumable biomass is at a maximum and easily accessible because the sea ice has broken up, and have finished their development by the onset of winter. Thus the cycles of these birds are so regulated as to coincide with the most favourable periods – from the point of view of food, and of climate in relation to the risks of being snowed under, which is the greatest danger for brooding birds. In contrast Giant Petrels have a semi-winter breeding cycle, beginning to nest in July – two months after the end of the breeding season of the former group. Longer periods of incubation (with an average of fifty-nine days) and raising (up to 128 days) no doubt explain the earlier breeding of this species.

The adaptation of these Procellariiformes to the polar environment is also shown by their modes of nesting, since they all seek the most favourable microclimates. Thus in Adelie Land the mean temperature of the sites of petrel colonies varies from $-3 \cdot 1°$C to $-2 \cdot 9°$C, whereas it is $-4°$C at the nearby meteorological station. Nesting sites generally are more or less sheltered from the wind, diminishing its cooling power by a quarter

(Mougin). The various species differ in choice of site, depending on their sensitivity to cold. Each species has to make a choice, depending on its size and resistance to cold. Giant Petrels – large birds which can resist intense cooling – are not afraid to settle on very exposed pebbly slopes, and so avoid any appreciable accumulation of snow, and can easily take off into the wind. Their wide-open pebble nests give only poor protection to their broods. Pintado Petrels habitually nest on wind-swept slopes, and sometimes on cliffs. Their nests are backed by rock, but are wide open in other directions and never dug in, so that they are poorly protected from the wind but in no danger of being snowed under. Fulmars and Antarctic Petrels build their nests on north-facing cliffs, facing the sun and well protected from the prevailing winds, and readily occupy any hollows in the rocky walls, so that they are in greater danger from snow. Snow Petrels and Wilson's Petrels also have to run this risk, being forced to concentrate on protection from cold since their small size is a serious handicap in thermal equilibrium (the adult Wilson's Petrel weighs only 40g and the chick at hatching 7g). Snow Petrels nest among boulders at the foot of a cliff, their pebble nests being hidden among the piles so that each is effectively a chamber with many openings. Though the shelter these give is incomplete they greatly reduce the force of the wind; but they are frequently snowed under. Wilson's Petrels nest in actual burrows and in fissures in the rocks, each with an opening just large enough for the bird leading to an entrance corridor and a brood chamber of variable size. Such nests are very easily snowed under, causing high mortality.

The success of polar birds

We have seen that species have succeeded in colonizing the polar environment, through marked morphological and physiological adaptations. Despite these however, the rigour of the climate and the irregularity of the ice cycle from year to year cause very severe losses.

The breeding cycles of these birds, and especially penguins, are strictly related to the cycle of the sea ice. Upsets in the timing of this cycle may cause catastrophes involving whole populations. Thus the hostile environmental factors, always very severe, are increased in abnormal seasons.

The fecundity of antarctic birds is generally low, the Adelie being the only penguin which regularly lays two eggs and whose female can replace one or both of them in case of loss. Mortality at the egg and chick stages is enormous, caused mainly by climatic factors which sometimes act directly, sometimes indirectly through feeding difficulties, and most often in both ways at once. The chicks cannot fully regulate their body temperature during their first stages of growth, since their thermoregulation only becomes established very gradually. Up to the ninth day, an Adelie Penguin chick cools like an inanimate body, and in a violent blizzard can survive for only a few minutes on its own. Despite improving individual and collective thermoregulation, this vulnerability persists even longer, and every blizzard causes increased mortality, as is shown by the corpses which litter the colony sites. Violent winds may carry the chicks away from their colonies to where their parents cannot recover them. Blizzard-blown ice may cover them completely, freezing them to the ground where they die in their tracks. The young are especially susceptible during certain critical phases of their development, particularly when they are becoming independent, and if the adults do not return to the colony at the proper time (for example because of the state of the ice) a high proportion of the young dies. This mortality is directly caused by starvation, at a time when a chick needs plenty of food in order to thermoregulate.

The flooding of Adelie Penguin nest sites and the snowing-over of petrels' nests, are further mortality factors which can kill a large proportion of the eggs or chicks in a colony, while predation by skuas and Giant Petrels may also be important. Inexperienced breeding adults behave inefficiently – especially in protecting their chicks against cold and against the raids of unoccupied rivals which dispute their possession, and their young suffer much higher mortalities than those of experienced adults. Forms of social maladaptation also militate against the success of the broods, and considerably increase mortality (Birr).

The mortality rate is generally very high. A mortality of 60 per cent among Adelie Penguins is considered good, and it may rise to eighty or even locally to 100 per cent where there has been a meteorological catastrophe. Of 100 eggs laid, on the average twenty are lost during incubation, five do not hatch, and 45-50 chicks disappear before reaching the immature stage, so that the mortality rate up to this stage is 70 to 75 per cent (Sapin-Jaloustre). The egg mortality among petrels may reach 50 per cent., especially as a result of abandonment by the adults. Their total mortality of eggs and chicks is generally between 50 and 60 per cent, but may reach 80 per cent especially in the Pintado and Wilson's Petrels (Mougin). However, mortality rates in some years are considerably lower, reaching only 25 per cent among Emperor Penguins in Adelie Land as a result of exceptionally favourable conditions. The annual growth of populations is thus highly irregular, as in all impoverished communities whose physical environment is subject to large fluctuations.

Adult mortality of such species is low – resulting from predation by Sea

Leopards *Hydrurga leptonyx* and Killer Whales *Orcinus orca*, by accidents (including those to females associated with laying), and by various illnesses and parasitosis – and penguins and petrels are potentially long-lived. Thus antarctic birds renew their population very slowly. As among other seabirds, the presence of non-breeding adults provides a kind of reserve, allowing the populations to maintain a certain equilibrium: it is known that some individuals of the King Penguin begin to nest only at eleven years old, and the same is probably true of the truly antarctic penguins. The success of the broods varies greatly from year to year, as a result of irregularities in the climatic cycle, and (in this unusually hostile environment, where abnormal circumstances are frequent) these populations must be able to compensate for such effects.

The Antarctic community

Despite its extremely hard conditions the Antarctic is thus occupied by many birds, which are highly adapted to exploit the food resources of this apparently desolate part of the world. We have seen that, while the continent itself is a true desert of virtually zero productivity, the arctic seas are in contrast very rich. The phytoplankton, consisting essentially of diatoms, is very abundant and supports flourishing food chains. Adelie Penguins usually feed almost exclusively on euphausids, though sometimes fish account for almost 40 per cent of their diet (Sladen *et al.*). Emperor Penguins have a more varied diet including fish, cephalopods and crustacea, all caught by swimming underwater. Among the Procellariiformes, the Snow Petrel feeds indifferently on crustacea, cephalopods and fish, while it is also to some extent a carrion-feeder. Pintado Petrels and Fulmars feed more strictly on plankton – especially crustacea but also cephalopods – while cephalopods plus crustacea are the main diet of Wilson's Petrels. The amounts taken by birds are considerable, though obviously less than those swallowed by the great antarctic whales for which euphausids are the main source of food.

The only available estimates of the biomass of antarctic vertebrates concern Adelie Land (Prévost 1963) – which it should be noted is a relatively favoured area. Other sectors, in which productivity is no doubt just as high, are completely devoid of birds because of the absence or inaccessibility of nesting sites (such as cliffs, the vertical snouts of glaciers, and areas of too-rigorous climate). Conclusions derived from Adelie Land are thus not necessarily valid for the whole of Antarctica. The biomass of birds and mammals is comparatively high throughout the year, varying from 300 to

1,920kg per hundred metres of coastline, with maxima at the beginning of the summer (late October to early December) and winter (April to May). Although at least two species of mammals and eight of birds take part in this community, penguins are in the majority. Represented in total by about 50,000 individuals, they form a biomass of around 515 tons, while petrels (totalling 3,000 individuals) contribute only 1·4 tons and seals 21 tons, so that Penguins are the principal consumers. The two maxima recorded in the annual fluctuations correspond to the arrivals of the Adelie and Emperor Penguins, so that the latter's inverted cycle allows an almost continuous

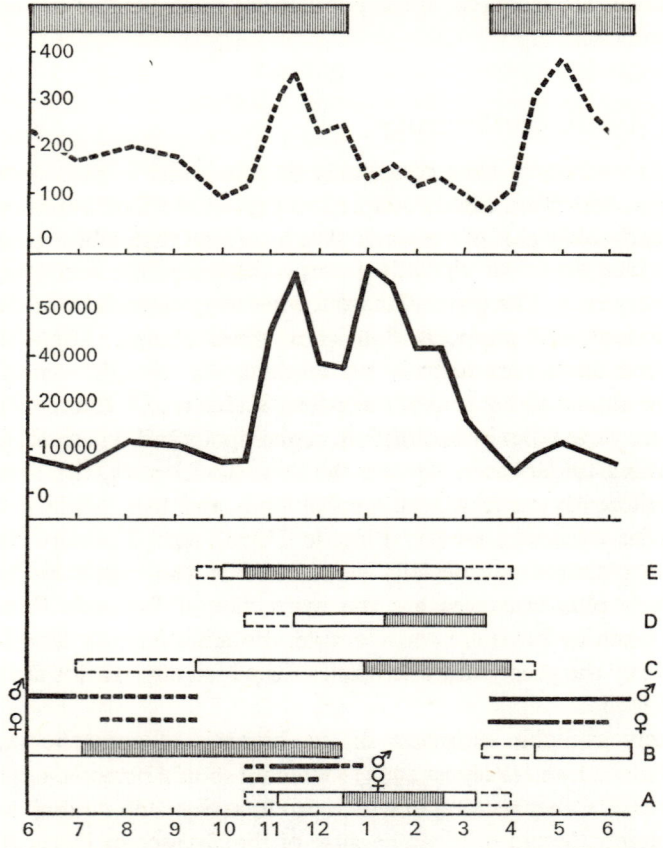

Figure 80. Seasonal differences in numbers of birds (and a mammal) to be found on the antarctic coast. Above, the biomass of birds. Centre, the number of individual birds. Below, periods of residence of various species (shaded area indicates rearing young). A. Adelie Penguin. B. Emperor Penguin. C. Giant Petrel. D. Petrels and skuas. E. Weddell's Seal.

exploitation of the consumable biomass. There is in this way no competition during the annual cycle, but rather an alternation between these two species, whose dietary preferences also are somewhat different.

The waters near the coast are by no means the only ones to be exploited by birds. We have seen that during the pairing season the adults are never all together in the colonies, but that the sexes replace one another – one going to feed in the open ocean while the other looks after the egg and chick. Moreover, the brooding birds do not form a consuming biomass, since they maintain a prolonged physiological fast. In fact the greater part of the food is not collected near the coasts, but far out to sea, thanks to the fasting and travelling abilities of these principal consumers. These adaptations allow a large colony of gross feeders to exploit the resources of the wide area over which they disperse.[1]

The biomasses of birds at the coast are least just when the plankton is most abundant in the ocean waters.[2] There is a close parallel between the numbers and biomass of birds at sea and the annual variations in plankton density. Thus the various phases of the cycle of polar birds are remarkably well adjusted, not only to the seasonal changes in the physical environment, but also to the fluctuations of the biomass. It is especially notable that the young of two species of penguin have become independent of the season at which plankton is most abundant – the Emperor Penguin in December and the Adelie Penguin in February.

The great majority of the higher vertebrates, and especially the birds, of Antarctica are fish and plankton feeders, and these make up the dominant biomass of the region. However, two bird species – like one of the seals, the Sea Leopard – are at a higher trophic level. Giant Petrels, while they readily eat carrion (corpses, seal placentae and rubbish) and can fish in the sea, prey on young Emperor Penguins between August and December. Great Skuas similarly prey on the eggs and young of the Adelie Penguin. However, these birds necessarily make up only an insignificant biomass: the 90-odd

---

[1] These penguins thus differ profoundly from other species, such as *Eudyptula minor* and *Megadyptes antipodes* of New Zealand, which feed close to the coasts. In these mates replace one another at the nest at frequent intervals (1–3 days), and feed their young once or twice a day. These differences in breeding behaviour reflect adaptations to fundamentally unlike environments.

[2] It may be noted that these variations can be explained above all by the marked variation in the vertical distribution of zooplankton. During the summer, the latter is concentrated near the surface, while in winter it migrates in deep water towards warm currents. The quantity of plankton in the whole water column does not change markedly with the seasons (Foxton); but although the total biomass remains constant these variations are still important for the surface predators, since the accessibility of the biomass is as important as its presence.

Giant Petrels and 100 Great Skuas in Adelie Land account for only about 530kg. Although they account for a significant part of penguin mortality, their influence on the ecological community is slight.

The principal vertebrate consumers of this community are certainly the penguins, whose remarkable adaptations and the way in which their diets complement one another during the annual cycle deserve to be recalled. The antarctic community is thus well balanced, despite the extreme conditions of a most unfavourable environment to which few birds have succeeded in adapting themselves.

The polar environment of the Arctic seas

The marine environment of the Arctic is not lacking in parallels with what we have seen in the Antarctic. During the winter the seas are congealed in ice, which transforms the Arctic Ocean into a vast white expanse which breaks up in the spring. Nevertheless, the climate of the marine polar zones is not as cold as that of Antarctica whereas the arctic land suffers from a much more rigorous winter climate. However, the environmental conditions there are no less severe for animals.

The inhabitants of this zone get their living from marine resources. Virtually all the birds belong to systematic groups very clearly distinct from their antarctic equivalents. Because of the rigorous climate there are not many marine species. However, these few species make up enormous populations, since they are essentially gregarious and form huge colonies because of the rarity of favourable nesting sites and the abundance of nearby food. The dominant group is the Alcidae, represented by several species of guillemots *Uria* and *Cepphus*, razorbills *Alca*, little auks *Plautus*, and puffins. While many arctic birds have circumboreal distributions, some like the puffins of the north Pacific are confined to the great ocean basins. Another important group is the Laridae: while the Sterninae are represented only by the Arctic Tern *Sterna paradisea*, the Larinae are highly developed with several species proper to the Arctic – the Glaucous Gull *Larus hyperboreus*, Sabine's Gull *Larus sabini*, and Ivory Gull *Pagophila eburnea*. The Kittiwake *Larus tridactyla* belongs as much to the arctic world, although it also nests farther south, like the Great Black-backed Gull *Larus marinus* which breeds especially on the west coast of Greenland at a latitude of about 73°N. The Procellariiformes are much less developed than in the Antarctic, and are represented here only by the Fulmar *Fulmarus glacialis* – although the populations of this species are very large.

Certain arctic and antarctic birds belonging to distinct systematic groups show a quite remarkable degree of convergent evolution. Although able to fly (except for the Great Auk) the auks' bearing, gait and swimming ability are comparable to those of penguins, which are also approached by certain antarctic diving petrels.

Some birds which frequent floating ice and nest in the neighbourhood of snow are remarkably uniformly coloured, their plumage being almost devoid of pigment. This is especially true of the Ivory Gull, an entirely white bird whose chosen habitat is among the pack-ice. In pigmentation as well as in some other characteristics, it thus resembles the Snow Petrel.

As a general rule these arctic seabirds take all their food from the sea, but it should be noted that a few, such as the Glaucous Gull and some less polar gulls, make up their diets with a significant proportion of food taken on land. Fish, molluscs and crustacea are the principal prey. Fish are eaten by the majority of species – principally by the guillemots and razorbills which make them almost their exclusive diet (up to 95 per cent for Brünnich's Guillemot) but also by the Kittiwake, Arctic Tern and most large gulls. Molluscs are eaten by the Black Guillemot (up to 20 per cent of its diet), Glaucous Gull and Arctic Tern, and littoral species form the basic food of the Eider while the Fulmar takes a high proportion of pelagic ones. Pelagic crustacea are the favourite prey of Little Auks, and form a considerable proportion of the diets of Fulmars, Arctic Terns and Black Guillemots (up to 20 per cent), while littoral species are taken by the Eider and Glaucous Gull (up to 29 per cent). Annelids are not an important component, though all the birds (especially puffins) take them occasionally. Skuas and Glaucous Gulls are predators on eggs and young birds, while, of course, all the large gulls readily eat carrion.

A few of these birds, especially the Razorbill and Brünnich's Guillemot, have very restricted diets. All these are fish-eaters, since fish are an abundant resource which does not vary greatly with the seasons. Most of the other birds are distinctly polyphagous. Their diet varies with locality and with the time of year

These birds together exploit the whole expanse of the sea, while each specializes in a well-defined sector. Fulmars and Little Auks fish the open seas, even during the breeding season, while other species such as the Kittiwake become pelagic only when wintering. Still others fish near the coasts, at distances depending on the species and on the season, and disperse during the winter. They concentrate at certain favoured points, where productivity is especially high because of favourable oceanographic conditions. In the arctic seas, as in other sectors of the oceans, there are currents differing very much in temperature and in salinity, and the areas where these meet are always favourable for the formation of flourishing communities.

Study of oceanographic charts of this part of the world shows that the large colonies of seabirds are indeed established near these favoured areas. For the same reason, zones of high productivity occur at the snouts of glaciers, where the fresh water produced by the ice, or the streams which flow from it during the summer, meet the warmer salty seawater. Such conditions result in a proliferation of plankton and allow high productivity, and very many birds establish themselves in the neighbourhood and fish in these rich waters (Lövenskiold 1963, Belopolkii 1963). Thus more than 2,000 Kittiwakes, as well as other species, have been noted opposite the snout of a glacier in western Spitzbergen, feeding mainly on the crustacean *Thysanoessa inermis* which abounds in these waters (Hartley & Fisher 1936).

The great majority of arctic birds are migratory, entirely vacating the northern parts of their ranges at the coming of winter, when they disperse southwards and become more or less pelagic. The Arctic Tern is a long-range migrant, which winters at the edge of the antarctic ice (see Chapter 31). However, a few such as the Fulmar off Greenland haunt the arctic zones even in midwinter, their seasonal movements being determined more by the accessibility of marine food resources than by temperature. All arctic birds have well defined breeding seasons which coincide with the northern summer, and not one shows the remarkable inversion of the Emperor Penguin. They arrive earlier or later at the breeding territories depending on their ecological needs, Guillemots and Kittiwakes being among the first and Arctic Terns the last.

The breeding sites are selected so as to offer the greatest shelter to the broods, bearing in mind the thermal requirements of the various species. Cliffs cut up by narrow ledges are the favoured biotopes of many birds, especially auks, each of which chooses a particular microhabitat. The Little Auk and Black Guillemot nest in crevices and cracks, and the Razorbill mainly under blocks of stone or in winding passages. However, other auks nest in the open seeking no shelter from the cold, to which they seem less sensitive probably because of their greater size. Puffins nest in true burrows, which they dig in the soil by using their stout bills like pickaxes, reaching a depth of 2 to 3m or even more. These burrows provide an eminently favourable microclimate, the more so since they are usually dug in sunny sites exposed to the south.

The different sensitivities to cold of arctic birds are also shown in the quantities of nesting materials that they accumulate. Some, undemanding and able to maintain their own internal temperature and that of their young without external insulation, lay their eggs on bare rock. This is true of the guillemots which have lost all nest-building behaviour – their eggs are prevented from rolling off the cliff only by their highly pointed conical shape. Terns lay their eggs in plain shallow depressions in the soil. Other birds in contrast arrange grass, moss and seaweed to form real nests. The most elaborate constructions are those of Eiders, whose twig nests are thickly lined inside with layers of down. Built on the ground, they are hidden in low vegetation which shelters them from the wind.

Arctic birds thus turn favourable microclimates to good account, each species placing itself in the best possible conditions for nesting with regard to the vulnerability of its eggs and young, which themselves show undeniable adaptations to the cold and humidity of arctic habitats. This is especially clear in the thick shells of guillemots' eggs, the better protected from contact with cold rock; and in those of the Fulmar, laid on bare soil while the temperature is still between $-20°C$ and $-30°C$, and protected by strong calcareous layers. Similarly, the young on hatching are covered with a thick downy layer which effectively protects them from the cold, while their parents brood continuously and remain on the nest throughout their first stages of development. Many species grow rapidly, the continuous arctic day giving the adults a very long daily period of activity; and this is an undeniable advantage, allowing these birds like their relatives of the tundra to gain maximum benefit from the short favourable season of the Arctic.

Terrestrial polar environments

Terrestrial environments scarcely exist in the Antarctic, because of the extremely small area of exposed ground at high latitudes, Antarctica being almost entirely covered with ice. They are exceedingly impoverished, so that not a single community of terrestrial birds has been able to develop in the southern polar region. There are scarcely more than a few rails, two ducks (*Anas eatoni* on Kerguelen and Crozet Islands and *A. georgica* in South Georgia) and a single passerine, the South Georgia Pipit *Anthus antarcticus*. However, strangely enough, Macquarie Island was once occupied by a parrot, the Red-crowned Parakeet *Cyanoramphus novaezelandiae erythrotis*, until it was extinguished there through human interference. This fed, at least in part, on crustacea and other small animals from the heaps of seaweed accumulated on the shore. Of typically antarctic birds, the sheathbills (*Chionis*) are the best adapted to terrestrial life. Confined to the Antarctic and intermediate between waders and gulls, they avoid the water and live ashore though they are strong on the wing. They are robbers, feeding on the eggs and young of seabirds as well as on marine animals and various rubbish.

In contrast, the arctic polar regions include huge areas of exposed ground.

E

Between the coasts of the frozen sea and the northern forests (taiga), the wide expanses of *tundra* provide a very special environment in both the Old and New Worlds. The climate of this terrestrial polar habitat is extremely harsh. In winter the temperature is always below freezing, and heavy snow-falls deny access to any kind of food. However, there is an alternation between this very long unfavourable season and a much more favourable though shorter one. The tundra enjoys a true summer, during which the mean and maximum temperatures are quite high. Thus at Verkhoyansk in Siberia at 67°35′N, a locality long known as the 'pole of cold', the annual mean temperature is − 17°C with minima at about − 60°C; yet the summer maxima reach + 15°C. In Lapland, the mean for July lies around + 7°C to 9°C, with maxima of + 15°C to 20°C.

Despite the brevity of the arctic summer − often of only three ice-free months − it allows a fairly abundant vegetation to flourish. Very many lichens with some mosses, grasses and various other herbaceous plants form a dense multicoloured carpet. Tree-like species are absent, mainly because of the poverty of the soils and the fact that the summer thaw affects only their topmost layers, so that the tundra does not support a single tree and the few shrubs flourish only in the most sheltered places. The trees of temperate regions are represented in the tundra by dwarf and stunted forms, especially birches (*Betula*) and willows (*Salix*), which are important elements in the arctic flora.

This environment has allowed an arctic terrestrial avifauna to develop, most of whose members depend on freshwater resources. Marshes, peat-bogs and stretches of open water under rather diverse ecological conditions, offer the greatest supplies of food – their productivity being decidedly higher than that of the terrestrial habitats, which are poor in foods suitable for birds. As a result, the dominant bird groups are waders and ducks. Ruffs, plovers and godwits abound in various microhabitats in accordance with their ecological preferences. The ducks are represented by some very characteristic species, such as various dabbling ducks, diving ducks (Tufted and Scaup Ducks), eiders, scoters, Long-tailed Duck *Clangula hyemalis*, golden eyes and mergansers. There are many geese and swans in the tundras, which are their preferred nesting habitat (and that of divers *Gavia*).

Though fewer birds live off the terrestrial resources of the tundra, they still form reasonably flourishing communities. Wagtails, pipits, Shore Larks *Eremophila alpestris*, Wheatears *Oenanthe oenanthe*, Bluethroats *Luscinia svecica*, thrushes, specialized leaf warblers such as the Arctic Warbler *Phylloscopus borealis*, buntings including the Lapland Bunting *Calcarius lapponicus* and Snow Bunting *Plectrophenax nivalis*, and many finches

(*Carduelis*), exploit the vegetation and insects. Terrestrial arthropods play an important part in the ecosystem, and it is noteworthy that most of the waders feed on them while breeding – even those which take crustacea and molluscs on migration or in their winter quarters. Flies (chironomids and tipulids) and their larvae are especially important as food resources. Herbivorous birds are represented primarily by ptarmigan (*Lagopus*), an important group of game birds which apart from a few mountain species are confined to the Arctic. Finally, various raptors live on these birds or on the arctic mammals. Rodents, especially the lemmings characteristic of the tundra, constitute a most important food resource. Buzzards, falcons (the Gyr Falcon *Falco rusticus*) and owls (the Snowy Owl *Nyctea scandiaca*) have deeply penetrated the arctic world.

Arctic birds may be divided into two distinct groups, the first strictly confined to this environment, whereas the second are ubiquitous species whose ecological flexibility fits them for the conditions of the tundra. Thus one can see such highly specialized birds as the arctic buntings, Gyr Falcon and Snowy Owl living side by side with widespread ones like the Wheatear. In a district of Swedish Lapland the dominant bird, making up over 50 per cent of the total population, was the Meadow Pipit *Anthus pratensis*, which is very widely distributed in temperate Europe (Alm *et al* 1966). However, because of the severity of the environment this avifauna is poorly diversified.

As we have seen, the terrestrial arctic environment is characterized by the alternation of a long hard season with a short favourable one. As a result, physical and biotic variables fluctuate considerably and the standing crop biomass – both the vegetation and the aquatic and aerial invertebrates – varies enormously, with a relatively high but short-lived peak followed by an exceedingly low minimum in which the quantity available falls almost to zero. Furthermore, what food there is during the winter is denied to most birds by snow and ice-bound water. Birds have had to adapt to these conditions by compressing their breeding seasons to the minimum shortest possible periods. The various reproductive phases are accelerated, their timing being such that the young are reared during the most favourable period when the biomass is at its height. Thus it is that many arctic birds arrive at their breeding grounds already paired, with the females already fertilized. Their arrival is synchronized with the melting of the snows, following its fluctuations from year to year, and nesting begins at once. This timing has been observed especially for Ross' Goose *Anser rossi* in northern Canada (Ryder 1967). Brooding success varies greatly with the meteorological conditions, as has been shown for forty-odd species nesting in Iceland (Bengtson 1963).

This abridgement of the breeding season is undoubtedly related to the continuous daylight enjoyed throughout the summer at these high latitudes (Karplus 1952). The passerines show a special circadian rhythm, with a much longer period of activity than those of their relatives nesting at lower latitudes, whose activities are interrupted by the coming of night. Arctic birds show no more than a slowing down of their activity during the hours of minimal illumination. This prolonged activity allows them to feed their young much more often during the twenty-four hours, which results in accelerated growth and earlier fledging. The young of a pair of American Robins *Turdus migratorius* remain in the nest for only nine days in Alaska at a latitude of 69°23′N, whereas those of the temperate zone stay for an average of 13·2 days. During this shortened period, because of the increased period of daily activity, the young are fed about as many times as those of birds which nest farther south. The duration of daily illumination also allows a pair to rear a greater number of young, and as a result clutches are larger in the arctic (see Chapter 12). Conversely, many birds there rear only one brood a year, because the favourable period is so short.

The enormous flcutuations in standing crop biomass, and the fact that food is only available for part of the year, lead to the situation that almost all terrestrial arctic birds are long range migrants which entirely desert their nesting areas during the winter. The ducks and waders are among the most confirmed migrants. However, a few birds are sedentary during the winter. In particular, Snowy Owls and some other predators are only partially nomadic (except under exceptional circumstances) since stocks of mammals assure them of sufficient food during the winter. Ptarmigan are still more sedentary, since they do not leave the highest latitudes and can winter up to 75°N. They are well protected against cold by a very thick lining of feathers which are white in winter and thus highly cryptic in colour. These birds feed principally on the buds and twigs of dwarf willows which form up to 94 per cent of their winter diet (West & Meng 1966), and whose high calorific value and availability throughout the bad season allow the ptarmigan to winter in most of their range. Ptarmigan shelter in small chambers scooped out beneath the snow, and move around along tunnels leading to the vegetation on which they feed, so that they take advantage of a microclimate which is distinctly more favourable than the surrounding environment. They also seek out high ground free of snow, where they often gather in flocks which disperse only in the spring.

Some arctic birds undergo great fluctuations in numbers, just like those of some mammals belonging to the same community. These fluctuations are most often cyclical, and are still poorly understood, but their great amplitude is in accordance with the simple structure of arctic terrestrial communities (see Chapter 14).

Despite the faunal impoverishment caused by severe environmental conditions, arctic terrestrial birds thus form fairly flourishing communities, more diversified than those of the mammals. While there are in the tundra mammals which feed on the vegetation (notably reindeer or caribou and specialized rodents such as the lemmings), and predators which live on them, there are neither mammalian insectivores nor aquatic mammals. By their ability to travel, birds are especially well adapted to exploit such resources, temporary but abundant during one phase of the annual cycle. The abundance and relative diversity of arctic birds is in strong contrast to the paucity of arctic mammals in individuals and species – characteristics which are closely linked to the great differences in mobility between these two vertebrate classes.

Chapter 21

Coastal Environments

COASTS present very varied ecological conditions, depending on the nature of the substrate, on the relief, and on a great number of other factors both physical and biological. As a result they include strongly differentiated habitats, very varied in productivity, and supporting diverse avifaunas whose composition and variety depend on the ecological conditions.

Rocky coasts are ecologically the poorest, but are nevertheless colonized by seaweeds of various species, distributed in successive zones, while the rocky pavements are scattered at low tide with small pools and basins supporting a specialized invertebrate fauna. Sandy coasts are much richer, with a highly adapted faunule occupying the interstices between the grains of sand and varying in composition and density with the grain size. Mudflats are still richer, and are the most favourable environment in the intertidal zone. In Europe these favoured areas are found especially on the east coasts of England in the sector south-east of the North Sea, along the Dutch coast, and in France (on the Bay of Mont St Michel, the Gulf of Morbihan, the Vendée and especially the Bay of Aiguillon). They can of course develop only on coasts subject to tides, and the salt-water swamps elsewhere (as on the Mediterranean) are very different.

Two primary physical factors ensure high productivity in littoral habitats. The first is the play of tides, which twice a day introduce mineral and organic substances which are then laid down in the intertidal region, and make mud-living organisms available to predators at regular intervals. The second is the zone of contact which these habitats provide, between the aquatic and terrestrial phases and between marine and freshwater environments.

Every high tide brings in a considerable amount of matter in suspension: in the Waddenzee in Holland this is estimated at 9mg per litre at the entrance from the open sea, and 20mg at the centre. Its most important components are mineral salts containing nitrogen and phosphorus, and organic substances. Waters which are shallow even at high tide are relatively warm and richly irradiated by the sun. Thus this environment shelters a phytoplankton of uncommon density and is the site of very intense photosynthesis.

The Waddenzee is a natural community of exceptional richness. This stunning abundance is characteristic of all similar intertidal environments, in which microphagous animals (especially molluscs) can proliferate and form vital links in the food chains to carnivorous fish and birds.

Mudflats and birds

The considerable biomass and very high productivity of mudflats have attracted birds to exploit their resources, many of them specialized for such habitats. This is especially true of the waders (Charadriiformes), most of which are dependent upon this environment for at least a part of their annual cycles, and also of certain specialized Anatidae such as the eiders and scoters. Many gulls and terns also exploit the seashore and readily feed on mudflats, while some Ardeidae are dependent on the shore and use its resources either temporarily or permanently. Some birds are found there throughout the year, and others for only parts of their cycles, and many waders show cyclical changes in their ecological preferences during the course of the year. Many plovers, sandpipers and their allies nest in terrestrial habitats, near the stretches of fresh water with which the arctic tundras are sprinkled. Their diets include an extensive range of aquatic organisms, notably molluscs and especially aquatic insects and larvae. Thus Dunlin *Calidris alpina* feed almost exclusively on larval and adult tipulid and chironimid flies, according to studies undertaken in Alaska (Holmes). At the end of the breeding season these birds leave on migration and then spread into the intertidal environments which provide them with resting places and their chosen winter quarters. They thus become marine and littoral, and their diets are now almost exclusively based on molluscs, crustacea and polychaetes. The great resources and high productivity of the mudflats allow them to maintain very large populations, which the rigour of the arctic climate and the shortage of food force out of their breeding areas for a large part of the year.

The intertidal zone, and especially the wide mudflats, are sought out also for the quietness and security which they offer to birds – some of which are temporarily flightless during the moult – since the mudflats are inaccessible to terrestrial predators for which the soil is much too yielding. Thus the birds find at the same time abundant food – very welcome when they are facing the expenditure of energy involved in the moult – and a sure refuge. The whole European population of Shelduck *Tadorna tadorna* gathers in the south-eastern part of the North Sea, along the Danish and German coasts, in a moult migration immediately after they have bred.

Of all the higher vertebrates, birds are indisputably the best adapted to exploit mudflats and other intertidal environments, since only they can cope with a substrate which is both solid and liquid. Furthermore, the play of currents, and especially of tides, forces any animal exploiting this environment to make constant movements, which the birds can do thanks to their powers of flight. Seals are the only mammals to have colonized mudflats, thanks to their amphibious adaptations, but they are of minor importance compared to the birds, which occupy a dominant place in these ecosystems.

Herbivorous birds

Besides the phytoplankton, the primary production of the intertidal zone consists of many seaweeds, and a few highly specialized higher plants such as the eel-grasses *Zostera*. On these plants graze birds with a specialized herbivorous diet: in western Europe Brent Geese *Branta*, which are distinctly marine in winter. They wait around during high tides and as the sea goes out resort to the coasts, where they begin to graze – swimming at first and then on foot upon the mud. They stop feeding when the rising tide again covers their pasturage, so that their feeding rhythm is linked to that of the tides. Solely vegetarian, they feed principally on eel-grass but also on seaweeds, especially *Enteromorpha*, which they take increasingly towards the end of the winter, probably as a result of impoverishment of the beds of eel-grass. They also graze on halophytes such as the glassworts *Salicornia* (Salsolaceae) and on grasses (*Puccinella*).

The Wigeon *Anas penelope* is another anatid which nests in the north of the Palaearctic and winters along the coasts of western Europe, living like the Brent Goose. These ducks, which at this time are entirely marine, characteristically gather at the heads of sheltered shallow bays. They feed for preference on eel-grass and Tassel Pondweed (*Ruppia*) and later on seaweeds, taking scarcely any food of animal origin. In contrast to Brent Geese which favour rhizomes and leaves especially, Wigeon mainly eat seeds. Their circadian rhythms are less closely regulated than those of the geese, and they feed mainly at night. While the diet of these birds is distinctly specialized, their preference for eel-grass and secondarily for seaweeds may not be marked as it appears, since this behaviour has been accentuated by human interference. In areas where these birds are effectively protected they come as much to the land and graze on salt meadows, whereas the effect of man has been to force them seawards. None the less, under all circumstances intertidal vegetation does form an important part of their diet.

Some other birds share this diet, at least while wintering. In Europe the

Coot *Fulica atra* frequents coastal waters during the winter, while the coasts of other parts of the world are similarly settled by birds occupying the same ecological niche. In South America this is particularly true of the Kelp Goose *Chloephaga hybrida*, a species belonging to a genus of Patagonian origin all of whose other species are distinctly terrestrial. This goose, found along the coasts of Chile from 40° South latitude to Cape Horn, is strictly confined to the intertidal zone, avoiding dry land and rarely going on to the water. When pursued, it runs along beside the sea rather than fly or swim. It prefers rocky coasts where it feeds on seaweeds, especially *Porphyra umbicalis*. The narrow localization of this bird is very reminiscent of that of a reptile, the Marine Iguana *Amblyrhynchus cristatus* of the Galapagos, also an algal feeder which never leaves the intertidal zone and coastal rocks.

Carnivorous birds

Those birds which live on the secondary production of coastal environments are more numerous than the herbivorous ones. Most are opportunists with a varied diet, but nevertheless all show confirmed dietary preferences which divide them into several feeding groups and thus avoid undue competition. Many feed especially on molluscs, while others catch crustacea or insects. No fish-eating bird is strictly dependent on the intertidal environment, though it is frequented by cormorants, gannets, pelicans, terns, gulls and various herons and egrets, while fishing raptors (sea eagles and ospreys) are by no means absent.

Certain Anatidae are among the mollusc-eating birds. The Eider *Somateria mollissima* and its relatives feed principally on molluscs, which they swallow whole under water. Their gizzards break up the shells, which are thrown out in the form of grit, so that they play a part in a geological process by contributing towards the formation of calcareous sands. Their depredations can be very significant: every day an eider of 2·5kg eats about 475g of the flesh of cockles or 330g of mussels, representing about 250 cockles or 150 young mussels. A colony of 16,000 eiders, such as that on the Waddenzee, annually eats 1,100 million cockles, 220 million mussels, and about thirty million crabs. These figures allow the impact of these birds on the biocenose to be measured, while the productivity necessary to maintain them can be judged.

The scoters *Melanitta* also feed almost exclusively on molluscs – oysters, mussels and various other shellfish – although they supplement them with crabs and even small fish. The steamer-ducks (*Tachyeres*) of Patagonia share this diet and feed in the same way. Their prey consists of various molluscs, especially mussels, chitons, limpets and other gastropods, taken mainly from populations among the seaweeds, for which they have to dive to no great depths. Their diet is completed by smaller proportions of crustacea, echinoderms and even fish. These ducks too are gross feeders, for no less than 450 mussel shells of various sizes have been found in a single stomach. This mollusc-based diet is shared by many of the Charadriiformes, which are specialized to seek them out. The Knot *Calidris canutus* feeds above all on tellins, cockles, winkles and other gastropods, all proportionate to the bird's size. An individual has been seen to take up to 980 of these in twenty minutes. According to analyses undertaken at Mellum in Germany, 77 per cent of stomach contents included molluscs (*Littorina*, *Macoma* and *Hydrobia*) – sometimes forming the whole of the remnants of prey but sometimes far from the largest part (Ehlert). Relatives such as the sandpipers too are great mollusc feeders, and also eat polychaete worms. Oyster-catchers (*Haematopus*), which owe their name to their dietary preference, unlike the foregoing do not swallow shells. An Oystercatcher's bill is transversely compressed but remarkably stiff, so that it can be neatly slid between the valves of a shell, which the bird can thus force open in order to swallow the flesh. An Oystercatcher, weighing an average of 460g, eats more than half its own weight of molluscan flesh per day, made up of 140 two-year-old cockles, 200 one year old, or 2,000 of less than one year. In England (at Morecambe Bay on the Dee Estuary) the intake of an Oystercatcher has been estimated as 500 cockles a day, equivalent to 325g of flesh (Davidson 1968). Other birds with more generalized diets also take molluscs, especially many gulls, some of which open the shells by carrying the molluscs into the air and dropping them on rocks in order to break them open.

Other birds live on crustacea, especially curlews *Numenius* which feed most eagerly on crabs, taking up to fifteen every day. The crabs are swallowed whole and broken up in the stomach of the bird, which then regurgitates the fragments of carapace. However, the diet of curlews is far from constant throughout the year. In Holland they feed on crabs during the summer, while these are at their most abundant and frequent the sandbanks; but from November to March, when the crabs retreat to deeper waters, they take mainly molluscs (cockles and tellins). This change of diet, seen also in many other birds, reflects their adaptation to the amounts and kinds of food available.

Other waders forsake molluscs living in the mud for small animals swimming near the surface, especially crustacea which form another important resource of the intertidal zones. This is true of certain sandpipers, particularly the Redshank *Tringa totanus*, whose activities are thus complementary to those of the foregoing birds. Members of other groups also feed regularly on crustacea, the sole diet of some such as the Grey Gull *Larus modestus* which is confined to the Peruvian coasts bathed by the Humboldt Current. It haunts sandy beaches where it behaves like a wader, with the same quick movements, and also mixes with flocks of sandpipers and plovers. Here it feeds almost entirely on a decapod crustacean of the family Hippidae, *Emerita analoga*, which it extracts from the sand in the zone of breaking waves. In this it shows a fundamental difference from other gull species, which mostly have remarkably wide diets. The Dusky Gull *L. fuliginosus* of the Galapagos shares this diet of crustacea taken on the beaches to some extent, but complements it with other foods and especially rubbish.

Polychaete worms form part of the diets of many waders, and are the principal prey of some: their remains were found in 96·7 per cent of the stomachs of Dunlin *Calidris alpina* analysed at Mellum in Germany. While nearly 40,000 worms were counted, there were only 2,000 molluscs, while crustacea and insects accounted for a minute proportion (Ehlert).

The wide range of available prey has thus allowed a well-diversified avifauna to establish itself in intertidal environments, and especially on mudflats. The enormous biomass formed by the populations of molluscs and crustacea provides enough for the birds which live on them, which are very numerous and like all birds have large appetites. Flocks of waders sometimes take flight in tens or even hundreds of thousands in favourable biotopes. A concentration of 500,000 waders has been reported in Holland on the Waddenzee, and may be greatly exceeded elsewhere. According to estimates made in this area (Verwey), the 40,000 Oystercatchers which concentrate there eat three thousand million cockles a year. The 8,000 Curlews each eat fifty crabs a day for eight months, and 600 molluscs a day for the remaining four months, accounting for a total of 100 million crabs, 300 million tellins and 300 million cockles. The sandpipers take at least thirty-five thousand million molluscs and more than a thousand million worms. The dominant species of waders thus take in a year three thousand one-year-old cockles, one thousand million younger ones, 2·5 thousand million tellins, thirty-five thousand million gastropods, 500 million winkles, several thousand million worms, and about a hundred million crabs. To these amounts must be added those taken by less abundant species, whose impact on the biotype is nevertheless far from negligible. Thus gulls take at least fifty million crabs a year. Similarly high values have been calculated for the Bay of Aiguillon in the Vendée, an area where very many waders – especially Knots, Dunlin and Black-tailed Godwits – winter or stop on

417

passage. The Knot population eats 2·7 million small gastropods (*Hydrobia*) a day, and that of Dunlin more than five million polychaetes.

There is no accurate information on the sizes of the prey populations, so that the effects of these predators cannot be estimated, but the total figures seem to be astronomical. The quantities of food necessary to maintain such populations of waders can only be found in environments which are exceptionally productive, especially at the secondary level. This is true of mud-flats. The birds are strictly dependent on the delicate equilibrium which is necessary at the high level of productivity of this environment, to which they are closely adapted. Thus the impact of birds on this biotope is considerable. It is estimated that in the Waddenzee they take 10 per cent of the cockle populations, whereas men fish only 1 per cent, while their predation on tellins amounts to about 10 per cent of the stock. However, despite their magnitude, these values are perfectly compatible with equilibrium in the populations of prey on which the birds live, though in certain instances the predation is still greater. It has been calculated that the Oystercatchers wintering in England at Burry Inlet annually take between 240 and 860 million cockles. Depending on the breeding success of the molluscs, this represents between 20 and 70 per cent of the population of cockles which have attained their second winter. As a result there is conflict with the cockle fishers, who consider the Oystercatchers as pests (Davidson 1968).

Insects form another important resource of the intertidal zones, which are populated by a specialized fauna. Beetles and flies are easily the dominant groups – the principal prey of Sanderlings *Crocethia alba* whose stomach contents have been analysed at Mellum in Germany were various beetles (*Dyschirius*, *Bledius*, *Phyllotreta* and *Heterocerus*) and flies including *Scatophaga*. However, Sanderlings also take various crustacea. Insects gather in the masses of seawrack and rubbish left by the waves, which attract terrestrial birds – especially passerines narrowly dependent on this environment, which also hunt small crustacea and worms. In Europe Rock Pipits *Anthus spinoletta* (subspecies *petrosus* and *littoralis*) haunt beaches to this end, while in North America certain buntings occupy the same ecological niche, especially near sand dunes. Along the South American coasts bathed by the Humboldt Current, the ovenbird *Cinclodes taczanowskii* is found only in this environment. The Yellow Warbler *Dendroica petechia* of American mangroves may be met on the beaches, picking up the characteristic small animals. These principal predators may be joined by other passerines on their migrations, which are often coastwise, while waders also catch littoral insects.

The waves also carry a considerable amount of other rubbish, including the carcases of many marine animals such as the large mammals, whales, dolphins and seals. This organic waste forms a food source exploited by numerous birds, including all omnivorous gulls and certain raptors, especially kites and vultures. South American Black Vultures are especially common along the beaches while condors – which were believed to be mainly restricted to the high Andes – regularly descend, and are to be found on the coasts of Peru and Chile feeding on rubbish carried by the tides.

The ecological distribution of littoral birds

Thus littoral birds and especially waders – while they make up a single community and are seldom narrow in their diets – divide up into several groups depending on their preferences. Thus, they are distributed across their habitat in accordance with its variety. They are profoundly influenced by the availability of food, and its variation in time and space produced by the play of tides, which control the daily rhythm of activity and the movements of the birds. Thus the whole biology of waders is marked by shifting and always highly adaptive characteristics (Spitz 1964, Recher 1966).

Their ecological distribution is dependent especially on the nature of the substrate. Reefs and rocky shores are the least densely populated habitat, with the least-diversified avifauna, because of the violent action of the waves, the small area uncovered at each tide, and the smallest productivity in foods usable by birds. In the New World the Surfbird *Aphriza virgata* and the Polynesian Tattler *Heteroscelus incanus* are scarcely ever found away from stony beaches, so that their populations are limited by the small extent of this intertidal habitat. Turnstones *Arenaria interpes* are more or less characteristic of rocky shores, where they turn over pebbles by levering at them with their bills, in order to catch the animals hidden beneath. Sanderlings are also to be found in this habitat. Sandy beaches are frequented by larger and more diverse groups of birds, and are the optimal habitat for Whimbrel *Numenius phaeopus*, Ringed Plover *Charadrius hiaticula*, Kentish Plover *C. alexandrinus*, and Sanderling *Crocethia alba*. Still more birds live on the mud, among them ducks (Shelduck, Pintail and Wigeon), Grey Plover *Charadrius squatarola*, Ringed Plover, Curlew *Numenius arquata*, godwits, various sandpipers, and Avocets. Channels with muddy banks through which the tidal currents run out, often through associations of halophytic plants, are the chosen biotope of various sandpipers (*Tringa ochropus*, *T. glareola* and *T. hypoleucos*) and curlews. Obviously, these are only the optimal locations, since most birds have a wider ecological distribution and are also found, though at lower densities, in secondary habitats.

Within the same biotope the birds form a single community, with an 'altitudinal' distribution in relation to the water level. A series of successive bands may be distinguished as one leaves the water-line, in accordance with their degree of drying-out and their physical characteristics, such as humidity of the substrate, and whether the interstitial water does or does not reach to the surface to form a superficial film. Biotic characteristics are equally important, since the composition of the interstitial microfauna varies considerably in relation to these physical factors of the environment, to which the birds respond by way of their dietary preferences.

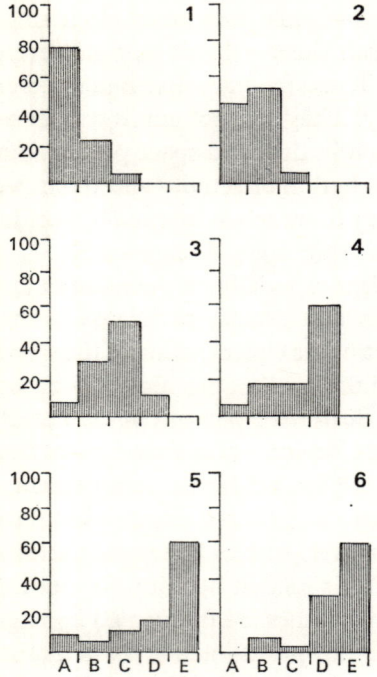

Figure 81. The beach zonation of migratory waders on the American coast (*Ereunetes mauri* in New Jersey, others in California). A. the area not emerging from the water. B. the area covered by a water film. C. the water's edge. D. the landward zone within 30 cm of the edge. E. the landward zone beyond this. Ordinate, percentage representation of the birds in the various zones. 1. *Erolia minutilla*, 2 *Ereunetes mauri*, 3 *Erolia alpina*, 4 *Limmodromus* species, 5 *Limosa fedoa*, 6 *Recurvirostra americana*.

The birds do not by any means confine their search for food to the exposed zones of the shore, and many take their prey from ground covered by a layer of water – whose depth is crucial for purely physical reasons, since wading birds can fish only in water of a depth proportional to the lengths of their legs and bills. This results in another zonation, this time below the water-line, according to the depth to which the substrate is covered. However, food gathering underwater is possible only in calm areas. Along exposed reefs and rocky zones the agitation of the water, together with the always steeper gradients, prevents birds from exploiting the food resources.

Thus a segregation may be demonstrated into zones parallel to the water-line, which are especially obvious at low tide. According to studies in California (Recher), zones which are wet but not covered with a film of water are the domain of the Least Sandpiper *Erolia minutilla*, Semipalmated Plover *Charadrius semipalmatus* and Grey Plover *Pluvialis squatarola*. Zones permanently covered with a water film, and the water's edge itself, are occupied by the Western Sandpiper *Ereunetes mauri* and Dunlin *Erolia alpina*. Beyond the water-line, the very shallow zone is exploited by dowitchers of the genus *Limnodromus*, the Greater Yellowlegs *Tringa melanoleucus* and Willet *Catoptrophorus semipalmatus*. The deeper zone is searched by the Godwit *Limosa fedoa* and American Avocet *Recurvirostra americana* – whose greater size, the fact that some of them take food in mid-water or at the surface rather than at the bottom, and the swimming abilities of phalaropes and avocets, allow them to exploit feeding resources denied to the others. A comparable distribution can be shown in Europe (Spitz). On mudflats exposed by the ebbing tide, Ringed and Grey Plovers haunt the dried mud, Dunlin wet mud at the water's edge, and Redshank and Black-tailed Godwit the shallow water. On sandy beaches, Kentish Plover explore the dry sand, Sanderling wet sand, and Bar-tailed Godwit and Whimbrel the edge of the water.

This distribution in zones roughly parallel with the water's edge can vary greatly in relation to the slope and physical characteristics of the shore – the more gradual the slope, the more distinct are the expanded zones. Zonation is especially obvious on mudflats, where the water moves without marked turbulence. Nevertheless it can be seen even along wave-pounded beaches. This distribution is clearly seen only at low tide, since the birds return towards the land pursued by the rising sea, and the zonation tends to be blurred and to disappear at high tide.

Waders show marked gregariousness outside the breeding season, especially when they are gathering food at low tide. Some, such as Grey Plovers, distribute themselves at random across the habitat, evenly spaced out. Others, such as Redshanks, show higher densities here and there within a continuous distribution, or even form loose assemblages like the Ringed Plover and Dunlin. Others again, such as the Knot *Calidris canutus* and Black-tailed Godwit *Limosa limosa*, carrying this gregarious tendency further, form dense flocks from which individuals are only exceptionally isolated. These variations in grouping, and in the co-ordinated manner in which all the members of a flock move in the same direction, reflect differences in gregariousness and social tendencies.

Zonal distribution parallel to the water's edge is not rigid, and does not prevent the formation of mixed flocks. While Oystercatchers and Common Sandpipers have no constant associates, Green and Wood Sandpipers *Tringa ochropus* and *T. glareola* are almost always together, while Spotted Redshanks *T. erythropus* associate freely with Knots *Calidris canutus*. Dunlin *Erolia alpina* and Ringed Plover *Charadrius hiaticula* form mixed flocks, in which the proportions of each species depends upon the nature of the substrate owing to their slightly different preferred zones, mentioned above. Different species take part in multispecific associations to very different degrees, while there is much variation in relation to environmental conditions, and to the state of the tide at a given place. Certain associations loosen at high tide and later re-form. This is true of Bar-tailed Godwits *Limosa lapponica* with Grey Plovers *Pluvialis squatarola*, and of Knots with Grey Plovers (in which the latter usually lead the flights).

The broad ecological distributions of waders may give the impression of being static, but this is by no means true. The play of tides means that the map of areas open to the birds is endlessly redrawn, with a periodicity of a little more than twelve hours, and waders occupy and leave the various sectors of the intertidal zone in a regular cycle. Furthermore, the rhythm of spring and neap tides leads to considerable changes in the extent of drying and submerged zones during each lunar cycle. Independently of these regular cycles controlled by astronomical phenomena, the strength and direction of the winds can affect the distribution and depth of the water. Wind also has a great direct influence on birds in such exposed zones as mudflats and sandbanks, forcing them to seek refuge in more sheltered areas such as estuaries and salt meadows.

Obviously, the rhythm of the tides determines the waders' food-gathering cycles. At low tide these birds spread out widely across the mud-flats, following the ebb, and this is the principal feeding period of the cycle. As the tide rises they move back with the flood, gradually reassembling and banding together on higher ground, where they form large dense flocks. Only a few waders such as avocets join the ducks and geese in seeking safety offshore, where they swim about in flocks while waiting for the next low water. Sometimes, at exceptionally high tides, waders find no exposed ground and have to take wing, circling over the sea as they wait for the tide to turn. Thus the distribution of birds in the intertidal zone fluctuates end-lessly, and the most notable feature of their biology is certainly this mobility of their populations.

Since the principal food-gathering area lies in the tidal zone, the capacity of a particular region to support a given feeding population (for example of wintering birds) is determined by the greatest area uncovered during neap tides (which may be much less than half the area exposed at extreme low water). It is important to bear this fact in mind when estimating the carrying capacity of a habitat – which depends much more on this fact of topography, controlling the accessibility of food, than on the quantity of food itself, which is always much more than is needed. This control of carrying capacity through the maximum area uncovered at a particular time, also no doubt explains the nature of wader migrations. Since tides are of different amplitudes in spring and autumn, the course of migrations cannot be the same in both seasons, and the greater densities during the spring are related to the greater expanses uncovered at that season. Furthermore, waders always migrate in successive waves, replacing one another at a given place. In the autumn the adults go before the young birds, thus avoiding competition which would be harmful to the latter. Finally, similar species which occupy very close feeding niches show different modes of migrations, so that they do not compete in the staging areas where they would otherwise arrive simultaneously. For example, in North America the Least Sandpiper *Erolia minutilla* avoids domination by the Western Sandpiper *Ereunetes mauri* by migrating earlier.

Thus waders are remarkably diverse in their biological specializations, so that competition between them is reduced. Their variety of diets and feeding sites allow the members of a community to deploy over a wide ecological range, but never involve unduly narrow specializations. Waders are the most opportunistic of birds, adapting to the fluctuations of the intertidal environment on which they are dependent – an ability essential to birds living in such a mobile habitat.

The breeding of shore birds

Some shore birds, such as gulls and terns, nest on the coast or in its immediate neighbourhood, whereas others completely change their habitats and resort to freshwater inland. This is true of the great majority of waders (only avocets nest near brackish water, and only oystercatchers near the coast). Most of their breeding areas are in the arctic tundras. We have seen that such a change of habitat involves switching diets, and most waders which feed on molluscs and polychaetes in their winter quarters are predominantly insectivorous while breeding.

The most marine and the largest birds, such as the gulls, are gregarious, nesting in colonies not far from the shore. They defend their nesting sites from predators, relying on their own aggressiveness and on collective defence. Their eggs, unlike those of pelagic birds, are cryptically coloured. Their young are nidifugous or at least partly so – although they do not have to find their own food, they can very soon travel far from their nests. These are usually hidden in vegetation, or consist of mere depressions in the ground. The chicks too are cryptically coloured, and thus escape more easily from their enemies.

In contrast the Charadriiformes (as defined here) are solitary nesters which carefully hide their isolated nests in vegetation. The nests are usually scanty, made of material collected in the neighbourhood so as to blend with the surroundings. The eggs and young are markedly cryptic in coloration, and the young are nidifugous, allowing them to escape the more easily from their many enemies. The dispersal of the nests also allows the birds to exploit the habitat more effectively. However, avocets and stilts, which nest in less-accessible sites (often on islets), do so in colonies, though very loose ones, while godwits occasionally do the same.

It is noteworthy that many waders begin to nest only when relatively old. Oystercatchers nest only after their fourth year, and no doubt many small species wait until their second year. Non-breeding birds vary very much in behaviour, and a sizeable proportion remains in the winter quarters or in favourable areas along their migration routes. This summer residence means that flocks of waders can be seen throughout the year in some places, and this complicates the study of their migration.

Coastal lagoons

Behind a line of dunes along a very low-lying coast, there are often shallow lagoons, which are strongly saline. Such lagoons are characteristic of the Camargue in southern France, of the Marismas of the Guadalquivir in Spain, and of certain flat islands in the Bahamas. The open water is surrounded by pans of crystallized salt and a zone of strongly salty mud. Thus this environment is characterized by a concentration of salts (especially sodium chloride) which often approaches saturation. As a result it is biologically poor, since only highly specialized organisms can survive in it. Besides a very specialized phytoplankton confined to waters of high salinity, only certain halophile plants such as *Ruppia* and *Chara* grow there. Among the few animals are some molluscs, the branchiopod crustacean *Artemia salina* whose adaptations ensure that it swarms in such waters, and the larvae of certain insects especially ephydrid flies. The bottoms of these stretches of water are liquid mud with a very high proportion of organic matter, which often stinks as it decomposes.

These coastal lagoons have inland equivalents, even at great altitudes, since certain lakes (such as those in the high Andes of Peru and northern Chile at altitudes of more than 4,000m) are highly saline, with similar ecological conditions. Few birds are established in this environment, because of its low productivity and the high degree of specialization needed for success. Besides a few waders, especially avocets and stilts, the few truly characteristic birds are flamingos (*Phoenicopterus* and related genera), which are restricted to this environment. Highly specialized, they have been able to colonize the coasts as well as the high Chilean plateaux, and are found from sea-level to 4,000m wherever ecological conditions are suitable. Furthermore, they are very sensitive to variations in ambient salinity, and do not breed when the salt content falls below a certain threshold. They feed by taking up mud made liquid by the action of their feet, and filtering out the nutrients with a highly specialized buccal apparatus. The mud itself, with a high content of organic matter (up to 91 per cent) derived from bacteria and blue and green algae forms one component of this diet, which is completed by molluscs, crustacea, annelids and the seeds of certain plants, together with small fish. Flamingos nest in the same environment, establishing dense inaccessible colonies in isolated lagoons or on low islands, and are the only nidifugous birds which nest truly colonially. They are the sole users of an environment from which other birds are virtually absent, and in which the cycles of transformation of living matter and of the transfer of energy take place mainly at lower trophic levels.

Mangroves

Mangroves form a very special intertidal environment, confined to low-lying tropical regions. Although they consist of monospecific stands of two main types of mangrove plant, *Avicennia* and *Rhizophora*, which are very distinct, there is a certain similarity at least in physiognomy: these trees with rounded crowns are perched on 'stilts' thrust into the intertidal mud. Mangroves provide a markedly eutrophic environment which benefits from the loads of nutrients brought in by the tides. The aquatic habitat of mangroves is marked by a regular alternation of submergence and drying-out, and is sheltered from waves and strong currents by the network of stilt roots and pneumatophores so that mud can be deposited in thick layers. Aquatic invertebrates, insects and fish are numerous. Mangroves also provide a terrestrial habitat: whereas at root level there is an intertidal swamp, at foliage level there is a true (though monospecific) forest. Thus mangroves shelter an arboreal fauna, living side by side with the aquatic one.

Mangrove birds reflect this double opportunity for colonization very precisely. This habitat is the more attractive to them since their flying ability enables them to get there easily, and since mammalian predators are absent because the loose ground is unsuitable for terrestrial locomotion. The aquatic environment is exploited by many rails, sedentary waders and other small wading birds, which are joined by wintering migrants. Skimmers (*Rhynchops*) find suitable fishing grounds in the still waters, as do many species of terns and kingfishers. The tree environment is occupied by a great number of passerines among which insectivorous species are dominant, and flycatchers are well represented, especially in the mangroves of Asia and the Pacific. In tropical America the Yellow Warbler *Dendroica petechia* is a mangrove bird, whereas its North American races occupy very different habitats. Parrots and swallows complete the fauna, while various other birds come from neighbouring forests.

The double ecological role of the mangroves has been put to good use by many ardeids, for which this is a preferred habitat. These birds are dependent on the aquatic environment since they need a rich diet based on molluscs, crustacea and fish, whereas they build their nests in trees. Egrets, herons ibises and spoonbills are correspondingly numerous and diverse in this habitat. The Scarlet Ibis *Eudocimus ruber* of tropical America is rather characteristic there, while cormorants, frigate birds and pelicans are equally numerous and have established flourishing colonies and roosting places in mangroves. Nevertheless, this special environment includes scarcely any endemic elements, since the species are also found in other very different habitats, either aquatic or forested.

Chapter 22

Continental Waters

FRESHWATER habitats offer abundant resources of all kinds for many birds. Continental waters are very diverse in physical and biological characteristics, and can be divided into many though intergrading categories, whose sole common feature is the presence of water which (with some exceptions – Dussart 1966) is of low salinity. A primary division separates *running* waters (lotic habitats) from *standing* or *still* waters (lentic habitats). Streams always begin with a fast current, whose speed gradually falls, until on reaching the lowest level it is almost imperceptible or even nil. Thus such biotopes pass insensibly into lakes and pools, thence into swamps in which there are more and more solids, and so into zones which are merely wet. Another distinction can be drawn between permanent stretches of water such as lakes, and zones of temporary flooding – particularly important in the tropics.

Apart from faster and slower currents (or none), the principal variables of fresh waters are their temperatures, transparencies, contents of dissolved gases (oxygen and carbon dioxide), and salinity – especially their contents of nitrates and phosphates, which are of such prime importance that they are often the limiting factor. These physical characteristics profoundly affect the aquatic ecosystems, modifying them both qualitatively in their faunistic composition and quantitatively in the biomass they carry.

Running waters are distinguished above all by their speed of flow and the volume of water transported. Torrents – which for obvious reasons are confined to hilly terrain – consist of turbulent, cold and well-oxygenated water. In the lowlands the flow of water, in streams and then rivers, is slower and less confused. These waters vary in temperature and chemical content, but they are characterized especially by their average rates of flow, and by their often very complex flow patterns in which floods alternate with low water levels. Physical conditions can change very markedly from one reach to another – especially in relation to their gradients, which control the speed of flow and correlated characteristics.

Still waters form *lakes* situated in natural depressions closed on all sides and filled with water, or *pools* where running water accumulates when held up by a change in gradient or by a geological barrier. However, these two terms are often used interchangeably. Such stretches of still water are of very varied mixed origins, and biocenoses are markedly influenced by their sizes and depths and the gradients of their edges, as well as by the various physical characteristics of their waters – especially temperature and content of dissolved gases and salts.

In fresh water the primary producers are microscopic algae (especially Chlorophyceae) and diatoms forming the phytoplankton, and higher plants including mosses, and many phanerogams, which may be fixed or floating and are highly adapted to aquatic life. Bacteria and fungi take part in the breakdown of organic substances and in the nitrogen, sulphur and carbon cycles. The primary production may reach very high values, so that a considerable biomass of plant foods is available to the animals. However, part of this cannot be used and forms a real trophic block, since some shore plants including reeds and rushes are of extremely high productivity but are protected by hard indigestible envelopes, while others are more or less poisonous, so that they often serve to support only organisms which provide animals with food. However, their elements are returned to the cycle when they decompose, and the proportion which is directly usable by animals remains large.

Without entering into details of the fauna which occupies continental waters, we may note the diversity of protozoa, molluscs, worms, crustacea and insects. Freshwater biocenoses in Germany include at least 2,180 species of insect, 1,450 worms, 1,320 of protista, 430 of water-mites and other arachnids, 310 of crustacea and 127 of molluscs (Illies). These figures are undoubtedly under-estimates and tropical biocenoses are still richer. In general the dominant groups are oligochaete worms, crustacea and insects – including mayflies, dragonflies, stone-flies, bugs, beetles, caddis-flies and true flies. The dominant vertebrates are of course fish, of which there are at least 1,300 freshwater species across the world, belonging to more than seventy families (Sterba). Amphibia often play an important part, especially in warmer regions where terrapins and water-snakes are also found.

On the whole these animals form very flourishing communities. However, their aquatic environments are very variable in richness, with biomass and especially productivity varying greatly from one to another. Marshes and areas of temporary flooding are particularly productive, whereas fast-running water and deep steep-sided lakes are oligotrophic. There are also often wide periodic fluctuations within a single habitat, according to an annual cycle controlled by the changing insolation and by very complex physical and biological cycles including rises and falls in water-level. These rhythms make it necessary for most of the water birds to undergo regular migrations. Some are among the widest-ranging migrants, whereas others make only local movements whose regularity is no less striking.

Thus the habitats provided by continental waters are very varied, and most are extremely complex because of the interaction of many physical and biological factors.

The birds of continental waters

Continental waters offer important food resources, which the birds have not failed to exploit. As in marine and intertidal environments their mode of locomotion gives them the advantage over mammals, only a few swimming species of which are specialized for the exploitation of these aquatic resources. The difficulties of access met with by most mammalian predators are a further advantage to the birds.

The birds which have colonized these habitats belong to various, mostly non-passerine, systematic groups. Grebes and divers are restricted to fresh water, at least during the nesting season. The Pelecaniformes are represented there by pelicans, cormorants and darters. The Ardeiformes are especially abundant, fresh waters being the habitat preferred by most herons, egrets, bitterns, ibises and spoonbills, by the Boatbill, Shoebill and Hammerhead, and by most storks. Apart from a few marine species, almost all the Anseriformes – ducks, geese and swans – are confined to fresh water, though some freshwater species frequent the seashore while wintering. Among the Ralliformes, the cranes, limpkins, rails, moorhens, coots and finfoots are almost all confined to these habitats, the few exceptions being mainly among the rails. The Charadriiformes are very well represented there by jacanas, painted snipe and most waders – although the latter become marine outside the breeding season – while some Lariformes (gulls, terns, and skuas) frequent fresh waters.

All these birds show more or less obvious adaptations to the aquatic environment. The plumage of swimming and diving birds is remarkably water-resistant. Some birds have webbed feet and others have elongated toes, giving enough supporting area for them to move about on yielding ground or aquatic vegetation. Other aquatic birds, have elongated distal segments in their legs which allow them to wade in deep water without preventing them from perching. Continental aquatic habitats are preferred too by kingfishers, most species of which are confined to the neighbourhood of water, although some in the tropics have become forest birds. These habitats are also occupied by raptors – both diurnal such as fish-eagles, ospreys and buzzards, and nocturnal such as fish-owls and the Short-eared Owl – many of which show specializations for the capture of fish. While passerines are also present, the dippers (Cinclidae) are the only ones which are truly aquatic. Many others live among the vegetation of river banks and lake shores, and get their food either directly or indirectly from the water. This is true of certain wrens, some thrushes (the Bluethroat *Luscinia svecica*), tits (the Penduline Tit *Pemiz pendulinus*) and babblers (the Bearded Tit *Panarus biarmicus*), many warblers (including reed warblers, grass-hopper warblers and grass warblers), the wagtails, some wood-warblers, many icterids and some buntings. Swallows (including martins) come to catch insects over stretches of water, and like starlings they gather to roost in reed-beds on migration.

All these birds exploit the wide range of food resources provided by aquatic environments. Some are herbivorous and others carnivorous – mostly taking worms, molluscs, insects or fish – while some are micro-phagous and exploit the plankton. Although most are polyphagous, merely showing a preference for one category of food, some – especially the strictly fish-eating birds – are very much more specialized feeders. Water birds are usually tied to a particular habitat, which they seek out for the cover it provides and for certain kinds of food. Many aquatic birds such as herons and marsh terns, which nest colonially, are highly gregarious. In contrast, some others such as rails and bitterns are solitary, which is explicable in behavioural rather than ecological terms.

Birds of fast-flowing water

Fast-flowing, often torrential, streams are characterized by water which is clear, relatively cold, highly oxygenated, but poor in organic substances. Their bottoms are usually gravelly or pebbly, and the speed of the current greatly reduces the density of aquatic vegetation. A microfauna highly adapted to this environment hides under the stones, consisting of insect larvae (midges, mayflies and caddis flies), worms and some molluscs, and a richer one in the mosses which cover the stones and banks.

This habitat, offering few types of food, is occupied by very few character-istic birds. Prominent among these in the temperate zone of Eurasia, North America and the Andes are dippers of the family Cinclidae – blackbird-sized passerines, whose short tails give them something of the stance of wrens. No other bird shares their ecological niche. They swim on the surface by paddling with their feet, and wade half-submerged in shallow water in order to feed by probing the bottom; but more particularly they can submerge entirely and swim adeptly by means of their wings. Though they appear to move about by walking on the bottom, all the time they are keeping down and progressing by using their wings. In faster streams they stay submerged by turning their bills into the current and leaning forward, while helping themselves with wings opening and closing at a fairly rapid rate. On the average a dive lasts up to half a minute. Peculiarities which may be considered as adaptations to their mode of life are their thick dense plumage, relatively much larger uropygial glands for waterproofing the plumage than those of other passerines, and opercula protecting their nostrils (Penot, Goodge 1959). Dippers feed especially on insects and their larvae (caddis-flies, stone-flies and mayflies) which they seek on the banks and under torrents, and also on spawn, fry and small fishes taken in the same way. They build their nests beside streams, as large balls of dry grass and moss with side entrances. Their territories extend in strips along the stream.

In the temperate zone, the only other birds which haunt fast streams are wagtails (*Motacilla*), which merely exploit the food resources offered them by the overgrown margins. Ducks in general are absent from streams however, the small group of torrent-ducks *Merganetta* – a highly specialized branch of the dabbling ducks remarkably well adapted to this environment – is confined to torrents at the temperate and subtropical levels of the Andes. These small ducks are of an elongated shape very suitable for swimming in fast waters. They swim easily among the eddies, upstream as well as downstream, and leave the water by jumping on to rocks where they rest leaning on their long stiff tails. They fish either by diving from boulders at regular intervals, or while swimming in the calmer water under the shelter of a rock. Torrent-ducks feed almost exclusively on the larvae of stone-flies *Rheophila*, which they seek out under stones and in cracks with their fine narrow bills, whose remarkable softness no doubt protects them from damage. However, they seem to supplement this diet with molluscs and fish. Thus these birds are notably well adapted to a highly specialized ecological niche (Niethammer, Scott, Johnsgard 1966). It is remarkable that such an adaptation has been achieved to this degree only in South America, where many torrents pour down both slopes of the Andes. Although this environment is available in other countries, the ecological niche is practically empty. Few other ducks frequent torrents, though the Harlequin Duck *Histrionicus histrionicus* does so in the mountainous regions of the arctic, the Black Duck *Anas sparsa* those of Ethiopia, and Salvadori's Duck *Anas waigiuensis* those of New Guinea.

Anhingas sometimes fish in very turbulent waters, while Collared

Pratincoles *Galachrysia nuchalis* of western and central Africa cling in extraordinarily dense flocks to rocks, in the midst of the tumultuous waves of the great rivers where they hunt insects.

Slow-flowing waters

Slow streams present richer faunas and more feeding opportunities than torrents. Conditions vary considerably especially in relation to the speed of the current. Since this is always much slower than in torrents, the bottom is often sandy or muddy. Oxygen content is usually high, which allows flourishing biocenoses to be established – the more so since mineral and organic constituents are constantly renewed in this open system, which receives contributions from the surrounding areas whether terrestrial or aquatic (as from the drainage of swamps). The more placid reaches and areas of still water carry a plankton of significant density, while there is much vegetation underwater and along the banks. Animals too are numerous, the composition and density of the fauna varying greatly in relation to the nature of the bottom and the speed of the current. There are fewer invertebrates attached to the bottom than in fast-flowing streams, but more swimming animals (the nekton) and burrowing ones such as molluscs and the larvae of dragonflies and mayflies. The numerous fishes include species adapted to various ecological niches, depending on the force of the current and the properties of the water.

Streams carry a fairly diverse avifauna. Wherever the riverside vegetation is dense enough to give certain reaches the appearance of still water, moorhens, coots, other rails and grebes as well as some ducks are to be found. Sand and shingle banks are occupied by several small waders – plovers and sandpipers – and more rarely by terns and gulls. This river habitat is equally occupied by herons of various species, some of which build their nests in the large trees of the riverside forests; but these are also found in the lake environment, and are no more typical of streams than the cormorants and fishing raptors which are to be seen on large rivers.

Kingfishers, however, are more or less characteristic of running water, though they may be met near lakes where conditions are suitable especially in the tropics. These birds are remarkably well adapted to catching fish, which form their principal food. Analysis of stomach contents of European Kingfishers *Alcedo atthis* have shown that fish account for 65 per cent of their diet, especially small species and individuals such as minnows, sticklebacks and fry, but they also take aquatic insects such as dytiscus beetles, dragonflies and bugs, amphibian tadpoles, worms and molluscs.

Other aquatic species take similar diets. Kingfishers are only found near sites suitable for their nesting, which involves the digging of deep burrows in the soft soil of steep banks.

Lakes and pools

In contrast to the environments discussed above, these are characterized by still waters. However, this is not to say that these waters are entirely motion-less, since currents of very diverse origins often cause a stirring of the water mass which is of great biological importance.

Lakes are very diverse, in relation especially to their sources, sizes, the shape of their basins, the gradients of their banks, the relationships between their areas and depths, and the physical characteristics of their waters. Lakes may be divided according to their productivity, especially at the primary level. This depends especially on the amounts of mineral nutrients brought in by the inflowing waters, but also on the depth and geographical characteristics of the basin. Deep lakes, with a relatively low total surface/ mass relationship, are oligotrophic; the productivity of plankton is low, while their steep and sharply contoured sides do not allow colonization by aquatic or amphibious vegetation. On the other hand, shallow lakes which receive a sizeable contribution of minerals are eutrophic, since in them the production of plankton is high (although subject to considerable seasonal fluctuations), and their waters are rich in suspended organic matter. Further-more their shores are flat, shallow and swampy, rich in submerged vegeta-tion.[1] Such lakes grade into smaller bodies of water, which have no pelagic zone. Their shallowness makes them very susceptible to external influences, especially to heating by the sun, and results in large fluctuations in their conditions. Some are oligotrophic and others eutrophic, sometimes with very high productivities.

Lakes are much more complex biological systems than running waters. The upper layers of their open zones are populated by a rich plankton of algae and phytoflagellates, on which lives a correspondingly abundant zoo-plankton which allows fishes to flourish. In contrast the inshore waters are rich in aquatic phanerogams – reeds at the edges, and submerged and then floating plants as one passes towards the open water. These support an abundant fauna, the more so since they provide it with shelter and protection

[1] Oligotrophy and eutrophy represent stages in the evolution of lakes. Young ones are oligotrophic and then evolve under the influence of contributions from outside and become richer in nutrients, which allows population growth and consequent changes in the faunal composition. In theory this process accelerates, though the rate at which successive stages are reached varies. The final stage leads to the filling-up and death of the lake.

from waves and currents. This fauna, more diverse than that of the open water, includes animals living both on the plants (the *periphyton*) and on the bottom. Aquatic birds have taken advantage of this many-levelled productivity. Their variety and representation in the fauna depends on the diversity of food and habitats, while their abundance is naturally linked to the productivity of the environment.

Lakes are characterized by a concentric zonation all round their shores, passing from dry land to open water. The distribution of the birds follows this zonation precisely.

The *littoral zone* is essentially characterized by its shallowness, which allows an abundant vegetation to flourish. It follows immediately on the coastal zone, which is occupied by sub-terrestrial plants including *Carex*, grasses and a few trees such as willows. A first belt is formed by reeds (*Phragmites* and *Arundo*) and catstails (*Typha*), and a second one in deeper water by rushes and club-rushes. This is the special domain of the rails, birds of secretive habits whose general morphology seems to be well adapted to moving about in dense vegetation. They very seldom fly and move by swimming and especially on foot (which seems contradictory in view of the wide distribution of the family and some of its member species). Some, the almost wingless rails of the Pacific, have entirely lost the power of flight, while most of them moult all their remiges at once which makes them totally flightless for a time. Their best safeguard is thus their habitat, impenetrable and inaccessible to predators.

This is also the zone of grebes and free-swimming rails such as coots and moorhens, while certain small herons such as bitterns and night herons are dependent on this environment, in which they freely meet the Purple Heron. These birds hide their nests in the midst of dense vegetation, in which they are admirably camouflaged. Many of these are floating nests, merely made fast to aquatic plants, so that they are not flooded when the water-level rises. This is also the preferred habitat of many ducks, especially the Mallard, and of swans. It is occupied by many passerines, especially warblers. Reed warblers and their allies (*Acrocephalus*) are highly adapted to marsh vegetation, through which they move with ease and sling their nests between stems or among the foliage. Reed Buntings and many icterids also frequent this type of habitat.

Beyond these belts, lies a *zone of submerged vegetation*, forming several successive belts which penetrate within the reed association where these are not too dense. These plants include those with floating leaves like the water-lilies, and farther out ones which are entirely submerged. The Potamogeto-naceae (or Naidaceae) are often dominant, notably the pondweeds and

434

Ruppia, and also *Myriophyllum, Elodea,* wild celery *Vallisneria* and *Chara,* forming a carpet over the bottom or a network throughout the body of water.

In this zone are found almost the same birds as in the belts of reeds, many coming to feed either on the plants themselves or on the molluscs, crustacea and insects which hide in the network of vegetation. The zone is especially important for the diving ducks, lovers of *Ruppia* seeds and *Vallismeria* roots. Most of these birds do not stay here permanently and only swimming birds can exploit this zone, where the vegetation is not firm enough to support those which mainly walk about. The sole exception (confined to warm regions), is provided by jacanas. This zone is also frequented by marsh terns (*Chlidonias*), which are adapted to life on freshwater and come here to fish and to capture insects on the wing. Dragonflies, flies, beetles and their larvae are their most important food but they also eat worms, small amphibians and fish, and search the fields and meadows for terrestrial insects (especially grasshoppers). Marsh terns also nest in this environment of lagoons, marshes and shallow pools, where their nests of heaped up plant debris float in the open. They form colonies of tens or sometimes even hundreds of pairs, safe from terrestrial predators. Although they can swim, the young live on the rafts which their parents have maintained and strengthened throughout the incubation period.

Here one passes into a much deeper zone, which may at the scale of a lake be termed pelagic. In the upper layers of these open waters, there is a real plankton in which are found cyanophytes, algae (especially Chlorophycae), protozoa, and larger animals such as rotifers, crustacea (copepods, ostracods and cladocera), and even certain insects or their larvae. Microphagous fishes exploit these planktonic resources, and in turn serve as prey for carnivorous species, while the lower waters are occupied by benthic organisms. The open water is but little frequented by birds. Birds merely fly over it, or seek refuge there in flocks – like the diving ducks, which are poorly adapted to small sheets of water by their flying characteristics, since their small wing-areas make take-off difficult.

Such a zonation is shown by all lakes throughout the world, their zones being occupied by different but ecologically equivalent faunas. Thus on the high-altitude lakes of the Andes the White-tufted Grebe *Podiceps rolland* is characteristic of the zone of submerged vegetation, which it never leaves, in contrast with the Silvery Grebe *P. occipitalis* which frequents sheets of open water. The coots are also characteristic of this zone, especially the Giant Coot *Fulica gigantea* which builds there its huge nest, forming a kind of floating platform, endlessly repaired by the addition of fresh material to compensate for the loss due to rotting underneath. In contrast bulrushes

(*Scirpus totora*), which form an abundant belt around most Andean lakes, are the chosen habitat of glossy ibis, night herons, and many ducks, some of which scarcely ever venture away from this vegetation or from the sheets of open water surrounded by it. Some however, especially the Puna Teal *Anus puna*, remain by preference beyond the plants, towards the open water where the most characteristic species is the Crested Duck *Lophonetta specularoides*.

Thus birds show an ecological specialization for different zones, even within a single systematic group such as the ducks, whose ecological demands vary greatly in such a way as to distribute them among all the habitats. In North America, deep lakes surrounded by forests are the domain of the Bufflehead *Bucephala albeola*, Goldeneye *B. clangula*, Harlequin *Histrionicus histrionicus*, and mergansers. In contrast, shallower eutrophic lakes are occupied by the Canvasback *Aythya valisneria*, Redhead *A. americana* and Ruddy Duck *Oxyura jamaicensis*. This ecological distribution arises in part from dietary preferences, in part from the density of plant cover and the presence of nest sites suitable for the species. Some birds build their nests on solid ground, sometimes quite far from water – up to more than a kilometre for ducks of the genus *Anas* – whereas others like the diving ducks (and divers, which avoid walking because of the difficulties it presents to them) float theirs on the surface of the water. Most, however, build amongst tall reeds and rushes – not at the centre of clumps but rather at the edges, on the border between two plant associations such as rushes and floating plants. Ducks need both cover for their nests and open water for take-off, so that the most favourable environment for them is the most varied, where plant associations meet. Such an environment is also the most productive, as a result of edge-effects between the different habitats, which is an important consideration in managing stretches of water so as to build up maximal stocks of waterfowl.

Diets of lake birds

The distribution of lake birds is also affected by their diets. As in all communities, coexisting species take complementary diets, and so avoid undue competition and are able to exploit all the resources of a single habitat.

Three main sources of food are available to aquatic birds, the first being plants. The very prolific submerged vegetation is taken despite its low calorific content, which makes it necessary for the birds to take large amounts and thus to weigh themselves down. Some of these birds are large like the swans, which browse on the shore plants and grub up submerged plants and roots by plunging their heads and necks (though they supplement this diet with some molluscs and insects, and even fish). Others, especially some species on tropical lakes, tend to be flightless, so that they are not limited by weight. The constancy of their physical environment and of plant productivity allows them to be sedentary and more or less to lose the use of their wings, since they need to fly neither to change habitats during the annual cycle, nor to flee from enemies which are absent from their lake habitats. Among the coots, the European Coot *Fulica atra*, 80 per cent of whose diet consists of the stems of submerged plants (*Chara, Myriophyllum* and *Potamogeton*) gathered by diving, has kept the use of its wings and can fly, although clumsily. In contrast the Giant Coot *F. gigantea*, of lakes at high altitudes in Peru, has lost the power of flight and uses its wings only to help it over the water. It too feeds mainly on submerged plants, especially *Myriophyllum*, which it also uses to build its nest. Seeds are a vegetable food source with a much higher calorific content, and are present in the lake environment although not available to all birds. Many plants have heavy seeds – like those of *Ruppia* which in Europe play an important part in the economies of some stretches of water – which quickly fall to the bottom, and are thus accessible only to diving birds.

The second main resource is provided by insects, which occupy various ecological niches in the lake environment and whose productivity at certain levels may be very high. Many birds also take other small aquatic animals, especially crustacea, worms and molluscs. Rails, for example, catch many insects and their larvae, and also spiders, worms, molluscs, crustacea, and even small fishes and frogs, and indeed young birds. Many waders too are primarily insectivorous, as are the marsh terns *Chlidonias*. Insects are the almost exclusive food of those passerines which occupy the belts of reeds and rushes, especially the reed warblers. These birds, which can cling in acrobatic attitudes to thin vertical stems, feed on insects mainly taken there but also in flight over the water, and complement this diet (shared by certain American icterids) with crustacea, small molluscs and even tadpoles and tiny fish. Aerial insects are preyed upon by swifts and swallows, for which stretches of water are excellent hunting grounds.

The third dietary resource is provided by fish. Many birds take appreciable amounts of them, and some even make them their exclusive food. Fish are caught mainly in the littoral zones where they are common, but are the sole prey available to the birds of the pelagic zone. They are almost the sole food of cormorants, which catch them underwater. The common Cormorant (which is also a marine species) eats 400 to 750g of eels, pike-perch, perch and roach a day, swallowing eels 65cm long. Divers, which can reach a depth

437

of up to 25m and stay submerged for two minutes, are great fish-eaters, scarcely complementing this diet with other animal prey. Among ducks the mergansers (*Mergus*) are specialized fish-eaters, with narrow beaks ending in hooks and in particular with rows of teeth adapted to holding fish. They fish by diving, staying underwater for almost two minutes and reaching a depth of 4m.

Large grebes too feed primarily on fish. The Great Crested Grebe *Podiceps cristatus* eats 200g of fish a day – mainly roach, gudgeon and bleak, but also perch up to 20cm long. They complete this diet with the small prey – insects, crustacea and molluscs – which are the essential food of small species such as the Little Grebe *P. ruficollis*. Grebes show a tendency towards flightlessness, parallel to that of the coots. They spend all their lives on the water and their adaptations to aquatic life are obvious: spindle-shaped bodies, remarkably water-resistant plumage resembling fur, feet carried to the back of the body and thus ensuring efficiency in aquatic propulsion while forfeiting it on land, laterally compressed tarsi which reduce water resistance to forward motion, and webs on the toes. As a result grebes swim and dive efficiently, remaining underwater for up to a minute while travelling more than a hundred metres. They can control their specific gravity by expelling air trapped between their feathers, and also no doubt from their air-sacs, and float entirely submerged beneath the water with only their necks protruding. The European species have retained the power of flight, which enables them to make irregular migrations in order to avoid unfavourable winter conditions, especially the freezing of stretches of water. In contrast certain American grebes have lost the use of their wings, as is especially true of the giant Atitlan Grebe *Podylimbus gigas*, which is incapable of true flight through the reduction of its wings and the muscles which work them (Bowes), as is the Short-winged Grebe of Lake Titicaca *Centropelma micropterum*. This evolution cannot be explained in terms of diet, since fish have a high energy content, but is related to the constancy of tropical lake environments, which do not undergo marked minima of productivity.

The Ardeidae too are great fish eaters. Herons and egrets make fish their essential food, though complementing this diet with molluscs, worms, insects, amphibia, snakes, and even small mammals and birds. The daily intake of a Heron is estimated at more than 300g of fish which is completely digested, bones and all, by its very intense chemical digestion. The fishing grounds of herons are shallow swampy areas. Some like the European common Heron fish by day, others such as the night herons and bitterns are active mainly at night, while small bitterns are by choice crepuscular.

Finally, the raptors of aquatic habitats are also mainly fish-eaters, such as sea-eagles and ospreys which fish especially in the pelagic zones. Harriers (*Circus*) prefer areas overgrown with vegetation, in the midst of which they build their nests on the ground, and where they hunt very varied prey and particularly amphibia.

Thus water birds exploit all ecological niches, especially in respect of diet, while concentrating on the most important levels. A single group may distribute itself through a series of niches, characterized as much by topographic localization and habitat as by the kinds of food taken. This is above all true of the ducks, a family dominant in aquatic habitats in numbers both of species and of individuals. Ducks have long been divided into two groups by their method of grasping food and by their diets, biological characteristics which are correlated with well-defined anatomical and morphological characters. Surface or dabbling ducks prefer to feed near marshy vegetation, on muddy bottoms in shallow water. They gather their food by skimming off the surface of the water and filtering it, through the horny filter plates with which their bills are equipped, or by upending themselves in the well-known way so as to reach the bottom without leaving the surface. These ducks are predominantly vegetarian. The food of the Mallard *Anas platyrhynchos* is made up of 90 per cent plants (seeds, shoots and leaves), the remainder being molluscs, worms, insects and small fish and amphibia. The Teal *A. crecca* feeds mainly (up to 70 per cent) on seeds and various aquatic plants, as well as on various animals such as the larvae of chironomid flies. The Gadwall *A. strepera* prefers leaves and roots. The Shoveller *A. clypeata* gathers its food mainly at the surface of the water, thanks to its flattened spatulate bill and filter of fine plates, taking minute prey which make up the surface plankton. Although the major part of this food is vegetable, almost a third is of animal origin, which is much more than that of any other dabbling duck.

In contrast, diving ducks keep to deeper water, clear and open. They dive completely under the water to seek their food, and can thus reach greater depths and exploit food resources which are inaccessible to dabbling ducks. Their food too is predominantly vegetable, but includes a higher proportion of animal origin, while those species which frequent the seashore on migration are distinctly carnivorous. The Red-crested Pochard *Netta rufina* browses on submerged water plants (water milfoils, pondweeds and *Chara*) which are almost its only foods. In contrast the Tufted Duck *Aythya fuligula* is largely carnivorous, feeding on molluscs, crustacea, insects and occasionally small amphibia and fish. One-quarter of the diets of 'stiff-tails' (*Oxyura*) is of animal origin, including insects and molluscs. Finally, the mergansers are strictly fish-eating ducks.

G

The depth of stretches of water often varies considerably during the year, and since ducks gather their food under precise conditions part of it may be made inaccessible by flooding, under too deep a layer of water. It is most important to take this availability into account, in considering the resources open to many aquatic birds and especially the ducks. It is noteworthy that few lake birds have very specialized diets – apart from some fish-eaters, insectivorous passerines, and herbivores whose foods are not subject to considerable fluctuations. Most will freely take a wide diet, though showing preferences for certain foods. The diet of a given species depends, among other things, on the locality within its range, and on the season. Thus the proportion of animal food increases during the summer, in agreement with the growth of animal biomass and the greater need for proteins during the breeding season. This great flexibility allows the birds to compensate for temporary deficiencies, and to vary their diets in accordance with local conditions and with variations of the available biomass in time and space.

Swamps

Small bodies of water are characterized by considerable development of their littoral zones, at the expense of the middle zone of open water which is greatly reduced or even suppressed. This marshy environment is very important, since it is generally of very high productivity because of its especially favourable physical and biotic factors. The water level there varies with the time of year and with the rhythm of rainfall, some marshes even drying out completely for a time. Marshes are very complex environments, in which the phytoplankton is very abundant, with cyclical fluctuations which are usually marked, while rooted plants are equally numerous and form a dense cover. Insects and other aquatic invertebrates swarm, as do fish.

This eminently eutrophic environment is a preferred habitat for birds, including many which we have mentioned in relation to lakes – notably Ardeidae of all species, and of course dabbling ducks which are favoured by the shallowness of the water, while waders are another dominant group. The essential food of many of these birds is made up of insects – orthoptera, beetles and bugs – although they also feed on molluscs, worms and other aquatic invertebrates. Thus the Redshank *Tringa totanus* has an 88·5 per cent carnivorous diet, including 39·5 per cent of insects, 15·5 per cent of molluscs, 10 per cent of crustacea, 9·5 per cent of oligochaete worms, and 9·5 per cent of arachnids. Almost the same proportions are found in the diets of other waders, especially sandpipers and plovers. Amphibia provide an equally important trophic level in marshes which are exploited by birds, especially by herons which also feed on rodents and insectivores. Raptors are present, notably harriers (*Circus*) which find their optimal habitats in marshes.

Most marsh birds are of rather solitary habits, which is explicable in terms of the nature of the environment – favourable over considerable extents and thus allowing dispersion of populations and considerably reducing the risks of predation. The density of vegetation further favours it by partitioning the mating pairs, which often nest secretively with markedly cryptic eggs (such as those of waders). However, some species such as herons and spoonbills are gregarious, probably because their size protects them from predators.

Marshes grade into less aquatic environments such as peat-bogs and wetlands, from which birds dependent on open water are absent, but where certain waders are to be found which choose such terrain for breeding.

Tropical fresh waters

There is a variety of wet areas in the tropics, from the low margins of rivers to lakes and swamps, which are often very rich ecologically and therefore occupied by great numbers of birds. Most of such environments are subject to great fluctuations – even in forested areas, whose rivers undergo cycles of rising and falling water level as a result of the alternation of wet and dry seasons. Productivity fluctuates in amplitude annually with changing conditions of the environment: sandbanks appear and disappear, and nesting sites and sources of food become inaccessible for part of the year. These changes result in local but regular migrations by cormorants, whistling ducks, pratincoles and many other birds.

Such fluctuations are particularly marked in the wide flood-plains of northern Africa. High water levels in the rivers result in the flooding of large low-lying areas in Senegal, the inner Niger Delta, Tchad and the Sudan, transforming what during the dry season are savannahs sprinkled with lagoons into shallow lakes during the wet. The advance and retreat of the waters cause changes in the physical conditions of the environment and large fluctuations in the kinds and amounts of the standing crop biomass, and accurately control the birds' annual cycles.

Tropical freshwater environments are occupied by avifaunas distinctly different from those of similar temperate area. Waders, so characteristic of the arctic regions, are almost absent from the tropics as breeding birds, apart from a few plovers (though the lapwings are more diversified than elsewhere) and their chosen habitats are often far from stretches of water, on grassy plains of more or less marshy ground. Many more species and individuals of waders winter in the tropics, and one may wonder whether they have not eliminated many autochthonous species during the course of evolution (as palaearctic ducks may have done especially in West Africa). In contrast, many tropical aquatic birds are unknown in temperate regions. The Rallidae – gallinules, moorhens and crakes – are well represented, as are cormorants and pelicans, and especially the Ardeidae, which are much richer in species than elsewhere. Tropical herons, egrets, night-herons and bitterns are very diverse. The Hammerheads are a notable component of the African aquatic communities to which they are confined. There are many species of ibis, several of them sympatric, and of jabirus, wood ibises, openbills and spoonbills. Ducks are represented by specialized forms, occupying ecological niches which are often markedly different from those of temperate dabbling ducks.

Many ecological specializations have arisen in these aquatic communities so as to avoid unduly intense competition, mainly involving the choice of nesting sites. Some of these birds nest in the midst of reeds and other aquatic plants, while others seek out tall trees forming the riverine forests which are such a widespread environment in the tropics.

Other specializations concern diet. The three feeding classes already noted in temperate lakes – insectivores, fish-eaters and herbivores – are found here too though herbivores seem to be less numerous, in accordance with the greater complexity of this environment and the much greater biomass of invertebrates. This has also allowed birds to specialize in the capture of certain types of invertebrate which play only minor parts in temperate communities, as is especially true of mollusc eaters. Openbills feed almost exclusively on molluscs, as does the neotropical raptor the Everglade Kite *Rostrhamus sociabilis*, which winkles snails from their shells with its hooked bill. Amphibians provide another abundant source of food for wading birds, especially storks, jabirus and ibises. Fishes too are very numerous, allowing populations of cormorants, pelicans and herons to establish themselves on watercourses and beside lakes, together with fish-eagles and fishing owls. The abundance of plankton in the marshes and sheltered bays of shallow lakes allows other birds to specialize in microphagy. This is especially true of the skimmers (*Rhynchops*) and of the spoonbills.

Chapter 23

Temperate Forests

THE temperate zones as a whole are characterized by large climatic fluctuations during the annual cycle. Alternation of a hot and a cold season controls the ecosystems' rhythms of activity, causing large fluctuations in the standing crop biomass available to the birds. The most favourable season is relatively short, especially at higher latitudes, and is clearly defined, coinciding with the beginning of the summer. Breeding takes place at a closely controlled time and almost simultaneously for the whole community of birds at a given place, in contrast to the staggered timing to be seen in the tropics.

In contrast to this season, the most favourable for physical conditions of the environment and for available biomass at various trophic levels, is the opposite one. Low temperatures and inadequate sunshine slow down the tempo of life. The biomass available to birds suffers a marked decline, passing through a minimum whose acuteness varies between localities and trophic levels. The birds must therefore either face bad conditions, or leave those parts of their ranges which are uninhabitable for part of the year. It is not surprising that a large fraction of the temperate avifauna is migratory, only fluctuating populations being able to exploit varying resources.

Winter conditions are most severe at high latitudes, making inaccessible a large proportion even of that food which does survive. At lower latitudes in contrast the winter is mild and often wet, and the trophic minimum correspondingly less marked. A greater variety of resident birds is able to get a livelihood, and these places often actually receive migrants from harsher regions. This is notably true of the Mediterranean area, where many of the winter consumers are birds on migration from northern zones (Blondel). Thus in the north temperate regions the climatic gradient from north to south is matched by a gradient in the behaviour of birds, and in their responses to fluctuations in the physical and biotic factors of the environment.

Temperate habitats are distributed especially across southern Europe, northern Europe and a large part of North America, with equivalents in South

America (Argentina and Chile) and Australia. We shall consider here those of the northern hemisphere, which are in every respect the best known, and shall ignore habitats of the Mediterranean type despite their great biological interest. Apart from the aquatic habitats already considered in Chapters 19 and 20, low altitudes in the temperate zone support two contrasting environments, open habitats and forests.

Man has profoundly transformed the original steppes by cultivating them, and has considerably extended the area of open habitats by felling and clearing the forests. This is especially clear in Europe where the climax vegetation was largely forest, and in eastern North America, once covered in huge deciduous forests. Now the forests are replaced by woodlands, while the open plains have been transformed by cultivation. These profound modifications have led to a shift in the balance of avian communities, favouring the species of open habitats at the expense of forest birds. A varied environment, of cultivation mixed with woods and hedgerows, is favourable for many birds, which have therefore proliferated except where kept down by shooting or by modern agricultural techniques. Among the birds most characteristic of such open habitats are game birds such as the partridges of the Old World and the quails and grouse of North America. Bustards, lapwings, larks, wheatears, chats and many pipits are characteristic of this environment in Eurasia, with certain buntings *Emberiza* where it is broken up by hedges and bushes. In America certain icterids, such as the blackbirds *Sturnella*, are equally characteristic of the plains.

While open habitats are occupied by strict insectivores such as the swifts, martins, swallows and flycatchers, and by omnivores such as the thrushes and crows, it is noteworthy that many of the birds such as the finches and larks are predominantly graminivorous, at least for a good part of the year. This is related to the high productivity of the plains in certain types of food, whether natural from wild grasses and other herbs or as a result of cultivation. Raptors too find ideal hunting grounds in these open habitats, where prey are not only more conspicuous than in closed habitats, but can be pursued and captured through a wider range of possible flight patterns. As a result, although eagles, buzzards, falcons, kites and most owls usually nest in the forest or among rocks, their hunting grounds are in open country. Rodents provide them with an important dietary resource, and thus play a considerable part in these ecosystems.

Temperate forests form a series of closed habitats which have retained much more of their original natures than have open ones. Although reduced in area and modified by sylviculture, they have preserved a characteristic avifauna whose specializations are very interesting. It should be noted that the natural climax formation of much of the temperate regions, and especially of Europe, is largely forest. Although only fragments remain of the original forests – which stretched across the continent, almost uninterruptedly, though with physiognomies differing from area to area – waste ground tends to revert to this condition, and it is not surprising that most European birds are forest species though some are secondarily adapted to open environments. Few terrestrial species can do without trees or hedgerows – hardly any except the Skylark, Meadow Pipit, Corn Bunting, bustards and the Lapwing, which are also ground-nesters, and were originally restricted to clearings of edaphic or climatic origin.

The richness of the forest avifauna is also explained by the microclimates of closed habitats. These physical conditions of the forest habitat, which are equally favourable to other animals and notably to potential prey species, have certainly helped to attract birds during their colonization of the temperate countries.

Forests are divided into two well-known types, coniferous and broad-leaved. Conifer forests extend through the northern parts of the north temperate region, forming a huge belt across North America, Scandinavia and northern Asia, as well as in mountainous regions farther south. The sub-arctic climate is cold for a long season, and markedly dry. Many tree species are distributed according to the climate and the kind of terrain. The Norway Spruce *Picea excelsa* is northerly and mountainous, belonging to the subalpine zone. The Fir *Abies pectinata* is mainly a mountain tree of central Europe. The pines *Pinus* include many species found at various levels in the mountains, some in the Mediterranean region and others such as the Scots Pine *P. sylvestris* farther north. The Siberian taiga is made up of various conifers such as *Abies sibirica*, larches *Larix*, pines *Pinus sylvestris* and *P. cembra* and Norway Spruce *Picea excelsa*. These evergreen forests, with scanty undergrowth, provide shelter for animals throughout the year.

Deciduous forests, extending farther south than the conifers, occupy regions of moderate temperatures, less rigorous winters, and plentiful rainfall (between 750 and 1,500mm a year) which is evenly distributed. The undergrowth, itself deciduous, is generally luxurious, consisting of a herbaceous layer and a layer of bushes and shrubby trees important for birds, while the ground itself shelters more flourishing communities than that under conifer forests. The deciduous forests are made up of diverse species, including birches *Betula*, oaks *Quercus*, beeches *Fagus*, hornbeams *Carpinus*, chestnuts *Castanea*, ashes *Fraxinus*, and limes *Tilia*. In North America other species of beeches, oaks and chestnuts, many maples *Acer* and walnuts or 'hickory' *Carya* are the main constituents of the tree flora. These

445

large trees are accompanied by many smaller ones, such as wild apples and pears and mountain ash.

In contrast to the situation in tropical forests, and especially rainforests, the tree species are not much mixed and there is always one extensively dominant species, so that one can speak for example of a beech-wood or an oak-wood, while conifer forests are often monospecific over huge areas. However, even under natural conditions several species do mix, for example firs and beeches at middle altitudes. Man has of course profoundly changed the balance between tree species, and has introduced exotic ones to which the native birds are better or worse adapted.

Every type of forest presents markedly different ecological conditions, as much in the density of cover as in the kinds of vegetable and animal foods available to the birds. Insects are in general specifically tied to their food-plants. Furthermore a single type of forest can show many variations in height, density, floristic composition and physiognomy, in response to innumerable factors – edaphic, climatic or consequent on human interference. Such diversity explains the differences between the avifaunas established in these environments, and their local variations. While faunistic composition is often almost uniform within a single type of forest, the balance of the species is often very diverse.

The primary production of forests is often large. The net plant production in England has been estimated at 3,180g/m² for pine forests and 1,560 for deciduous forest, which may be compared with values of 1,250 for a field of wheat and an average of 1,725 for fields of sugar-cane. However, a considerable part of this production is a trophic dead-end, being stockpiled in the form of wood which few animals can use, and which only becomes available to them after being re-cycled by decomposing organisms. Furthermore the foliage, another large part, can be used by birds only as secondary consumers, acting through insects which are an important element in the forest ecosystem. In contrast, birds can use the primary production represented by fruits and seeds, though this is not as great as in tropical forests.

The birds of temperate forests

These forests are occupied by a very characteristic series of birds which, at least in Europe, form the essence of the temperate terrestrial avifauna. Some belong to families almost all of whose members are birds of closed habitats, like the woodpeckers represented by nine species in Europe and by a greater number in the USA and Canada. This is true also of the creepers (Certhiidae) and to some extent of the tits (Paridae), although some of the latter occupy marshy areas studded with trees. Other forest birds belong to families which are distributed through a greater variety of biotopes, but which have very characteristic forest representatives. Such in Europe are several owls such as the Tawny Owl *Strix aluco*, grouse such as the Capercaillie *Tetrao urogallus* and Hazel Hen *Tetrastes bonasia*, raptors such as the Buzzard *Buteo buteo*, the Goshawk *Accipiter gentilis* and the Sparrow Hawk *A. nisus*, the Cuckoo *Cuculus canorus*, the Wood Pigeon *Columba palumbus* and Stock Dove *C. oenas*, the Woodcock *Scolopax rusticola* and many passerines such as the Turdidae (Robin *Erithacus rubecula*, Blackbird *Turdus merula*, Mistle Thrush *T. viscivorous*, Song Thrush *T. philomelos*, Redstart *Phoenicurus phoenicurus*, and Nightingale *Luscinia mergarhynchos*), the Wren *Troglodytes troglodytes*, the Sylviidae (leaf-warblers *Phylloscopus* and typical warblers *Sylvia*), the Nuthatch *Sitta europaea*, the Goldcrest *Regulus regulus* and Firecrest *R. ignicapilla*, many of the Fringillidae (Greenfinch *Chloris chloris*, Hawfinch *Coccothraustes coccothraustes*, Chaffinch *Fringilla coelebs*, Crossbill *Loxia curvirostra* and Bullfinch *Pyrrhula pyrrhula*), the Jay *Garrulus glandarius* and the Nutcracker *Nucifraga caryocatactes*.

In North America some of the forest birds are related to Old World species, as a result of ancient circumboreal distributions such as those of crossbills and tits. Others are characteristic of America, especially the woodpeckers, the many tyrant-flycatchers such as the Great Crested Flycatcher *Myiarchus crinitus* and Acadian Flycatcher *Empidonax virescens*, the Red-breasted Nuthatch *Sitta canadensis*, Turdidae such as several species of *Hylocichla* (among others the Wood Thrush *H. mustelina*, Swainson's Thrush *H. ustulata* and Veery *H. fuscescens*), the many Parulidae (especially the Black-and-White Warbler *Mniotilta varia*, Black-throated Green Warbler *Dendroica virens*, Blackburnian Warbler *D. fusca*, Pine Warbler *D. pinus* and Oven-Bird *Seiurus aurocapillus*) the Red-eyed Vireo *Vireo olivaceus*, and among the Fringillidae the Evening Grosbeak *Hesperiphona vespertina*.

Some of these birds never leave the dense forest, while others have a wider distribution. This is true for example of the Chaffinch in Europe and of the Red-eyed Vireo in America, whose ecological plasticity explains their success and the fact that both are among the commonest and most ubiquitous birds within their ranges.

Forest birds show a series of adaptations to their environment; among the most conspicuous of these are adaptations to life in the trees, which allow their possessor to perch and to move about with ease among the branches, often taking up the most varied attitudes. Among other things, flight within an environment studded with obstacles demands great manoeuvrability, even if this is acquired at the cost of spring speed. It is attained especially by a large lifting area of the wings, which are proportionally short but very wide, as in hawks which hunt within the forest.

Some at first sight unexpected birds are met with here, such as the Green Sandpiper *Tringa ochropus* which nests in the marshy forests of Scandinavia and northern Russia, often making use of old nests in trees. Some ducks too keep to the banks of rivers and pools within forests. The North American Wood Duck *Aix sponsa* nests up to twenty metres above the ground in hollow trees, from which the young drop to the ground. So do its representatives in Japan and northern China the Mandarin Duck *A. galericulata*, the goldeneyes (*Bucephala clangula* and related species) and the mergansers (*Mergus*).

Ecological specializations

The avian communities established in temperate forest are well diversified, sympatric species occupying very various ecological niches so as to reduce competition and allow the best use of the resources.

CHOICE OF HABITAT

This specialization is shown primarily in choice of habitat. Each type of forest has its own characteristic fauna. Apart from ubiquitous species such as the Chaffinch, most show a preference for a particular kind of forest. This specialization is explicable in terms both of diet, and of factors linked to the habitat itself, such as the density of vegetation and the microclimate.

Thus certain birds are peculiar to conifer forests. In Europe this is true of the Capercaillie, the Crested Tit *Parus cristatus*, the Goldcrest *Regulus regulus* (whereas its close relative the Firecrest *R. ignicapillus*, with which it overlaps widely, is less exclusive to conifers and also frequents broad-leaved forests), the crossbills, and the Siskin *Carduelis spinus*. In contrast, other species are never found there. Densities of birds among conifers are not very high compared with the corresponding figures for deciduous forests. A Czechoslovakian spruce forest supported 130 individuals (per 10 hectares, as in the following counts) representing sixty-three species of which the Coal Tit and Chaffinch were by far the commonest, followed by the Song Thrush, Goldcrest and Chiffchaff (Turcek 1956). A forest of firs at a height of 920m in Switzerland contained only 73·5 pairs representing twenty-seven species, of which the dominants were Chaffinches (thirteen pairs), Coal Tit (six pairs), Song Thrush (six pairs), Blackbird (six pairs), Goldcrest (four pairs), Firecrest (four pairs), Wood Pigeon (four pairs), Crested Tit (three pairs),

Robin (three pairs) and Wren (three pairs) – note the commonness of the ubiquitous species. A spruce forest at 1,850m sheltered seventy-two pairs representing only twenty species. These densities rise where conifers are mixed with broad-leaved trees, so that the avifauna of a beech-fir wood at 850m in Switzerland consisted of 121 pairs belonging to thirty-five species. Not only is the species range more open, but the greater productivity of such an environment allows it to support denser populations (Glutz von Blotzheim 1962). In contrast, plantations are distinctly poorer than natural conifer forests. A Dutch plantation of conifers sheltered only twenty-three pairs belonging to eight species, twelve years after it had been planted, and thirty-five pairs of sixteen species after fifty years (Tinbergen). Thus artificial forests are often no more than traps, for the birds and the whole animal community.

In North America the conifer forests are occupied by crossbills, tits such as the Brown-capped Chickadee *Parus hudsonicus* (though this is not strictly confined to conifers), and many wood-warblers such as the Magnolia Warbler *Dendroica magnolia* (especially in young forests), the Cape May Warbler *D. tigrina* (old open forests of spruce and fir), the Myrtle Warbler *D. coronata* (open forests), Audubon's Warbler *D. audoboni*, the Black-poll Warbler *D. striata* and the Pine Warbler *D. pinus*.

The avifaunas of deciduous forests are of entirely different composition, despite certain similarities due to the presence of more widespread species. Forests of the Pedunculate Oak *Quercus pedunculata*, so widespread in western Europe, with their dense undergrowth, are populated in summer by seven dominant species forming up to 71 per cent of the birds: the Chaffinch, Willow Warbler, Robin, Wren, Blackbird, Blue and Great Tits. Those of the Sessile Oak *Q. sessiflora* are occupied by an avifauna dominated by the Chaffinch, Pied Flycatcher, Wood Warbler, Robin, Willow Warbler, Wren, Tree Pipit, Coal and Great Tits, and Redstart. The dominant birds of oakwoods in winter are unquestionably the tits, which gather in mixed flocks. To these species must be added a series of larger ones, such as the Wood Pigeon, Cuckoo, Crow and several raptors such as the Buzzard, while woodpeckers are common and well represented (Yapp). The birch forest avifauna is dominated by the Chaffinch and Willow Warbler, alongside the less numerous Tree Pipit and Robin. Beechwoods are populated especially by the Chaffinch, Great Tit, Blackbird, Wren, Willow Warbler, Blue Tit and several other warblers. A parallel specialization is again found in America, where the species characteristic of deciduous forests are generally the Red-eyed Vireo, Wood Thrush, Tufted Titmouse *Parus bicolor*, Myrtle Warbler and several woodpeckers.

Bird densities in deciduous forests vary. The species are also often diverse, resulting from the dense undergrowth, allowing colonization by species which are strictly confined to this habitat. In a tall oak forest in Burgundy (consisting of 85 per cent Pedunculate Oaks and 15 per cent of beeches, with shrub layer formed of oak and beech seedlings, brambles and grasses), seventy-five pairs have been counted (per 10 hectares) representing forty-one species. This diversity is explicable by that of the environment, which allows birds of old forests (woodpeckers and Redstarts) to live side by side with those of bushes (typical warblers). The dominant species were the Blue Tit (12·9 pairs), Short-toed Treecreeper (6·2 pairs), Starling (6·1 pairs), Chiffchaff (6·0 pairs), Chaffinch (5·9 pairs), Tree Pipit (4·8 pairs), Great Tit (4·3 pairs) (Ferry & Frochot 1968). Similar figures have been obtained in German oak-hornbeam forests (79 pairs per 10 ha). The density rises towards the edge of the forest because of a greater variety of niches and higher productivity. Thus a Swiss oakwood with hornbeams at 350m was occupied by 138 pairs representing fifty-five species, the dominants being the Chaffinch (11 pairs), Blue Tit (11 pairs), Robin (11 pairs), Blackcap (10 pairs), Garden Warbler (9 pairs), Great Tit (8 pairs), Chiffchaff (7 pairs) and Redstart (6 pairs).

By comparison, man-made parkland provides a much more open environment. Although artificial it resembles true forests, but its very variable avifaunas are usually less diverse (though reaching sometimes about thirty species). In contrast, its densities of individual birds and its productivities are higher. Such a habitat near Lausanne in Switzerland was occupied by 170 pairs representing seventeen species, the dominants being the Blackbird (38-45 pairs), Greenfinch (30-38 pairs), Chaffinch (23 pairs), Serin 12-15 pairs), Great Tit (12-15 pairs), Goldfinch (10 pairs), and Redstart (9 pairs) (Chessex & Ribaut in Glutz von Blotzheim).

On the other hand, forests are much richer in birds than cultivated land. According to a study of a wide area in England (Williamson 1967), cultivated land in the Midlands carries from 11 pairs per 10 hectares of twenty-two species, to 50 pairs of thirty species. The dominants are the Blackbird (3·8 pairs), Dunnock (2·8 pairs), Skylark (2·3 pairs), Robin (2·1 pairs), Chaffinch (2·0 pairs), Whitethroat (1·41 pairs), Yellowhammer (1·3 pairs), Blue Tit (1·3 pairs), Song Thrush (1·3 pairs), Linnet (1·3 pairs) and Wren (1·0 pairs). Those areas with the greatest variety of habitats are more densely occupied than those devoted to a single crop, where the fauna is impoverished.

The distribution of the four species of leaf warblers (*Phylloscopus*) found in Burgundy (Ferry & Frochot 1958) is precisely related to the nature of the habitat: the association of Pedunculate Oak and Hornbeam on the plateau; beechwoods, especially on cool slopes; and the Hairy Oak(*Quercus pubescens*) association on the sunny ones. These distributions are shown by the following indices of abundance:

	Plant association		
Species	*Oak-hornbeam*	*Beech*	*Hairy Oak*
Chiffchaff	2.5	o	o
Willow Warbler	4	o	4
Bonelli's Warbler	o	1.5	9
Wood Warbler	o	5	o

Thus each of the very similar species in this genus establishes itself by choice in a particular habitat, which meets its specialized ecological needs.

The influence of habitat density is shown by the progressive changes in the avifauna, in its composition and the relative abundance of the various species, with the age of the forest areas concerned. Particularly striking researches have been undertaken in Burgundy, in tall cultivated forests with a brushwood understory (Ferry 1960). The association studied is made up of sessile Oak *Quercus sessiflora* forming 86 per cent of the tree layer, and Hornbeam *Carpinus betulus* forming 77 per cent of the shrub layer, while the herb layer includes a variety of small plants. This forest is cropped on a forty-year cycle, all the woody vegetation on a plot being felled to the ground except for a certain number of the tall trees. During the succeeding years a considerable herbaceous vegetation develops, followed by brushwood which redevelops in four to five years and smothers the herb layer. Bushy at first, this becomes shrubby with a high dense foliage, and progressively differentiates in height as it ages. While the tall trees change relatively little during a cycle, the understory passes through several distinct phases.

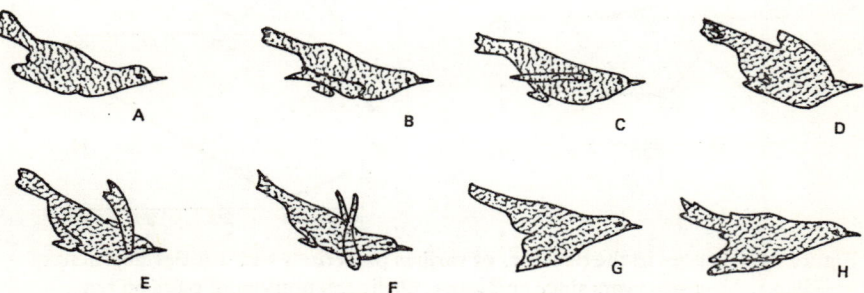

Figure 82. The successive wing positions of a Dipper *Cinclus* when flying underwater (from a film). A–D, phase of passive advance, E–H, phase of active progression.

As the habitat closes in, with the increasing density of vegetation, the avifauna changes. One can divide the birds of this forest association into four groups, according to their responses to the changing environment. The first group includes those birds which settle in the young underbrush, increase in density, and then disappear quickly since they need a low herbaceous or bushy layer with open spaces above it. These are the Linnet, Yellowhammer, Willow Tit, Nightingale, Whitethroat, Garden Warbler, Dunnock and Tree Pipit. The second group consists of species which arrive more or less quickly after the felling, increase in density, and then slowly decrease without disappearing completely (though if a plot were abandoned some of them might finally disappear as the forest attained its tall climax state). Among these are the Bullfinch, Marsh Tit, Blackbird, Chiffchaff, Willow Warbler

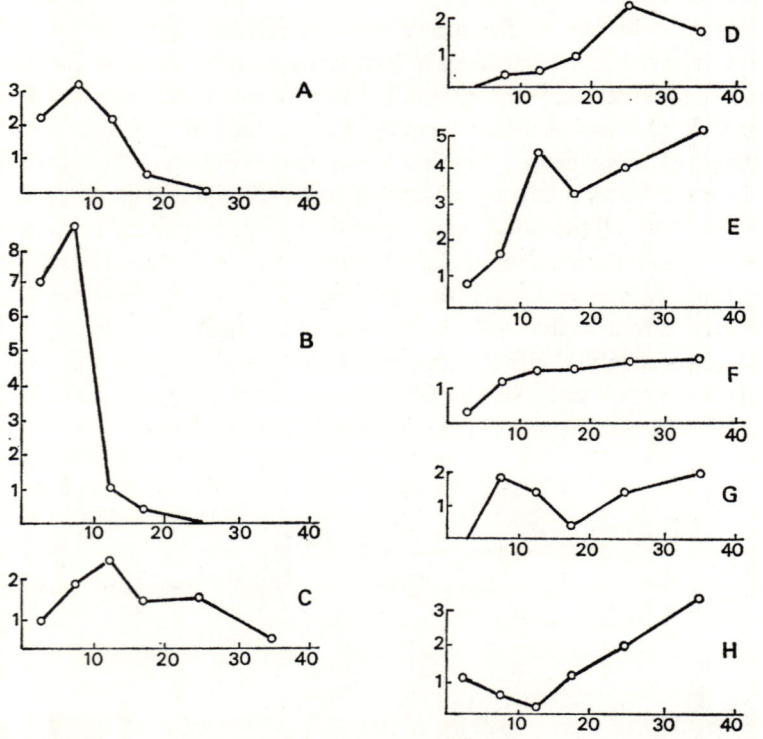

Figure 83. Changes in the densities of various passerines with the development of woodland. Abscissa, years since enclosure. Ordinate, number of pairs on ten hectares.
A. Yellowhammer. B. Whitethroat. C. Bullfinch. D. Marsh Tit. E. Robin. F. Great Tit. G. Song Thrush. H. Chaffinch.

Bonnelli's Warbler, Blackcap and Turtle Dove. The third group includes those birds which settle at a particular, more or less early stage, increase rapidly at first, and then more slowly as the popluations build up to their ceilings. This is true of the Jay, Great Tit, Blue Tit, Nuthatch, Robin and Wood Warbler. Finally, there are the birds whose populations pass through a maximum during the early stages, diminish or even disappear when the undergrowth becomes too dense, and later become more abundant again in mature plots. During a transitory period the undergrowth is too dense for these species which need open space between large trees, a condition fulfilled at the beginning and end of the development of the plots. This is true of the Chaffinch, Song Thrush, Short-toed Treecreeper and Great Spotted Woodpecker.

The proportion of migratory species tends to decrease in favour of sedentary ones, the maturing forest becoming more and more able to support birds throughout the year.

Similar changes have been shown by parallel studies undertaken in different environments, notably in pine forests in Britain (Lack 1937), spruce forests in Finland (Haapanen 1965) and American forests (Johnson & Odum 1956). These examples show how the balance between different forest species is established in relation to the nature and density of the plant cover. It also shows that many of the birds are not tied to forest itself as a canopy of vegetation, but much more to the understory and the bushy layer which accompany it.

VERTICAL STRATIFICATION
The avifauna shows a vertical stratification, distributing itself among the layers of forest vegetation which form a three-dimensional world (Colquhon & Morley 1949). In estimating the height above the ground which a particular bird prefers, it is necessary to consider its behaviour. During the breeding season many birds tend to establish their song posts at a higher level in the vegetation, in order to show themselves as conspicuously as possible, but it is necessary to determine as far as possible the layer in which the species gathers its food, which is ecologically the most important.

Temperate forests in general can be divided into five successive layers.
1   The canopy (above about 12 metres)
2   The tree layer (from 5 to 12m)
3   The shrub layer (from 1 to 5m)
4   The herbaceous layer (below 1m)
5   The ground.

453

The following table (taken, with most of the other data, from Colquhon & Morley) shows the percentage frequencies at which birds were encountered at different levels of a wood near Oxford. Since the observations were made in winter, only resident species are included.

Species	Layers				
	1	2	3	4	5
Wood Pigeon *Columba palumbus*	98	1	1	–	–
Nuthatch *Sitta europaea*	49	49	2	–	–
Blue Tit *Parus caeruleus*	25	44	26	4	1
Long-tailed Tit *Aegithalos caudatus*	26	39	29	4	2
Tree-Creeper *Certhia familiaris*	21	50	18	11	–
Coal Tit *Parus ater*	18	43	31	8	–
Marsh Tit *P. palustris*	4	30	42	22	2
Great Tit *P. major*	6	18	48	25	3
Goldcrest *Regulus regulus*	4	17	55	24	–
Blackbird *Turdus merula*	2	6	23	26	43
Robin *Erithacus rubecula*	–	–	36	40	24
Wren *Troglodytes troglodytes*	–	–	11	78	11

Some, such as the Wood Pigeon and Wren, are found only in one clearly-defined zone; whereas others, such as the Marsh and Great Tits, have wider vertical ranges.

The analysis may be carried further to take account of the population densities of the various species in successive zones, allowing the ecological

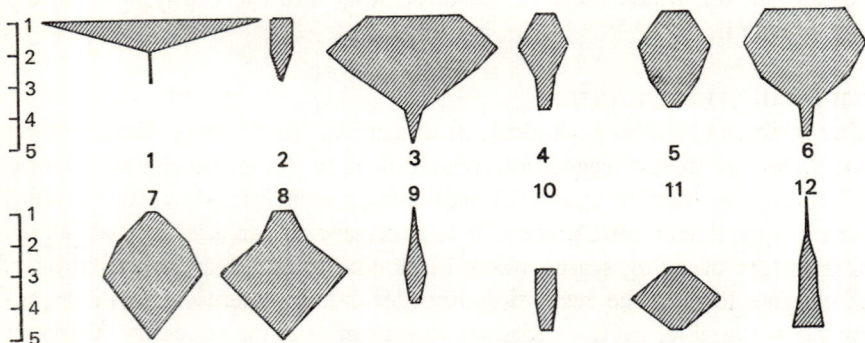

Figure 84. The distribution and abundance of birds among the layers of woodland vegetation in Great Britain. 1. the upper part of the canopy. 2. the tree layer. 3. the shrub layer. 4. the herbaceous layer. 5. the ground.

1	Wood Pigeon	7	Marsh Tit
2	Nuthatch	8	Great Tit
3	Blue Tit	9	Goldcrest
4	Tree Creeper	10	Robin
5	Coal Tit	11	Wren
6	Long-tailed Tit	12	Blackbird

dominance of each to be assessed. The Wood Pigeon is completely dominant in the canopy, followed at some distance by the Blue and Long-tailed Tits. In the branching layer the Blue Tit is dominant, followed by the Long-tailed, Marsh and Coal Tits. The Great Tit is the dominant species of the shrub layer followed by the Blue, Long-tailed and Marsh Tits. In the herbaceous layer the Wren is dominant, followed by the Great Tit and Robin, while the Blackbird is dominant on the ground.

Further species characteristic of the lowest layers include the Nightingale, Chiffchaff, Blackcap and Garden Warbler, while thrushes occupy the shrub and branching layers, and flycatchers the canopy. In contrast the ground is occupied by the Pheasant, which perches among the branches only at night, and by the Woodcock which hunts its food in the wet forests.

Temperate forest birds can thus be broadly divided into a number of communities. The middle and lower strata are by far the most populous, in numbers of individuals as much as of species. Birds of trunks and branches, which have a wide vertical range but keep mainly to the middle part of the trees, deserve a special category.

A similar vertical distribution has been shown for American forests, where it is especially clear among the Parulidae. Certain species such as the Canada Warbler *Wilsonia canadensis*, Oven-Bird *Seiurus aureocapillus*, and Yellow-throat *Geothlypis trichas* occupy the lower layer; others such as the Black and White Warbler *Mniotilta varia* and Magnolia Warbler *Dendroica magnolia* the middle layer; and others again – much less numerous but including the Blackburnian Warbler *Dendroica fusca* and Black-poll Warbler *D. striata* – keep to the crowns of the trees. In an aspen (*Populus tremuloides*) forest in Wyoming, the avifauna showed the following stratification (Salt 1957):

Foraging site		Species	Diet
Air – Soaring	Tree Swallow	*Iridoprocne bicolor*	Insects
Air – Perching	Alder Flycatcher	*Empidonax trailli*	Insects
	Western Wood Peewee	*Contopus sordidulus*	
Foliage	Calliope Hummingbird	*Stellula calliope*	Nectar
	Yellow Warbler	*Dendroica petechia*	Insects
	MacGillivray's Warbler	*Oporornis tolmiei*	
	Black-headed Grosbeak	*Pheucticus melanocephalus*	
	House Wren	*Troglodytes aedon*	
	Warbling Vireo	*Vireo gilvus*	
Trunks	Yellow-bellied Sapsucker	*Sphyrapicus varius*	
Ground	Red-shafted Flicker	*Colaptes afer*	Insects
	Mountain Bluebird	*Sialia currucoides*	
	American Robin	*Turdus migratorius*	
	Lincoln Sparrow	*Melospiza lincolni*	
	Chipping Sparrow	*Spizella passerina*	

H

Foraging site		Species	Diet
Ground	White-crowned Sparrow	*Zonotrichia leucophrys*	Seeds
	Oregon Junco	*Junco oreganus*	
	Song Sparrow	*Melospiza melodia*	
	American Goldfinch	*Spinus tristis*	

Thus temperate forest communities as a whole are dependent on bushes and shrubs, rather than on trees. The birds have a different distribution from those of tropical rainforests, in which a still clearer stratification of the avifauna can be distinguished (see Chapter 26) with the canopy occupied by an abundant and very varied fauna, which has no equivalent in the temperate zone. This lack probably arises from the fact that here the canopy is very poor in foodstuffs, especially fleshy fruits, while insects are also less numerous in this layer. Thus the birds of the temperate forests forage more in the undergrowth than in the high canopy, for essentially dietary reasons. Indeed, some birds of the canopy leave it to feed in other habitats, returning only because of the security or ethological facilities which it offers.

DIETARY SPECIALIZATIONS

Forest birds, which are thus preferentially distributed among the types of forest and layers of vegetation, are further specialized in diet, with each showing, at the least, marked preferences for certain foods. The following distribution, observed in British woodlands, may be considered as typical for other parts of Europe (Yapp 1962).

Diet	Dominant species	Other species
Vegetarian	Yellowhammer	Bullfinch
	Wood Pigeon	Blackcock
		Capercaillie
Omnivorous	Crow	Magpie
	Jackdaw	Great Spotted Woodpecker
	Chaffinch	
	Jay	
	Blue Tit	
	Great Tit	
	Marsh Tit	
	Nuthatch	
	Starling	
	Blackcap	
	Garden Warbler	
	Blackbird	
	Song Thrush	
	Mistle Thrush	
	Pheasant	
Carnivorous	Buzzard	Sparrow Hawk
		Tawny Owl

Diet	Dominant species	Other species
Insectivorous	Tree Pipit	Long-tailed Tit
	Chaffinch (in spring)	Tree Creeper
	Pied Flycatcher	Whitethroat
	Spotted Flycatcher	Green Woodpecker
	Coal-Tit	
	Redstart	
	Chiffchaff	
	Wood Warbler	
	Willow Warbler	
	Wren	
	Cuckoo	

This division of forest birds into dietary groups shows that – in contrast to forest mammals, most of which are vegetarian – few species live directly and solely on the vegetable production. No bird feeds on bark, and practically none on leaves. Only the Capercaillie eats the needles of conifers, and the Blackcock the buds of various shrubby or herbaceous plants.

Forests produce much fruit – acorns, beechmast and the seeds of conifers being especially abundant. Some rather large birds live solely on this highly nutritious vegetable biomass. The Wood Pigeon especially feeds on acorns throughout the year, though it also takes advantage of grain from cultivated land, sometimes causing considerable damage. Some birds of restricted diet are specialized to exploit the seeds of conifers. The Nutcracker *Nucifraga caryocatactes*, of conifer forests in the mountainous regions of Europe and the Siberian taiga, is almost entirely confined to these trees, and especially to the Arolla Pine *Pinus cembra*. This dietary specialization results in wide fluctuations in its populations, any failure in the fructification of the pine being followed by a massive emigration of Nutcrackers. The same is true of the crossbills, whose bills are highly adapted as tools for extracting the seeds from between the scales of pine-cones. Apart from these highly specialized species, the fruits of forest trees may make up a variable part of the diet of many birds. Thus in Czechoslovakia acorns are sought by twenty-nine species, including the Pheasant, Wood Pigeon, Stock Dove, Green Woodpecker, Great Spotted Woodpecker, Middle Spotted Woodpecker, Crow, Rook, Magpie, Jay, various tits, the Nuthatch, Blackbird, Starling, Hawfinch and Chaffinch. The seeds of various spruces (*Picea*) are eaten by thirty-nine species, from the Pheasant to tits, the Chaffinch and other finches. Naturally the method of taking these seeds differs from species to species. Large birds crack the woody shells with their massive bills, whereas small ones such as the tits and nuthatches fix the fruit in a crevice and hammer at it, or strip it with their strong bills in order to break up the shell. However, for most of these birds the fruits of forest trees are only a subsidiary food

457

taken occasionally, though some species take them more regularly. It is noteworthy that this dietary resource, however large, remains almost unexploited by birds.

The forest also contains other fruiting species, especially the bushes which form the undergrowth. Hazels *Corylus avellana*, hawthorns *Crataegus*, cornel-trees *Cornus*, crowberries *Empetrum nigrum*, ivy *Hedera helix*, mistletoe *Viscum album*, wild plums *Prunus*, buckthorn *Rhamnus*, brambles *Rubus*, bilberry *Vaccinium*, and viburnum *Viburnum* are sought out by many forest birds because of the fleshy fruits (berries or drupes) which they produce. Polyphagous birds often have a very varied vegetarian diet, including many of these fruits. The Hawfinch *Coccothraustes coccothraustes* in Czechoslovakia has been found to feed on the fruits of 112 plant species, from pines to oaks and from plums to brambles, its especially stout bill enabling it to tackle fruits with particularly hard woody shells (see Chapter 4). The Great spotted Woodpecker also takes a wide range of vegetable foods, including especially conifer seeds, nuts, acorns, plums and maple seeds. Smaller passerines – especially thrushes, including the Robin and Blackbird (Turcek) – show a very marked preference for blackberries, cornel berries and the berries of mistletoe and ivy. The amounts of these fruits taken by birds may be considerable.

The buds of trees are also sought by birds, especially grouse, various finches, crossbills, waxwings and tits (Turcek). The sap is another important resource for certain birds, especially woodpeckers. The Yellow-bellied Sapsucker *Sphyrapicus varius* of North America pierces tree-trunks to the cambial level and licks up the sap which oozes from the injuries, returning to take the daily yield. This behaviour is also shown less regularly by European species, especially the Great Spotted and Three-toed Woodpeckers. As many as forty-four plant species in Europe are attacked by these birds, especially pines, spruces, willows and maples.

Apart from some of those mentioned above, most birds use fruits as only part of their diets. Many take a mixed diet in which worms, molluscs, and especially insects and their larvae, are combined with fruit. This is true of thrushes including the Blackbird and Robin, and of some warblers such as the Whitethroat and Blackcap. The Robin, which forages on the ground, feeds especially on insects – including beetles, ants, caterpillars and other larvae – but also on spiders, woodlice, earthworms and small molluscs, while the vegetable part of its diet consists of berries – ivy, hawthorn, bilberry and elder. The Blackbird noisily turns over the carpet of dead leaves in searching for insects, molluscs, worms and also fruits and seeds. In the autumn it becomes more distinctly vegetarian, thus adapting to the seasonal abundance of berries and scarcity of insects. It also changes its habitat to some extent, leaving the forest and seeking the neighbourhood of man. Such changes are to be seen in all birds of mixed diet, and allow many of them to be sedentary or only partly migratory. The quantity of berries and other fruits, available to such partly sedentary birds as the Robin and other thrushes, is all the greater since during the winter there is no competition with other birds, the more delicate of which have left on migration.

Many other forest birds exploit the vegetable production through insects, which here as elsewhere form a considerable biomass. Insect larvae are especially appreciated by birds for their lack of a hard chitinous exoskeleton, and are therefore freely fed to the young. They and adult insects are sought at all levels in the forest, sometimes with narrow specializations in hunting technique. Tree-creepers and nuthatches seek their food on the tree-trunks, feeding on insects and spiders hidden in crevices of the bark. Woodpeckers exploit another group of insects, since they pierce the wood in search of wood-boring larvae, notably those of cerambycid beetles. Tits hunt on the branches and feed on eggs and coccoons hidden in the bark, which allows them to survive the winter in forests which are otherwise dormant. Finally others, such as the Chiffchaff, hunt especially in the foliage, seeking insects hidden under the leaves – a diet which allows them to return to their breeding areas earlier in the spring than species which take insects in flight, such as the flycatchers and certain other warblers. Ground-living insects are hunted by all the omnivorous birds already mentioned, and also by the green woodpecker – which is more or less specialized to catch ants on the ground, so that its distribution is to some extent determined by theirs.

Some forest birds are carnivores which feed on vertebrates – some of them even on other birds. Among the principal and commonest of these predators are the Tawny Owl, whose diet includes especially field-mice *Apodemus* and voles *Clethrionomys* and *Microtus*, and the Buzzard which feeds on small rodents and on rats and rabbits. However, it should be noted that this typically forest-nesting raptor usually gets its food outside the woods, and so only partly belongs to this ecosystem. These two species also take some birds, while others make them their habitual prey. The Sparrow Hawk hunts small birds, and the Goshawk larger ones especially Wood Pigeons. The Sparrow-Hawk is thus higher on the food chains than Tawny Owl and Buzzard, whose rodent prey are direct consumers of the vegetable production, since it lives on largely insectivorous birds which thus form an additional link between plants and the carnivore. In addition to true carnivores, temperate forests shelter several plunderers such as

Magpies, Jays and Crows, which visit nests in order to eat the eggs and young.

In general all forest birds are specialized in diet so that each occupies its own ecological niche. Most have very varied diets, and are not dependent on a single type of food, but each shows a preference for a well-defined category which differs from species to species. Despite this, there is some competition in a few cases. The same species of butterfly may be taken by one bird species as an egg hidden in a crevice in the bark, by another when it is a caterpillar on a leaf, by a third when it is a pupa buried in the ground, and by a fourth when it flies as an adult insect. At first sight these four birds, hunting at different levels, are after different prey, whereas in fact all are living on the population of a single insect which changes its habitats with its phases. Such matters must however be considered in relation to the abundance of forest insects. There is often a population surplus of prey, so that even in these cases competition is not as intense as one might think.

The majority of birds are consumers at a high trophic level, and exploit the primary production through intermediaries. Many forest birds serve to disseminate seeds, taking them to a distance and even 'sowing' them as the Nutcracker does by storing them away (see Chapter 5).

Nesting specializations

The nesting of forest birds shows adaptations to the environment and specializations which reduce competition. These are especially important since competition between birds is most severe during the breeding season, when territorial behaviour reaches its maximum intensity and the need for food is greatest.

In general, forest birds have powerful and highly specific songs. Among the leaf-warblers (*Phylloscopus*) for example, songs vary greatly from one species to another. The high intensity of these sound emissions is adapted to the closed environment, in which vocal recognition is more useful than visual behaviour. The latter can take place only at close quarters, and not in the long-range delimitation and defence of territory or the attraction of a mate.

Differentiation between the species is clearly apparent in their choice of nesting sites. Each species choses a well-defined layer, building its nests between certain limits of height above the ground and in a certain micro-habitat. The following table sets out the nesting sites of several European birds (Yapp 1962).

Branches of trees and bushes

Carrion Crow   Mistle Thrush
Jackdaw   Buzzard
Chaffinch   Wood Pigeon
Spotted Flycatcher   Long-tailed Tit
Blackcap   Jay
Blackbird   Magpie
Song Thrush

Holes

Jackdaw   Tree Creeper
Pied Flycatcher   Tawny Owl
Blue Tit   Great Spotted Woodpecker
Great Tit   Green Woodpecker
Marsh Tit
Redstart
Nuthatch
Starling

Near the ground

Yellowhammer
Chiffchaff
Garden Warbler

On the ground

Tree Pipit   Wood Warbler
Robin   Willow Warbler
Coal Tit   Wren
                                Pheasant

This stratification is not of course absolute, since the sites selected by most species vary considerably, especially in their height above the ground. None the less it reduces competition between related species, whose territories may thus be superimposed at different heights as is seen for example among the tits.

In contrast, competition is more intense between hole-nesting species. Woodpeckers hew their own cavities, which are afterwards used by other species which also occupy natural holes, especially in hollow rotting trunks. The number of usable holes is strictly limited, which leads to competition between the possible occupants. Great and Blue Tits often actually wrangle over them, and also enter into competition with Pied Flycatchers (although they have a clear advantage over these migrants and can occupy the nests before the latter arrive). Starlings are much more dangerous competitors because of their size and pugnacity. The increase in Starling populations throughout a large part of their breeding range has clearly taken place at the expense of other species. The number of holes suitable for nesting is evidently a factor limiting populations. For one thing, it has been noted that an artificial increase in their number (for example by providing nesting boxes) has resulted in at least a temporary increase in populations.

The staggering of nesting by different species also somewhat reduces the pressure on the habitat and interspecific competition. Although the birds of temperate countries nest at a particular time determined by the seasonal rhythm, some breed a little earlier than others, so that their maximum pressures on the habitat do not all coincide.

During the breeding season most forest birds are solitary, living in pairs and showing no gregarious behaviour. Only Rooks form colonies, and Starlings very loose associations. However, things are different in winter, when the migrants have departed and the avifauna is thus impoverished. Strictly insectivorous birds cannot endure the winter, whereas those of mixed diet can effect a 'conversion' by increasing their vegetarian tendencies. They are then more markedly gregarious: Wood Pigeons in winter form flocks numbering hundreds and sometimes thousands of individuals. This sociable behaviour is explained by the richness of the environment in foods which they can eat, such as acorns, beechmast and seeds. Passerines also gather in flocks numbering tens of individuals. These are sometimes monospecific but often include several species, as with tits which form very close interspecific social bonds. Only a few birds, such as the woodpeckers, raptors, Robin and Wren, continue to live solitarily or in pairs. Sometimes birds, notably Rooks and Starlings, gather only for the night, when they form real and often very large colonies. This is also true of small passerines – especially treecreepers, which gather in flocks in tree holes to pass the night huddled together. (See Chapter 6.)

Chapter 24

Deserts

BECAUSE of their rigorous environments, deserts are among the most difficult habitats for animals to colonize. Very marked aridity and consequent extreme poverty in food resources, exceedingly severe physical conditions, and an absence of shelter due to the lack of vegetation, ensure that birds like other animals find in them few opportunities favourable to life.

Deserts occupy a large area across the world, especially forming an almost unbroken belt in the subtropical zone of each hemisphere. The Sahara, and the deserts of the Near and Middle Easts, Arabia, central Asia, Mexico and the southwestern USA, are the most extensive in the northern hemisphere, while their equivalents in the south are the deserts of the Kalahari, central Australia and a part of Argentina, Chile and Peru.

The desert environment is characterized especially by its aridity. The humidity is always very low, and rain absent or very slight – in the form of brief and violent storms produced only by the arrival of moist air masses, often from a distance. The annual precipitation frequently amounts to no more than a few centimetres, and the rains are further characterized by their extreme irregularity, with an abnormally rainy year often followed by several very dry ones. Furthermore conditions vary considerably from one locality to another, even over short distances. In a large desert such as the Sahara it does rain every year, though not at the same spot. This great variability in time and space is extremely important for desert animals. Many deserts are also frequently swept by violent winds, which dry the atmosphere still further and have a considerable influence on living conditions. They often carry dust and sand.

Desert temperatures are very variable, and allow this environment to be characterized only in general terms. However, most subtropical deserts undergo high or very high temperatures, due to intense radiation not absorbed by any cloud layer, which of course further increases the aridity. The ground absorbs much heat which it then re-radiates, resulting in strong heating of the lower atmospheric layers. In areas at low altitudes the daily temperature varies between 30° C and 40° C, rising to 55° C, while the ground may reach 80° C. However, this heating involves only the

463

most superficial layers, and is no longer perceptible at heights of about a metre.

The absence of atmospheric humidity results in marked heat loss by the earth during the night, with a rapid and profound fall of night temperatures often to around freezing point or even lower – especially in the central Asian deserts at high latitudes, where on winter nights the thermometer may fall tens of degrees below zero whereas it rises to 35°C during summer days. Thus the desert climate is characterized by large daily fluctuations in temperature, a fact at least as important as the high day values.

Desert climates are affected by seasonal variations especially in temperature, which is higher (both on the average and in the extremes) during the summer and lower in winter. While the humidity varies only slightly, the likelihood of rain usually increases at certain well-defined seasons, depending on the rainfall regime of the particular zone (e.g. the spring rains of the northern Sahara).

Certain deserts have distinctive climates in accordance with their somewhat different and paradoxical origins. This is true of those which border the Pacific in northern Chile, Peru and Lower California, and the Atlantic in south-western Africa (the Namib). In these areas it rains only exceptionally, as a result of an atmospheric situation comparable to that of other subtropical desert regions. Near these particular areas, however, are masses of cold sea water brought from the Antarctic by the Humboldt and Benguela currents. The air takes up humidity and is carried inland, where it forms dense fogs (the *garuas* of the Peruvians). Thus these deserts are characterized by an almost complete absence of rain, low temperatures (15-18°C), very high atmospheric humidities (reaching almost 100 per cent), and an extreme lack of sunshine. Inconspicuous though sometimes abundant condensation in the form of dew compensates for the lack of rain, at any rate for part of the year. It should be noted that the marine influence affects only a layer a few tens of metres thick, above which extend layers of hot dry air isolated by a thermal inversion.

Thus deserts are rather variable though (apart from those last mentioned) they are all characterized especially by aridity. There are all the intermediate stages between 'absolute' deserts lacking any trace of rain, and arid steppes which are semi-desert for at least part of the year. We shall consider here only the truly desert environment. We exclude of course the polar zones, although in some of their essential aspects these belong to the desert environment. Desert soils vary very much: some areas are sandy (*ergs* in Arabic) and some rocky (*hamadas*), while others again are clayey. Some are salty, giving the environment very special characteristics which are reflected in the flora and to some extent in the fauna. The relief is of course also of considerable importance.

The conditions of the desert environment ensure that vegetation is exceedingly impoverished or absent. All the plants are highly adapted, in such a way as to reduce their water-losses and to take advantage of temporarily improved conditions. Trees – always small and often very thorny – are rare, while bushes are more common and form dense associations in the most favourable places. Some of the herbaceous plants are perennial, but most are annual and have only a brief life, surviving as seeds which germinate when conditions become favourable after rain. Thus the desert can quickly flourish, presenting a vegetated landscape which appears at first sight like a broad field of grass and flowers. This rhythm is very obvious even in those deserts with high atmospheric humidity such as those of the Peruvian coast. A highly specialized vegetation, capable of retaining the condensed fogs, develops to form temporary associations (*lomas*) as soon as the fogs thicken.

Desert plants on the one hand show very irregular rhythms from year to year (varying from place to place according to the rains), and on the other very local distributions. Plant associations form at the most favourable points, which are concentrated in certain places and leave other zones completely desert. The resulting discontinuity in time and place has a profound influence on the animals.

In general, conditions of the desert environment are very severe for birds. The most important factor is the lack of water, which poses problems of adaptation either directly or by way of the lack of food. The high temperatures recorded across vast areas provide another limiting factor, while they combine to make the environment particularly hostile. Desert birds have an annual cycle, adapted to local conditions and varying from region to region.

Desert avifaunas

These rigorous conditions impose severe restrictions on the birds capable of colonizing deserts. Birds are relatively well prepared to establish themselves in the desert – certainly better than mammals. Their 'preadaptations' include their high internal temperature, which assists in maintaining thermal equilibrium with the environment; their ability to tolerate a fairly considerable increase in this temperature; and their excretion of nitrogen in a concentrated form, with minimal water loss. Their great mobility, making it

465

possible for them to migrate, to wander, and to seek their daily food and drink, is also important.

It is necessary to distinguish among desert birds, between species characteristic of this habitat and those more widespread ones which can colonize arid zones as a result of their great ecologerca flexibility. Thus among the 300-odd species (including migrants and birds of passage) known from the desert zones of the USSR, only sixty are truly typical of deserts (Roustamov). In the Old World certain chats and larks are strictly confined to deserts, whereas raptors such as the Kestrel and pigeons such as the Rock Dove have much wider ecological ranges, and penetrate the most arid zones only by virtue of their remarkable adaptability. Correspondingly in the New World, forty of the 274 Californian breeding species nest in the desert scrub, but only seventeen of these are truly desert birds (Miller).

In the Old World, the dominant desert birds are the chats, a group of small Turdidae found in open habitats, several species of which are confined to the most arid zones (*Oenanthe deserti, O. moesta, O. isabellina, O. lugens* and *O. leucopyga*). The larks, another group of open country birds, are represented by characteristic desert forms such as sand larks (*Ammomanes deserti, A. cincturus,*) crested larks (*Galerida cristata, G. theklae*) and the Bifasciated Lark (*Alaemon alaudipes*). A raven (*Corvus ruficollis*), some shrikes, seed-eaters, (especially the Desert Sparrow *Passer simplex* and the Trumpeter Finch *Rhodopechys githaginea*), bustards, coursers (*Cursorius cursor*) and the Ostrich complete this desert-living avifauna. The sandgrouse *Pterocles*, all of which are characteristic of steppe habitats, have widely penetrated the true desert. It is noteworthy that all these birds are of terrestrial habits, perfectly capable of moving about and living on the ground.

Some birds, such as larks (*Calandrella cinerea, Ammomanes deserti*) and coursers, are characteristic of deserts with firm, especially stony, soils (*hamadas*). Others, such as the Pin-tailed Sand Grouse *Pterocles alchata*, Desert Warbler *Sylvia nana*, the Desert Sparrow *Passer simplex* and the Bar-tailed Desert Lark *Ammomanes cincturus,* are in contrast confined to those with sandy soils (*erg*). A greater variety of species is met in scrub on exposed slopes and in the beds of wadis, including especially thrushes and warblers such as the Blackstart *Cercomela melanura*, Fulvous Babbler *Turdoides fulva* and Desert Warbler *Sylvia nana*. Rocks and ravines attract still more species since they offer markedly more favourable conditions, being better sheltered from the sun, less dry and usually more densely vegetated. Rock-haunting species such as the Fan-tailed Raven *Corvus rhipidurus*, Streaked Scrub Warbler *Scotocerca inquieta* and various chats such as the Black Wheatear *Oenanthe leucura* have colonized these habitats.

Other more widespread birds have also populated the desert, notably the Rock Dove *Columba livia* and birds of prey such as falcons (e.g. the Lanner *Falco biarmicus*).

The desert faunas of the New World have evolved from very different components. North American deserts are occupied by birds such as quail (e.g. Gambel's Quail *Lophortyx gambeli*), pigeons (the White-winged Dove *Zenaida asiatica*), a fair number of hummingbirds, the Roadrunner *Geococcyx californianus* (a terrestrial cuckoo), a few tyrant flycatchers, wrens (the Cactus Wren *Campylorhynchus brunneicapillus*), mocking-birds (Le Conte's and Bendire's Thrashers *Toxostoma lecontei* and *T. bendirei*), finches (the House Finch *Carpodacus mexicanus*) and buntings (Albert's Towhee *Pipilo alberti*), apart from woodpeckers, nightjars, owls and diurnal raptors. In South America the deserts are occupied by characteristic species, especially several ovenbirds (the Coastal Miner *Geositta peruviana*, etc.) and seed-snipe (e.g. the Least Seedsnipe *Thinocorus rumicivorus*). The Blue-and-white Swallow *Pygochelidon cyanoleuca* nests in burrows dug under stones. Several other species occasionally penetrate into true deserts from less arid zones, especially from the lomas whose birds are distinctly more diverse.

For obvious reasons, aquatic birds are virtually absent from deserts, but in arid steppes subject to irregular but abundant rains they (and especially the ducks) play an important part. Alongside the more recent arrivals are a few relict species from a period when conditions (especially in the Sahara) were more favourable, which have survived near permanent waters.

Certain desert birds of the New and Old Worlds, belonging to clearly distinct lineages, show remarkable convergences. These resemblances involve both their external morphology and general appearance and their biology, notably their behaviour and ecological adaptations.

Desert avifaunas are distinctly less well developed in the New than in the Old World, especially in South America. This difference results from the relatively recent formation of certain American deserts, whereas those of the Old World are much older. The climate of Mongolia has been hot and dry since the end of the Eocene, while parts of the Sahara too have undoubtedly been desert for a very long time. This has given their faunas time to evolve and to form whole communities well adjusted to the environment, in contrast to the situation in America.

Characteristics of desert birds

Desert birds show few morphological adaptations. However, it is noteworthy that their locomotor apparatus is well developed. Forced to live on the ground

467

by the reduction in vegetation, their legs are strong. This is true of the larks and chats among the passerines, and the coursers, stone curlews and sand-grouse among the non-passerines. Other examples are the ground jays *Podoces* – curious corvids of the central Asian deserts which are related to magpies but have become terrestrial – and a terrestrial cuckoo, the Road Runner *Geococcyx californiamus* of south-western North America, which can run much faster than 20 km/hr and thus hunts reptiles on foot. Furthermore most desert birds are strong on the wing. Because their food is so thinly spread they have to gather it over a wide range, which drives them to large-scale nomadism and in particular to covering great distances (mostly every day) in order to reach their watering places.

Desert birds in general are smaller than related forms which live in moister habitats. However, this difference is related not to the aridity so much as to the high temperature. According to a general tendency (Bergman's Rule), warm-blooded vertebrates of the same type are larger in cold than in hot countries, as is confirmed by the example of desert birds. Possibly the poverty of the desert environment also plays a part, by curtailing the period of growth. Thus the Brown-necked Raven *Corvus ruficollis* of the Sahara is smaller than the Raven in Greenland *C. corax principalis*, and even than its other races including that of Mediterranean North Africa (*C. c. tingitanus*). The Shore Lark *Eremophila alpestris*, which is distributed across the major part of the holarctic region and includes a very distinct population *E. a. atlas* in the Moroccan mountains, is represented in the Sahara and in Arabia by a related species, *E. bilopha*, which is distinctly smaller.

Desert birds have very varied diets. Some are vegetarians and take leaves and buds but feed primarily on seeds – which are the sole food of others, such as the sandgrouse and pigeons, and less exclusively of larks and bustards. Many birds, such as the coursers, warblers, chats, shrikes and some larks, are either strictly insectivorous or include a large proportion of insects in their diets. Arthropods are common in deserts and since they contain a good deal of water they enable birds to meet at least a part of their water needs through this diet. Reptiles, which are very diverse in the arid zones, are also hunted even by larks, while the Roadrunner of the American deserts makes them its exclusive food. Mammals are the prey of diurnal and nocturnal raptors, while the fauna also includes omnivores such as the crows. Apart from a few strict vegetarians on one hand and the raptors on the other, desert birds are thus mainly polyphagous.

Densities are low in general and also very uneven, varying from place to place with the richness of the habitats. Some areas, especially sand deserts, are devoid of all life and particularly of birds, whereas stony deserts are generally less impoverished. The much moister beds of wadis and the bottoms of cliffs and of gorges are decidedly richer, as their vegetation shows, and there birds are more numerous and may locally reach quite high densities. This uneven density and the need to congregate in the most favourable place has led many desert birds to become gregarious, and some form large flocks outside the breeding season, and may be distinctly social even within it. Each pair of the Zebra Finch *Taeniopygia castanotis*, in the wetter parts of its Australian range, occupies a tree from which it excludes others even where they nest in loose colonies. In the desert, however, up to nine nests have been noted in a single tree. The White-winged Triller *Lalage sueurii* defends a large territory in coastal areas, but nests colonially in the interior of Australia. A series of social behaviour patterns, including even communal rearing of the young, is found there among birds which are not usually gregarious. This tendency is obviously a response to the rarity of favourable nesting places, which obliges the birds to gather at these points and therefore to be more markedly gregarious than in more favourable surroundings (Immelmann 1963).

The coloration of desert birds

The most striking characteristic of desert birds is their very pale coloration. According to Gloger's Rule discussed in Chapter 4, the birds of dry countries are lighter in colour than their representatives in wet regions. This difference is shown especially in their melanic pigmentation, while in addition patches coloured red or yellow by carotenoids are very poorly developed among desert birds. Though it is by no means universal, a few birds showing quite the opposite tendency, it does affect a large part of the avifauna: eighteen of forty-seven species nesting in the Sahara (Heim de Balsac) and twenty-seven of thirty-six nesting in the Karakoum desert of Turkestan (Dementiev 1958) are light in colour.

The coloration of these birds is characteristically dull and pale, varying from whitish to reddish, ochraceous or grey, while their underparts are often white. The pattern is generally of low contrast, apart from a few patches of reduced area. In the Old World deserts, notably the Sahara, such coloration is shown for example by the Courser *Cursorius cursor*, certain bustards and nocturnal raptors, the sandgrouse *Pterocles senegallus*, *P. coronata* and *P. lichtensteini*, the nightjar *Caprimulgus aegyptius*, the crested larks *Galerida cristata* and *G. theklae*, the sand larks *Ammomanes deserti* and *A. cincturus*, the warbler *Sylvia nana* and the sparrow *Passer simplex*. The same phenomenon is to be seen in South Africa, Australia and the New World deserts, and the universality of the trend shows that it is indeed related to desert conditions. This relationship between loss of pigmentation and environmental conditions is particularly obvious when one compares populations of the same species from humid and from desert areas. Crested larks *Galerida cristata* and *G. theklae* become paler and paler when they live in increasingly drier habitats. The Saharan race of the Stone Curlew, *Burhinus oedicnemus saharae*, is much paler than any other, and the same is true of certain diurnal and nocturnal raptors such as *Athene noctua* and *Bubo bubo*.

This desert type of coloration has been explained in many ways, especially in terms of camouflage and cryptic coloration. Certainly the coloration of most of these birds matches the habitats in which they live, and especially the colour of the substrate. The neutral dilute colours of the birds are close to those of the soil, and prevent their being noticed the more effectively since the bird instinctively reacts to a potential enemy by immobility, flattening itself to the ground so as to cast no shadow. The coloration of a given species evolves, not so much in relation to the humidity of the environment as according to the colour of the substrate, to the point where obvious correlations may be established across the various sectors of its range. This is especially true of the Sand Lark *Ammomanes deserti*: the races inhabiting deserts of black lava in Transjordania (*annae*), near Aden (*saturatus*) and in several mountain ranges of the southern Sahara (*geyri – bensoni* of Hoggar) are all dark in colour while in contrast those in areas with light-coloured soils such as the chalk hills of Arabia (*azizi*), are isabelline or beige (Houf and Salwa). The fragmentation of this species into a multitude of local races can be explained only in terms of the habitats in which they have evolved. It is noteworthy that the Sand Lark lives in very stony areas and among crags, whose colours vary greatly from place to place according to the terrain. It is thus differentiated into a multitude of local races, distributed in a mosaic. In contrast, a sympatric species the Bar-tailed Desert Lark *A. cincturus* shows a preference for sandy deserts. This habitat is much less subject to variation in colour, the sand usually being white more or less tinted with ochre. No doubt in relation to this, the species shows virtually no variation in colour throughout the areabetween Morocco and Arabia. Certain larks of the south-western African deserts (the Namib) have differentiated like the Sand Lark, while in the North American deserts the Shore Lark *Eremophila alpestris* is also differentiated into a series of local races, from the very pale form *utahensis* of Utah and Idaho, living on very light-coloured soils, to the reddish form *rubea* of the red-soiled areas of the Sacramento Valley, California.

Not only do these colours enable the birds to merge with their backgrounds, but the birds even behave as though they 'know' exactly how they are camouflaged. In a region where the soils make up a mosaic of very different colours, birds when alarmed often make for that part of the terrain which best matches their plumage. Thus there is, in some cases at least, a deliberate search for the most effective camouflage, through the closest possible match between plumage and background.

Thus the evolution of desert types of plumage may be explained in terms of the predatory action of carnivores, and especially of raptors. The latter, though only moderately diverse in the desert environment, nevertheless include a few eagles, buzzards and falcons (as well as kites, harriers and snake eagles), some of them present only during migration. Terrestrial carnivores are represented by a few small felines and by canids – especially foxes, which are abundant in the Sahara. Since potential prey cannot hide in dense vegetation, their only means of escape are to flee rapidly or to blend with their surroundings, so that cryptic coloration and reflex immobility are their best defence.

In accordance with their need for camouflage, most of these birds show no contrasting coloration, apart from a few reduced patches on the head or underparts, and some are dull and inconspicuous even in flight. However, some may need colours to function as sign-stimuli during certain types of behaviour. In that case these coloured and strongly contrasting patches are usually concentrated near the wings, as for example in the wing patterns of nightjars and bustards. The Bifasciated Lark *Alaemon alaudipes* has wings marked with black and white, which make it very conspicuous in flight and enhance its aerial nuptial displays which embody swoops and tumbles; but at rest the folded wings show the same dull colouring as the body, making the bird's camouflage very effective.

Taking all this into consideration, it is impossible to deny some adaptive value to cryptic coloration among desert birds. However, explanations relying wholly on these factors have been criticized on various valid grounds. One of these is that while the blending coloration seems adequate to our eyes, it may not be so to other animals whose vision differs from our own. This applies to mammalian predators, which search for prey largely by their sense of smell; but presumably not to birds of prey, whose colour vision is known to be very like our own.

The other objections are more serious. The loss of plumage pigment affects a high proportion of species to which it is not obviously advantageous. Desert raptors are themselves pale, though they have no natural enemies. They cannot benefit by passing unnoticed, since they hunt on the move and cryptic coloration is effective only when a bird is still. Furthermore this pallor effects even the nocturnal birds, especially the owls – the Saharan Eagle Owl

Bubo bubo desertorum has been described as 'the desert-coloured owl chasing a desert-coloured bat over desert-coloured soil under a jaundiced moon' (Buxton in Meinertzhagen 1954). It therefore seems that, since there is no direct connection between mode of life and coloration, the origin of the latter cannot be explained solely in terms of its protective value. It is found in many more birds than the few to which it is really useful, and so must have a more general cause, valid for birds which are potential prey and also for their avian or mammalian predators, and for birds living on the ground and also for rodents burrowing under it.

Some authors have therefore believed that the loss of pigment was a direct response of the animal to the physical conditions of the desert environment (Dementiev 1958). Since these conditions are exceptionally severe, it is in the interests of the animal to conserve its energy resources as carefully as possible. Desert animals in general are characterized by reduced vital activity and less intense metabolism. It is postulated that they economize in the taxing synthesis of pigments, which can only be indulged in by species living in richer environments where this 'luxury' is not harmful to the balance of energy. Furthermore, the pale colours of desert birds are produced by pigments which are only slightly oxidized, their synthesis having stopped short at the first stages in the transformation of the propigment tyrosin into the darker fully oxidized melanins. Thus the pallor may be the visible sign of reduced vital activity.

This effect, produced initially by the metabolic needs of desert birds, would be reinforced by secondary factors. Such lack of pigmentation is not by any means disadvantageous to birds in this environment. Though it deprives them of the variety of colours and patterns shown by their representatives in other habitats, and thus of visual signals for specific recognition, this is not important in any avifauna so poor in species.

To certain birds the loss of pigment is positively advantageous, so that although cryptic coloration is not the primary cause it has accentuated the trend by visual selection acting on some species of potential prey. The remarkable adaptation in colour of certain birds to their substrates could not otherwise be explained. The tendency is also beneficial in combating rising internal temperatures produced by solar irradiation. Pale plumage is a defence against undue heating, and helps to ensure the thermoregulation necessary to the physiological functioning and well-being of the bird. Incidentally, it has been noted that desert birds exposed to high temperatures have dense plumage sleeked against their bodies, whereas their relatives in cold countries have lax plumage. This is particularly obvious among the sandgrouse, coursers and certain larks. Plumage of this structure allows greater heat loss than an open lining, in which the entrapped air forms an efficient insulating layer.

Although much of the foregoing is mere hypothesis, such an explanation – based on the response of the bird's general metabolism to the hostile factors of the desert environment – does agree with the frequency of pigment loss in the animals of arid zones, whatever their modes of life. It does not exclude visual selection acting on some of them as a result of differential predation, leading to blending and cryptic coloration which certainly benefits those species which are the prey of visually-hunting carnivores.

Non-cryptic birds

Despite the above, not all desert birds show the pale coloration which may be considered as one of their characteristics, since some sport dark plumage or show strongly contrasting patterns. The proportions, given above, of cryptically coloured birds in the avifaunas of the Sahara and of the deserts of central Asia, show that a sizeable fraction does not show this characteristic, including certain raptors, the crows, and especially a large number of chats of the genus *Oenanthe*. In about ten species of the Old World deserts (especially *O. moesta*, *O. lugens*, *O. monacha*, *O. pleschanka* and *O. leucopyga*) the males at least show a strongly contrasting plumage in black and white. This conspicuousness is the more notable since related species living in the same environment provide some of the best examples of cryptic coloration. Though this type of coloration cannot be explained on ecological grounds, it is noteworthy that these conspicuous species generally do not occupy large open areas, but keep for preference to rocky biotopes. Taking advantage of this richer environment, they can dispose of greater energetic resources and avoid undue metabolic economies. Further, they can shelter under the rocks or in the denser vegetation offered by these habitats, where contrasting coloration may even be useful in breaking up their silhouettes and making them disappear in the play of light and shade. However, dark plumage absorbs incident radiation and causes an increase in internal temperature, against which the birds have to react by faster heat loss – an adverse feature because of the shortage of water.

Because of this coloration these birds are much more conspicuous to predators, and it is a remarkable fact that most of them are unpalatable. Thus in South West Africa two bustards live in the arid steppes (Meinertzhagen 1954). Ruppell's Korhaan *Eupodotis rupellii* is cryptic in coloration, whereas the Black Korhaan *Afrotis afra* is ornamented with strongly contrasting black-and-white plumage, which makes it very conspicuous. When alarmed, the first crouches or flees furtively, while in contrast the second makes its presence known by calls and by spreading its wings to attract attention. Now the flesh of the first is palatable (to man and probably to its natural enemies) whereas that of the second is quite loathsome. This statement is confirmed by studies on the palatability of many birds of contrasting coloration, according to the taste of man and of many carnivores, mammals and insects (Cott 1946). Birds with bright contrasting coloration which makes them clearly visible in their habitats are disagreeable to the taste of carnivores. This is notably true of the chats to which we have referred, and *Oenanthe lugens* and *O. leucopyga* actually proved to be the least palatable of all the birds tested. Thus it is clearly advantageous to these birds to show aposematic coloration, announcing their identity to possible enemies. The predators are thus put on their guard against the bad quality of prey of which they have had previous experience. Such coloration, in conjunction with behaviour which makes it conspicuous, forms an effective means of defence and should avoid attack by predators.

Thus the two tendencies observed among desert birds, far from being incompatible, complement one another remarkably. Like many other desert animals, the birds have evolved along two paths. This double tendency is not confined to desert animals, for it is expressed in different ways in other environments. Though the problem is the same everywhere, it is much more acute in a very poor environment, where the absence of plant cover obliges birds to rely on their coloration for concealment.

Desert birds and high temperatures

Almost all deserts are characterized by high temperatures, at least for a large part of the day, so that desert birds have to contend with this adverse factor, whose effects are aggravated by the lack of water. However, desert birds do not seem to differ notably from others in physiological peculiarities. A higher internal temperature is generally favourable to homoiothermic animals in hot surroundings. From this point of view birds are at an advantage over mammals since their temperatures are higher. However, the temperatures of those desert birds which have been studied are much the same as those in colder places. The same is true of their lethal temperatures: to survive, these birds have to keep their temperatures below the same thresholds as others.

It has been noted that in order to increase conduction birds tend to sleek their plumage – we have already seen that in some desert species it is characteristically very dense – and spread their wings so as to expose their thinly-feathered flanks, allowing them to act as radiating surfaces. The Ostrich behaves in the opposite way, since as the ambient temperature increases it fluffs out its plumage, increasing its thickness from 3 to 10cm. This behaviour, usually a defence against cooling, is paradoxical but explicable in the present case. The Ostrich, which because of its size can shelter from the direct sun only with difficulty, needs to increase the effectiveness of its insulation against solar heating. Sleeking the plumage and exposing poorly protected patches is only effective if the bird can get into the shade. Under these circumstances it increases the loss of body-heat, whereas the Ostrich acts to reduce the absorption of heat from its surroundings (Crawford & Schmidt-Nielsen).

Heat is mainly lost through evaporation within the respiratory apparatus. In general the loss of water through respiration increases with the ambient temperature by an average of four times between 34°C and 40°C. Beyond 40°C, most birds can no longer maintain thermoregulation, so that their internal temperatures tend to increase, but there are notable exceptions to this rule, and the Ostrich especially maintains its internal temperature at 39·3°C even when the ambient temperature reaches 51°C. In order to increase the loss of heat birds begin to pant by rapidly moving their throats, which can be seen to vibrate, and markedly increase their respiratory rates which must also speed up the loss.

In general the regulatory mechanisms of desert birds are the same as those used in other environments. Though in a few they do seem to be more effective, they can scarcely be considered as specific responses to the desert environment. It is noteworthy that the most effective heat loss involves the evaporation of a large amount of water. It is obvious that this process presents a very serious problem to birds, when aridity is the most critical factor of their desert surroundings. Finally, we may recall that the only bird known to be able to enter a seasonal state of torpor is a nightjar confined to the deserts of western America (See Chapter 6).

Desert birds and water

No bird can live without liquid, since thermoregulation involves the continual evaporation of water; and the evacuation of waste substances also makes appreciable demands. Although little information is yet available on the water metabolism of desert birds, it suggests that the amounts necessary for the maintenance of life are quite large, no bird being able to subsist solely on the water derived from the metabolic oxidation of its food.

However, in some ways birds have the advantage of mammals. They effectively excrete the waste products of proteins not as urea but as uric acid. As far as nitrogenous waste is concerned, birds' kidneys are more efficient than those of mammals: their secretions of uric acid are 3,000 times as concentrated as in the blood, whereas in the rat kangaroos *Dipodomys*, though this is a desert-adapted mammal, the concentration of urea in the urine is only twenty to thirty times higher than that in the blood (Smith). However, birds' kidneys are of a less highly developed and less effective structure than those of mammals, since the loop of Henlé (where so marked a concentration of urine is produced in mammals) is not developed. This hampers birds in the excretion of electrolytes, especially sodium salts, but the concentration of urine which cannot take place in the kidneys is compensated by a very effective reabsorption of water in the cloaca.

A much greater quantity of water is lost elsewhere, by evaporation, during defence against a rise in internal temperature. Many desert birds suffer water losses comparable to those of birds in humid environments but a few, such as certain graminivorous passerines and parrots and particularly the Budgerigar, appear to have physiological arrangements which allow them to thermoregulate while reducing their water losses to a minimum. The unavoidable losses, resulting from the absolute necessity of maintaining thermoregulation, are increased by the high temperatures of most deserts. The average losses of various small passerines are from 2·5 to 5 times greater at 40 to 44°C than at 34°C (thermal neutrality). In Abert's Towhee *Pipilo aberti* of North American deserts, five hours at 40°C results in a water loss equal to 10 per cent of body weight, while the loss reaches 30 per cent in the Zebra Finch *Taeniopygia guttata* under closely comparable conditions. Their consumption of water consequently increases in the same proportions. Thus the amount of water drunk daily by the Mourning Dove *Zenaidura macroura* rises from 6·5 per cent of body weight at 23°C to 24 per cent (four times as much) at 39°C.

In order to compensate for the water losses inherent in its normal physiological functioning, a bird thus needs regular and copious replenishment. The water resulting from the oxidation of metabolites represents only a trifling quantity when one considers the whole water balance of the organism. The bird must itself find the water it needs, in its drink or its food. Water sources are very widely spaced in the desert, and birds must be capable either of going without drink for extended periods, or of covering long distances to watering places.

It is remarkable that many desert birds can survive the great loss of weight associated with dehydration. This loss, under 30 per cent in the House Finch *Carpodacus mexicanus* and crossbill, may reach 37 per cent in the Mourning

Dove *Zenaidura macroura*, 40 per cent in the White-winged Dove *Z. asiatica*, and even 50 per cent in the California Valley Quail *Lophortyx californicus*. Whenever water is available the birds drink copiously and very rapidly compensate their losses. Mourning and White-winged Doves deprived of drink, and having lost up to 20 per cent of their weight, can drink in five to ten minutes a quantity of water amounting to 18 per cent of their own weight, so that rehydration is rapid. Thus, in a few birds at least, deprivation of water and the consequent dehydration seem not to be prejudicial to general well-being. It may be that certain desert birds have greater resistance to dehydration than those of other habitats.

The quantity of water necessary to a species varies greatly, depending of course on the size of the bird. Expressed as a percentage of the weight it varies inversely with weight, and thus in parallel with water losses. However, a few birds such as the Budgerigar have smaller needs than theory would predict.

Some desert species, especially strictly graminivorous birds whose diets include relatively little water, need water often and drink regularly. These include the Trumpeter Finch *Rhodopechys githagineus* in the Sahara (although this is a true desert species), the Striped Bunting *Fringilla striolata*, the Rock Dove *Columba livia*, and various sand-grouse in the Old World and Abert's Towhee and the Mourning Dove in North America. These birds, much commoner near sources of water than anywhere else, resort to them regularly (many no doubt daily) thus making true daily migrations. Sand-grouse, birds of strong and rapid flight, travel up to 50km to reach the shores of lakes, where they gather in large flocks to drink and bathe. Thus desert birds with large water requirements are forced either to keep to the immediate neighbourhood of water sources, or to be able to travel rapidly so that they can cover large distances to reach water. Those unable to do this have not been able to colonize the desert. Thus the California Valley Quail, though its resistance to dehydration is considerable, has not been able truly to colonize the desert because of its weak powers of movement.

The young of these species also need water, which is no doubt provided by adult pigeons in the form of 'pigeon's milk'. As for the sand-grouse, it has now been proved that the parents carry water to their young in the close and dense plumage of the breast, an observation made long since by the Saharians and already reported seventy years ago (Meade-Waldo). As has been established for *Pterocles namaqua* of South Africa and for *P. alchata* and *P. senegallus* of Iraq and Morocco, the male goes to drink and bathe abundantly in a pond, and then returns to the nest. It presents its belly to the young, stretching on tiptoe. They take the feathers between their bills and actually wring them out so as to extract the water. The feathers of sand-grouse retain water much better than those of other birds because of their special structure with barbules which are devoid of hamuli and form spiral flattened ribbons, ending in very fine components. From 25 to 40cm³ of water may be trapped in the belly-plumage of the bird. Even making allowances for the evaporation resulting from flight through dry air over a distance of 30km, from 10 to 25cm³ may thus be carried to the young, and no doubt enables them to maintain their water balance (Cade & Maclean 1967).

The fact that all these birds have to go daily to particular places where they gather in large flocks makes them especially vulnerable to predators, which as many observations show hunt them selectively in these places. Pigeons and sandgrouse show behavioural adaptations which serve to reduce the effects of predation, not flying directly to the water but perching in safety to examine the approaches before drinking. Also they can drink by sucking up liquid, which allows them to take considerable amounts in a very short time, and thus exposes them more briefly to the attacks of birds of prey (Cade). 1965). Similar behaviour is shown by certain weaver-finches of arid areas, such as the Australian Zebra Finch (Immelmann).

In contrast other desert birds seem able to do without water, and drink seldom and irregularly if at all. Among these are the Brown-necked Raven *Corvus ruficollis*, the Desert Warbler *Sylvia nana*, various chats (*Oenanthe*) and bustards, the Cream-coloured Courser *Cursorius cursor*, the European and Blue-cheeked Bee-eaters *Merops apiaster* and *M. superciliosus*, and most larks of the genera *Galerida*, *Calandrella*, *Ammomanes*, *Eremophila* and *Alaemon*, as well as the Barbary Partridge *Alectoris barbara* and the Ostrich. In the New World certain quail such as Gambell's Quail, and finches such as the House Finch, seem never to drink, and can certainly survive indefinitely in captivity without water. It is noteworthy that these birds are insectivores and omnivores which feed on buds, grass and fruit among other things, so that their normal diet contains a much higher proportion of preformed water, sometimes up to 90 per cent. The Thick-billed Lark *Rhamphocorys clotbey*, with a huge bill, feeds partly on lizards which it cuts up into fragments, while raptors similarly feed on water-filled prey. In some deserts succulent plants offer both food and water to the animals which feed on them. Cactus flesh is appreciated by many birds such as woodpeckers, while the hummingbirds of American deserts lick up sugary exudations and take nectar from flowers. In the same way the fruits of certain plants are sought by ostriches and bustards. Pallas' Sand-grouse *Syrrhaptes paradoxus* of central Asian deserts feeds largely on succulent Chenopodiaceae (*Agriophyllum*). Gambell's Quail feeds partly on the desert succulents of the USA and

Mexico. The fruit of saguaro is an important dietary resource for the White-winged Dove. It must also be remembered that concealed precipitation is abundant in most deserts. A considerable amount of water condenses as dew which wets the vegetation and the ground, from which it is drunk by birds.

The free water in deserts is often brackish or salty, but the birds which drink it seem to benefit. Others live on succulent plants whose tissues are rich in salts. In Australia certain parrots (*Neophema*) and the Zebra Finch *Taeniopygia castanotis* can gain weight while drinking seawater. Although the nasal glands of seabirds would be most useful to desert birds, they seem to be present only in the Sand Partridge *Ammoperdix heyi*, the Ostrich, and a few birds of prey. The secretions of the Ostrich's nasal glands have been shown to be especially rich in sodium, potassium and calcium. Other birds seem able to make use of saline water and eliminate the excess salt by a specialized renal mechanism, their urine apparently being richer in chlorides than that of other birds, while their serum itself may have a higher salt content which would allow the whole organism to function at a distinctly higher osmotic level. Such mechanisms, though still very poorly known (the few studies having been made mainly on the Savannah Sparrow *Passerculus sandwichensis* which lives in salt swamps), would be an especially valuable adaptation for desert birds.

Behavioural adaptations

Besides their physiological specializations, desert birds show ecological and ethological adaptations to this habitat. These no doubt play the most important part in the lives they lead in this extreme environment, and allow them to colonize and to survive in it.

The principal enemy of desert birds is indubitably the sun, which causes the desiccation and high temperatures of most deserts. The direct effects of insolation are especially striking, since a unit area of ground in the shadow of a tuft of vegetation receives only one-third as many calories per unit time as one exposed to the sun. The following different temperatures have been recorded at a single time and place: 31°C in the shade, 33°C in the shadow of a bush, and 59°C at the surface of the soil. At the latter level the temperature can be still higher, reaching 80°C on sand, and many desert birds are strictly terrestrial, if only because there is no vegetation.

Under these conditions birds tend to behave so as to protect themselves from direct insolation, though this is far from being the general rule. In the Sahara larks (and especially the Bifasciated Lark *Alaemon alaudipes*),

sandgrouse, bustards and coursers are often normally active even during the
heat of the day without seeking out the shade, and move about on soil whose
temperature exceeds 70°C. The thermal and water balances of such birds
are indeed truly astonishing. However, many desert birds do try to screen
themselves as much as possible from direct insolation. They are most active
during the cool of the morning and evening (though this tendency is also seen
in many other environments and is not confined to desert birds). Around
mid-day these birds rest in the most sheltered places.

As characteristic as this daily rhythm is the search for the most favourable
– least arid and least insolated – surroundings. Some desert birds are rock-
haunting, seeking out rocky gorges whose fallen boulders and xerophytic
vegetation provide shelter during the heat of the day. This search for shade is
especially evident during breeding. The brooding bird has to keep to its nest
throughout the day, if only to protect its eggs and young from direct insolation, and seeks out the most favourable nesting sites. Some species such as
chats arrange their nests in jagged fissures in the rocks. Others nest on the
ground itself but reduce the effect of the sun by leaving their nests in the
shadow of boulders or even of pebbles, such as larks which scoop out hollows
at the foot of rocks or under tufts of grass. The nest is usually placed to the
east of such a 'parasol', so that it is protected during the hottest part of the
day.

The temperature is lower at a certain height above the ground, and this is

Figure 85. The nuptial flight of the Hoopoe-Lark *Alaemon alaudipes*.

no doubt part of the reason for certain birds to build their nests in bushes, thus benefiting not only from the shelter of the vegetation but also from a more favourable microclimate. This is done in the Sahara by the Desert Warbler *Sylvia nana*, the Streaked Scrub Warbler *Scotocerca inquieta*, desert babblers and even sometimes by the Bifasciated Lark. In the American deserts cacti provide very favourable nesting sites: woodpeckers tunnel in the pulpy flesh of saguaro (*Cereus*), and tiny Elf Owls *Micrathene whitney* then use the resulting cavities. It must be emphasized that few desert birds nest underground – in contrast to those rodents which have colonized deserts, and have there accentuated their inherent burrowing tendencies. A few birds of arid regions do nest in tunnels, such as bee-eaters – but since the whole of this family nests in the same way, this cannot be interpreted as a response to desert conditions. In the Sahara there is only the Red-rumped Wheatear *Oenanthe moesta*, and even this uses rodent burrows for nesting. Indeed, no typically desert bird digs burrows, although subterranean life would shelter them from the effects of heat (with temperatures reduced by 30°C or more at a depth of 50cm) and dryness, giving their young a much more favourable microclimate at the same time as it protected them from enemies.

Figure 86. A Wheatear's nest built in a hole in a crag. A. the nest cup. B. a low wall of pebbles.

The annual cycles of desert birds

Most deserts show fairly well-marked annual climatic cycles. The rains, scanty as they are, are concentrated into a short season, which is thus the most favourable for breeding since the vegetation is then densest, insects

481

most abundant, and temperatures lowest. In contrast the summer is the least favourable season, because of its aridity and heat.

In most deserts, birds therefore breed at a certain time of year, depending on the rainfall cycle. In the United States breeding takes place at the end of winter in deserts characterized by winter rains, and later in those where spring is the rainy season. In the northern Sahara birds nest during the spring, the coolest and wettest season. As temperatures increase the birds stop breeding, with the graminivorous species generally the last to do so in accordance with the late ripening of seeds. Many then move away from the most inhospitable parts of their range, which are habitable for only part of the year. They gather in the most favourable biotopes, even in the arid steppes around the true desert, and thus take part in true migrations with a regular rhythm, although the distances they cover are often small. One may thus speak of some Saharan birds making journeys to summer away from their breeding grounds. The Cream-coloured Courser and the Bar-tailed Desert Lark *Ammomanes cincturus* nest in the northern Sahara at the beginning of the spring, and then leave to pass the summer in Barbary or even farther north, the desert areas where they bred having become uninhabitable because of the excessive temperatures and resultant desiccation. Movements of this type, elsewhere than in the Sahara, are much more general than had for a long time been believed. Such regular migrations are known especially in the Australian deserts, and express the adaptation of a consuming population to a standing crop biomass which fluctuates wildly at any given place.

Climatic conditions also vary greatly from year to year at the same time and place, according to a wholly irregular rhythm. The amount of precipitation may vary enormously, a year of what passes for abundant rain in a desert being followed by several very dry ones. The cycles of the vegetation and insects are profoundly altered, so that the standing crop biomass at what should be the favourable season varies greatly from year to year. The birds follow this irregular rhythm precisely, their fecundity varying with the richness of the environment. During decidedly unfavourable years, sedentary populations do not breed. Thus in Tunisia partridges do not nest in years of intense dryness, showing no reproductive behaviour while their gonads remain dormant. In the arid zones of the United States, the breeding success of California Valley Quail depends upon the rainfall. When it is markedly dry they do not breed and losses (especially among the young birds) are enormous. In contrast, desert birds show markedly greater fecundity during favourable years, with increased numbers of eggs per clutch and distinctly higher breeding success, so that their populations increase – only to diminish greatly if the following years should be unfavourable. Thus desert birds show great fluctuations in density, whose regulation depends precisely on the amount of available food, and thus on rainfall which is the ultimate limiting factor.

These great fluctuations in time are matched by corresponding variability in space, since the rainfall regime is as irregular from place to place as it is at one locality. The rains are mostly very local, falling in one region or another in different years. This phenomenon is well known to all desert dwellers, especially the Saharians who lead their flocks to different pastures depending on the local rainfall. The same is true of the birds, which have the advantage of exceptional mobility; and in fact a very high proportion of desert birds are nomads, capable of moving on a grand scale to concentrate at the most favourable places. Thus the total range of a given species is very large.

The Australian avifauna includes no less than 30 per cent of nomadic birds, in accordance with the great extent of desert zones on this continent, and the extreme irregularity of their rainfall regimes. It is here that nomadism has been most strikingly studied (Frith 1959, Keast 1960, Immelmann 1963). The birds are remarkably well adapted to the very special climatic conditions which obtain over the vast deserts of central Australia. This area lies between the northern zone of summer rains and the southern zone of winter rains, and so is extremely arid. However, the occasional local rain may fall any-where, so that at a particular place it may rain at any time. To be able to survive in such an environment, birds must be able to take advantage of all these showers, despite their extremely irregular distribution in time and space, and Australian desert birds are therefore characterized by lack of any annual breeding periodicity and by very marked nomadic habits. They are in breeding condition throughout the year, as a result of permanent pituitary and gonadial activity, and do not show the resting periods characteristic of birds in environments with regular seasonal fluctuations. In most of the species, pairs remain mated for life. Furthermore, moulting does not seem to inhibit reproduction, as it does in most birds, and they have been observed to nest while in full moult. They are also precocious, being capable of breeding while still very young.

The immediate stimulus for breeding is provided by the rain itself. Other factors – especially temperature, photoperiod and even the amount of available food – cannot be responsible, since the first nuptial displays take place within a few minutes when it unexpectedly begins to rain after long drought. Copulation and the beginning of nest building are to be seen a few hours after the first drops of rain in various birds such as the Black-faced Wood-Swallow *Artamus cinereus* and the Zebra Finch *Taeniopy giacastanotis*. The rain is thus the trigger which releases breeding behaviour, in birds which are always physiologically ready. This is an extremely important adaptation, since it allows the birds to avoid any waste of time. Brief but relatively abundant rain is followed, after a latent period which is often short, by an explosive increase in the vegetation and the insect populations. By beginning the breeding cycle from the first rains, the birds behave as though they foresaw this abundance, and gain time so that their young hatch when they can take advantage of the most favourable conditions.

However, this favourable period is always short. The brief rains are not repeated, and the vegetation and insects flourish within a short space so that their breeding is completed by the time conditions deteriorate again. The various phases of the birds' reproduction are also accelerated. In central Australia, both sexes of the Zebra Finch take part in collecting materials for the nest, whereas elsewhere only the male does this while the female waits at the nest and arranges the materials which he brings. This behaviour allows the nest-building period, thirteen days in less arid districts of New South Wales, to be shortened to seven to eleven days. The other phases of reproduction are similarly accelerated, so that the young may be reared before environmental conditions become unfavourable again.

The second adaptation of Australian desert birds is their strongly developed nomadism. Since the rain falls unpredictably at one place or another, the populations of most species are highly mobile and opportunistic. Some move quite irregularly while others such as the Masked Wood-Swallow *Artamus personatus* may follow the movements of areas of low atmospheric pressure so as to take advantage of the resulting rainfall. Nomadism is especially obvious among the ducks, whose behaviour varies greatly with their preferred habitats and diets (Frith 1959). Some such as the Black Duck *Anas superciliosa* are sometimes migratory and sometimes sedentary, but always with a regular cycle including a relatively fixed breeding season, in accordance with their preferred habitat on permanent stretches of water. Others in contrast are opportunists, in their exploitation of food resources as well as in their reproduction. These ducks are largely dependent on temporary stretches of water, whose ecological conditions change quickly and very irregularly in areas without a well defined climatic cycle. Some are of generalized diet and can feed on very varied foodstuffs, using different ways of taking and gathering them. Thus the Grey Teal *Anas gibberifrons* moves about remarkably erratically, individuals which have nested in one spot being apt to travel in any direction across the continent, and only a fraction of the population is sedentary. In contrast others have highly specialized diets which oblige them to seek out their indispensable conditions. Thus the Pink-eared Duck *Malacorhynchus membranaceus* feeds on nothing but planktonic organisms, which it collects by filtration and not at the bottom or edges of bodies of water like other species. Its food requirements thus make it narrowly dependent on precise ecological conditions, and it has to travel widely. It is still more nomadic than the Grey Teal, leaving no sedentary individuals. These ducks have no defined breeding season and nest wherever and whenever they meet favourable conditions. Like the terrestrial birds mentioned above, their responses are immediate. Any rise in water-level releases breeding behaviour in the Grey Teal, the eggs being laid seven to ten days later, whereas the Pink-eared Duck begins to breed only when the water reaches areas where it will remain shallow, or when it begins to retreat. This lack of synchronization between the two species can be explained by their dietary preferences: the Grey Teal feeds on aquatic insects (Corixidae and Dytiscidae) whose biomass reaches its maximum at high water, whereas the Pink-eared Duck feeds on planktonic organisms which proliferate in water when it is shallow or about to retreat.

Thus the birds of the Australian arid zones show remarkable adaptations to the peculiarities of the environment, which one might expect to be paralleled in America and Africa. However, in the Sahara for example, the seasons are much more regular than in Australia, with a decidedly higher probability of spring rain. This results in a well defined breeding season, though without altogether excluding nomadism among the birds, since a given population will nest at one place one year and another the next, depending upon local conditions. Thus for example in some years when there has been much rain, the Draa Valley in southern Morocco is covered in vegetation and nesting birds such as the Thick Billed Lark reach high densities there, while the neighbouring hamadas are dry and unpopulated, whereas in other years the opposite is true. Thus there is an adjustment of the consuming populations to the standing crop biomass, an important ecological adaptation. Only fluctuating populations can exploit fleeting and irregular resources from year to year. The range which they occupy is huge, and only the most favourable parts are colonized in a given year.

The reproductive rates of desert birds are poorly known, though in accordance with Hesse's rule (see Chapter 12) they are generally lower than those of their relatives in temperate habitats. Thus the normal clutch of Saharan chats (*Oenanthe moesta* and *O. leucopyga*) is four eggs, whereas that of *O. hispanica* in the Mediterranean region is five and that of *O. oenanthe* in temperate or cool countries is six. Clutches of three or even two eggs are as common among the desert species as they are rare elsewhere, while warblers and shrikes provide similar examples. The number of eggs per clutch, though invariable in some species (two for pigeons, coursers, stone-curlews and nightjars), varies greatly in others depending on environmental conditions. The amount of food available plays some part in these variations, which is easily explained in terms of the metabolic demands on the female, and similarly affects the success of the broods. Thus the populations of desert birds are subject to large fluctuations, with high peaks followed by massive reductions, resulting in very close adjustment to the environmental factors.

Chapter 25

Tropical Savannas

SAVANNAS of very diverse physiognomy extend across the tropical regions of both hemispheres, to latitudes which vary locally according to conditions. They form grass associations more or less densely interspersed with trees, and despite their diversity they all have the character of open biotopes. Savannas occupy especially large areas in Africa, crossing the continent from west to east between the edges of the Sahara and the rainforests of Guinea and the Congo, and then across eastern Africa. They are less extensive in south-east Asia, but occupy a large part of Australia, while in tropical America they stretch across the Orinoco basin (*llanos*) and over the Brazilian plateaus (*campos cerrados*).

Savannas as a whole require a warm and more or less humid climate. Their rainfall varies greatly though it may reach 1800mm per year; but the rains, far from falling throughout the year are confined to a limited period which may be as little as three months. This seasonal cycle of dry periods alternating with rainy ones is the most obvious characteristic of savannas. Variations in photoperiod are never of great amplitude, though they may reach two hours in Senegal. The temperature is almost as constant, though its variations are strongly influenced by the humidity. Whatever the conditions, all savanna regions show an alternation of dry and wet seasons.

The rainfall varies greatly from place to place. At low latitudes the rains are generally plentiful and prolonged, often with two rainy seasons. With increasing latitude the shorter dry season disappears so that the two rainy seasons become one, which then tends to shorten and produce less rain. Furthermore the amount of rainfall often varies widely from year to year at the same spot, by a ratio of as much as two to one. The dates when the various seasons begin also varies, so that the seasonal cycle may be very irregular at a given place, and this has considerable effect. In contrast to the plants of equatorial associations, those of the savanna show well-defined cycles. The trees and bushes are deciduous, losing their leaves during the dry season. The herbaceous plants are either annuals, passing part of their cycle as seeds, or perennials whose aerial parts wither and take on the yellow colour characteristic of dry savannas. The vegetation as a whole is resistant to droughts and fires. As a result savanna vegetation is highly specialized, with proportionally few trees and bushes and lower strata formed by herbaceous plants, among which the grasses are dominant. In contrast, tree-like and herbaceous species are extensively dominant in woody moisture-loving formations, so that the characteristic plant associations may be defined.

While all savannas are open habitats, they present very diverse aspects: in certain very dry zones homogeneous grass associations extend without any treelike layer, the grass cover being either continuous or in separate tufts giving the appearance of steppes. The *sahelian* types of savanna, of areas where the rainfall is scanty, consists of a very scattered herbaceous covering and small, spiny, widely-spaced trees. The wetter *sudanian* type, is more densely wooded, with still many spiny trees. Finally the *guinean* type is well wooded and carries a much richer flora, with palm trees and others of considerable size. There are numerous intermediates between these various types of savanna, depending on the rainfall and edaphic factors.

The contrast between even densely wooded guinean savannas and humid forest is often very distinct, forming a well-defined boundary, but sometimes there is an intermediate zone with a mosaic of forested islands scattered among savanna or of clearings in the forest. The wet forests penetrate the savannas along the rivers to form gallery forest. This is especially obvious in tropical America, where the Amazonian forest winds far to the south along the watercourses, allowing a truly Amazonian forest fauna to penetrate to the centre of Brazil.

The annual cycle causes large fluctuations in the standing crop biomass, including both plants and animal prey. The birds have to adapt to the annual rhythm of these conditions, by which their entire biology is profoundly marked.

The African savannas are much better known than those of other parts of the world. Furthermore, African biotopes show more obvious zonation and are more homogeneous over large areas than their Asiatic counterparts, in which there are more complex mixtures of habitats. So in this chapter we shall especially consider Africa – principally the sahelo-sudanian type of savanna (studied in Senegal by Morel 1968) and only incidentally the guinean type. Arid steppes resemble deserts, and their birds show the same adaptations as desert species.

Many existing savannas are not climax vegetational formations, but result from human interference. For a long time bush fires have opened wide clearings in closed habitats, converting stretches of forest into savannas, which are maintained in this state by regular burning during every dry season. These open environments are thus much more extensive than they were originally, and some authors maintain that the greater part of the African savannas is artificial, having mainly replaced dry forests. Savanna animals, and especially the birds, have profited from this extension of their preferred habitat.

Savanna birds

In Africa and Asia the dominant groups are pigeons, hornbills, bustards, certain game birds such as francolins and guinea-fowl, rollers, larks, certain shrikes, many weavers, starlings, and many thrushes and warblers including grass warblers. Weavers are without any doubt the most abundant of all these birds, and are the more conspicuous since they live in large flocks. Savannas with the densest tree cover support the most species, including certain forest birds. In Australia, terrestrial parrots are abundant in open environments. In America the faunistic composition is very different, with tinamous replacing the game-birds and various passerines such as icterids and finches occupying the place of the Old World weavers.

Raptors are common and diverse in open habitats, where they can hunt and find their prey much more easily than in forest environments, so that savannas are the domain of eagles, snake eagles, buzzards, kites and falcons. These environments are also occupied by most of the running birds belonging to the Ratites. The Australian Emu, the African Ostrich and the American Rhea of the Argentinian pampas are all steppe and savanna birds, whose terrestrial locomotion demands hard ground and sufficient open space for them to detect their enemies and escape from pursuit by fast running.

Savanna faunas are also distinguished by a certain number of negative characters. Typically forest birds do not penetrate them, and certain families which are well diversified in the great rainforests are represented at most by a few specialized species. This is true of the African turacos, most of which are forest birds except that the species of *Corythaix* are confined to savannas.

Savanna birds are very generally much paler, duller and less contrasting in colour than those of forest, while the plumage of many of the terrestrial species shows very marked cryptic coloration. These differences agree with the general rule that intensity of colour increases with the humidity of the environment. Further, there is a very clear gradient in coloration, parallel to that of environmental humidity, in passing from dry steppe to relatively humid guinean savanna.

Savanna birds may be grouped in many categories, depending on the vegetational strata they occupy and especially the manner in which they gather their food. Some – such as guinea-fowl, pigeons, rollers, bustards, weavers, grass-warblers, chats, larks, hoopoes and raptors – are terrestrial and live mostly on the ground. Others – such as barbets, colies, cuckoos, shrikes, sunbirds, parrots and most warblers – live in trees and never descend to the ground. Others again such as swifts, swallows, bee-eaters and night-jars, are entirely aerial and feed on insects which they catch in flight. Savannas are also inhabited by many aquatic birds. During the rains low-lying areas fill with water and occasionally form temporary flood zones of considerable extent. These are of the first importance to the birds, which find in them a greatly increased biomass. During the dry season these waters progressively regress, while their standing crop biomass tends towards a very low minimum. Despite this limiting factor the tropical aquatic avifauna is very diverse, which is explained by the multiplicity of ecological niches to be found in this environment. Forests are much less favourable to most of these birds (apart from some specialized forms), because in them stretches of open water are rare so that they are very poor in water birds compared with the neighbouring savannas.

Savanna birds as a whole are less diverse than forest ones. On the other hand, certain species such as weavers are very numerous: one has only to recall the extraordinary proliferation of the Red-billed Dioch *Quelea quelea* in Africa, due in part to artificial factors resulting from the cultivation of savannas which were in better equilibrium in their natural state. Savanna avifaunas always contain a few key species, whose biomass is large in comparison with that of the remainder.

Such abundance, together with the openness of the biotope which facilitates social contacts between members of a species, results in widespread gregariousness among savanna birds. Many, especially among the weavers and starlings, live in large assemblages, a tendency which is carried to extremes by the Social Weaver *Philetarius socius* of Africa which builds huge communal nests. The need to concentrate in the areas which are most favourable at a given time, taking part in large-scale nomadism and even true migrations, is in accordance with this gregarious characteristic.

Finally we may note that, in contrast to forest birds which are often narrowly localized by their precise ecological requirements, savanna birds tend to be very widely distributed. Thus in Africa many birds of these open biotopes are found from Senegal to the Cape. Such distributions are explicable by the uniformity of the environment over enormous areas.

Diet

Savannas offer terrestrial birds a wide variety of food resources. As everywhere they can specialize on a herbivorous or an insectivorous diet. Among the herbivorous birds, the small number of frugivorous species is striking. In Africa only the barbets, parrots, colies and turacos feed exclusively on fruits, though many omnivores take fruit as part of their diets. This negative characteristic of the avifauna is explained by the rarity of fruits in savannas, and still more by the seasonal character of the fruiting. Further, because of the relatively low nutritive value of fruit, birds have to eat a lot of them, which in savannas they could regularly do only by making long journeys. On the other hand, fruits do provide them with a considerable amount of water, which means that frugivorous birds are not so strictly tied to watering places as others.

Graminivorous birds are extremely numerous and diverse across savannas in Africa and Asia, including many pigeons and doves, sand-grouse, whydahs, weavers, waxbills and other small ploceids. In Senegal there are twenty-five graminivores as against only five frugivorous species (Morel). This proportion is repeated in other types of savanna, though there are proportionately more frugivores in the guinean savannas because of the more favourable conditions of that environment (in which the green pigeons *Treron* are characteristic). In America, graminivorous ploceids are replaced by innumerable finches such as *Sporophila*.

These birds have at their disposal a very wide range of seeds, which are especially abundant when they ripen at the beginning of the dry season. Despite dietary preferences differing from species to species, which prevent undue interspecific competition, most feed on a wide variety of seeds of very different sizes, without showing obvious feeding specializations. The choice of food depends above all on its abundance and availability. Grasses (even species of *Cenchrus* although these are not at first sight appetizing) provide a large proportion of the seeds. This lack of feeding specialization probably results from the fact that the food is abundant and widely distributed, providing an extremely favourable environmental factor. Rather than be dependent on a particular plant, whose fortunes they would share in case of poor fructification, the birds are able to compensate for the poverty of one type of food by another which is more abundant at a particular time. The success of *Quelea* is partly explicable by its lack of specialization. Wild plants, and especially grasses, produce even more seeds per unit area than plants cultivated by man in the same environment, which gives some idea of the standing crop biomass which they represent. Furthermore, despite fluctuations this source of food is almost continuously available to the birds throughout the year.

Because of the low water content of their food, seed-eating birds are very closely tied to water, which they need to drink frequently and copiously. This absolute necessity affects their distribution, for at least in the dry season they cannot go more than a certain distance from water sources, depending on their strength of flight. In Africa turtle doves such as the Vinaceous Dove *Streptopelia vinacea* and the African Collared Dove *S. roseogrisea* can travel tens of kilometres daily to reach water, and they are therefore widely distributed even across very dry savannas. Although less strong, weavers fly fast and easily, so that their field of action is still relatively wide. In contrast fire-finches and other waxbills are weaker, and being very susceptible to heat they live in the shelter of plant cover and cannot make prolonged flights. Thus they cannot leave the immediate vicinity of water sources, and keep to the surrounding vegetation. These more or less imperative demands explain the different distributions of various types of bird, which allow better utilization of habitat resources and reduce competition.

A greater number of savanna species are carnivores. Of 100 resident and migratory species in the savannas of Senegal, eighty feed on animal foods – fifty-six exclusively, while the remainder are polyphagous (Morel). Insects are the chief animal food. Far from forming a constant source throughout the year they show large fluctuations from season to season, correlated with the cycles of plant life and of rainfall. According to surveys the soil entomofauna, which mainly occupies the plant litter under trees and bushes, varies in quantity in the ratio of one to eight. The leaf fauna varies much more still, as a result of defoliation and despite the flowering and fruiting which often take place during the dry season. Variations in abundance of one to sixty have been demonstrated, so that birds of the tree layer have to face the largest fluctuations in their food resources.

All the feeding opportunities of the habitat are exploited by various classes of consumers. Plovers of the genus *Sarciophorus* and coursers, which do not shun the heat, hunt on the bare ground taking mainly tenebrionid beetles and ants. During the night they are replaced by stone curlews and the Bronze-Wing Courser *Rhinoptilus chalcopterus* which feed on nocturnally active prey – crickets and arachnids, especially scorpions and solifugids. Incidentally, this example shows how two birds may hunt over the same ground without competing, by inverting their cycles of activity and exploiting prey which show parallel inversion of their day and night cycles. Other birds, some hunting in thickets and others among the foliage, divide into many categories according to the layers in which they are found. Certain small and very active warblers keep to the foliage and there catch small insects. Cuckoos take a great many caterpillars. Woodpeckers, as everywhere, explore the bark and pierce the wood in their search for wood-boring larvae. Wood-hoopoes *Phoeniculus*, with long sickle-shaped bills, can extract insects hidden in cracks. The last category feeds on aerial insects caught in flight: swallows, swifts and bee-eaters during the day, and nightjars by night which prey on many tropical insects and can take prey several centimetres long in their widely gaping bills. In general ants, termites (including the winged forms which swarm at certain times), tenebrionid beetles and locusts (together with flies, as everywhere) form the principal food of the insectivorous birds. On the whole few savanna birds – whether terrestrial or the small species living in trees in accordance with the size of their prey – are strictly insectivorous. Many of them are migratory or nomadic. Further, many migrants from temperate regions, especially in Eurasia, come to swell the African populations of insectivores for part of the year, so that the seasonal surplus of production is exploited by fluctuating populations. The very low minimum reached by the insect populations explains the low density of strictly insectivorous autochthons.

Polyphagous birds are much commoner and more highly diversified. Their diet gives them the advantage over stenophagous ones, since they can more easily compensate for the lack of one class of food by switching to another. This is true of francolins and guinea-fowl, whose diets include almost everything edible on the ground – seeds, fruit, insects, and other small animal prey – and of ground hornbills *Bucorvus* which sometimes take minute insects, seemingly out of proportion to their own size. The Red-billed Dioch *Quelea quelea* itself takes a mixed diet, which for at least part of the year includes many insects.

Birds of prey are more numerous here than in any other environment. Each is adapted to a particular diet and lives on a certain class of prey. Some, such as the Martial Eagle *Polemaetus bellicosus*, hunt guinea-fowl and young ungulates. Many, such as the chanting goshawks *Melierax*, snake eagles *Circaetus* and sparrow-hawks *Accipiter*, take reptiles. Others are bird-catchers, notably the Harrier-Hawk *Gymnogenis typicus* which attacks weaver colonies. Vultures, which are absent from forests, are very common here since the savannas suit their soaring flight, thanks to the columns of warm rising air which are so frequent when the sun has heated partly bare ground. Further, their method of finding animal carcases by sight can be carried out only in open environments.

The annual cycles of savanna birds

The birds have to adjust the various phases of their cycles very closely, so as to find themselves in optimal conditions during the breeding season when their need for food is greatest. Most savanna birds therefore have precise annual cycles, adapted to local conditions and varying with them.

There is certainly no rule which rigorously links climatic fluctuations and the breeding season of a particular species. Most savanna birds have their breeding spread over a long period, while some appear to breed unseasonably when the season is at first sight unfavourable. Further, since ecological demands vary very much from species to species, there is considerable lack of synchronization between them, some birds nesting in the dry season and others during the rains according to their specific dietary needs. As a result there is no one breeding season for all the birds of a savanna region, but staggered breeding of the various feeding groups within the single biocenose. Many birds even breed during the height of the dry season, as long as their diets allow them to find sufficient food. This spreading-out of breeding benefits all the birds by reducing competition.

None the less, there are immediate links with environmental fluctuations. The beginning of breeding is sometimes actually determined by the arrival of the first rains. This is notably true in the most arid areas where rain is irregular, as is the almost immediate response of the birds to the beginning of the rains in the sub-desert regions of Australia. As a very general rule, the insectivorous birds nest during the rainy season and the seed-eating ones and raptors during the dry. In Senegal, the dry-season nesters are essentially the vultures, stone-curlews, rollers, raptors, bee-eaters (except the Blue-cheeked Bee-Eaters), sand-grouse, ringed parakeets and some larks. The rainy-season nesters are the cuckoos, shrikes, bustards, guinea-fowl, weavers, glossy starlings and warblers. Other birds, especially the fire-finches, swifts, francolins and whydahs, nest for much longer, straddling both seasons.

However, one must note the differences within a single feeding group, especially the seed-eaters. In Senegal the doves (except the Black-billed Wood Dove *Streptopelia abyssinica*) nest practically throughout the year, though some of the species do so less frequently during the rains. For this the relative poverty of the wet season in seeds, as well as the storms which destroy many nests, are responsible. Certain waxbills, especially the Senegal Fire-Finch *Lagonosticta senegala*, breed during a very long season which lasts ten months and begins at the same time as the rains. Although these birds are strictly graminivorous they are not restrained by the shortage of seeds, whose biomass is then at its minimum. The sand-grouse breed only during the dry season, whereas the weavers (Ploceinae) do so only during the rains. The latters' timing is explained by a change of diet, which may as in *Quelea* come to include a large proportion of insects, or even in other species become exclusively insectivorous. Thanks to this dietary change these birds can thus profit from the massive increase in insect populations. At the end of their breeding period, which coincides with the arrival of the dry season, seeds are at their maximum and can once more satisfy the dietary needs of these birds, which revert to seed-eating when breeding has brought their populations to a maximum. The dietary shifts of the Ploceinae, and their adaptations to resources available at different times, thus allow them to benefit from the annual trophic maxima in both insects and seeds (M. Y. Morel 1969).

The breeding chronology of some birds is also in harmony with the supply of vegetable material suitable for nest-building. This is especially true of the Red-billed Dioch *Quelea quelea*, which in order to build its nests takes 190kg of grass per hectare from the savanna around its colonies, from a total production of 700 kg/ha. In the dry season the plants cannot supply this considerable amount of material and the establishment of such dense colonies is possible only if the plant cover of the savanna is sufficiently luxuriant, so that the birds do not have to travel too far to collect the material indispensable to their breeding.

Thus it seems that there is no one preferred breeding season for graminivorous birds as a whole. The shortage of seeds at the beginning of the rainy season, when the old crop is becoming exhausted and the new one only just germinating, does not seem to have very profound effects. Sufficient seeds are probably left from the previous season and certain plants no doubt ripen their new seeds rather early, germination being desynchronized between the species. The variety of breeding demands and their spreading through the annual cycle reduce dietary competition between the various graminivorous birds. Population maxima in Senegal occur in October (the end of the rains) for the weavers, in February for the weaver-finches, and in May for the sand-grouse, whereas the stocks of pigeons are more or less constant throughout the year because of their uninterrupted breeding.

In contrast most of the insectivorous birds nest during the rainy season, together with the polyphagous species which need large amounts of insects for feeding their young. However, some species do nest during the dry season – surprisingly in view of the lack of insects at that time. Thus in Senegal the Cape Thick-knee *Burhinus capensis*, the Abyssinian Roller *Coracias abyssinica*, the Bronzed Starling *Spreo pulcher* and others nest in the

495

dry season, while the Black Bush-Robin *Cercotricas podobe*, the Grey-backed Eremomela *Eremomela griseoflava* and Crombec *Sylvietta brachyura* do so indifferently during both seasons. Except for the last-mentioned, most of these are small moisture-loving birds which find the greater part of their food by searching the leaf litter and other plant debris for beetles and arachnids – precisely the arthropods which show least tendency to disappear during the dry season – so that the supply of insects living near the ground is sufficient to support the breeding of these birds.

There is certainly an optimum humidity, acting no doubt through the kind of food necessary to each species, which causes a particular type of bird to nest at one season or another in a given region. This is clearly true of the West African nightjars. The Standard-winged Nightjar *Macrodipteryx longipennis* nests, in a broad band of fairly humid country along the edge of the tall forest from Senegal to Kenya, from February to April – that is during the dry season – and then migrates northwards while the wet season sets in. In contrast the Plain Nightjar *Caprimulgus inornatus* nests in the driest savannas from Mali to the Red Sea, to the north of the previous species' breeding range, from July to September – that is during the wet season – and goes south to pass the dry season. Thus these two species migrate simultaneously and in the same direction; but the one does so to nest while the other returns to its 'wintering' grounds, each in search of its ecological optimum.

Thus the chosen breeding seasons of the tropical savanna birds seem to be determined by their ecological needs. The active phases of reproduction – nest building, display and laying – are under the influence of external stimuli. This appears especially clearly among the birds which nest in the wet season, which begin to show breeding behaviour immediately after the first rains. Although (as has been shown experimentally for many species) tropical birds are sensitive to variations in the cycle of illumination, these are of minimal amplitude in the tropics and do not have the same importance for savanna birds as for those in temperate countries.

However, the rains are not wholly favourable to the birds which nest during the wet season, since in the tropics they are most often accompanied by violent storms and tornados which dislodge nests from the trees. The vulnerability of nests varies according to their construction. Those of doves being formed of loosely interlaced twigs in pliant trees, are especially fragile, whereas weaver nests are reliably solid and made fast to supple branches, so that they withstand the wind while many are also waterproof.

As a very general rule the birds of prey nest during the dry season, which is the more favourable to them, since during the rainy season their prey disappears into the dense vegetation. This makes hunting difficult, and would prevent them from ensuring a regular and plentiful supply of food for their young.

Thus the breeding seasons in tropical savannas have been established in a very special manner and differ from those in other habitats. There is no single favourable season as in temperate countries, but every species chooses the one which suits it best, so that birds can be found in the course of breeding almost throughout the year. Birds which breed at different times belong to very different feeding classes, and a particular seasonal cycle applies only to a given species or group, characterized by diet or other ecological demands.

The modified seasonal rhythm also has a profound influence on moult. Many have two annual moults, one before and one after breeding, so that they present two very distinct appearances. This is especially true of certain weavers, which show very marked sexual dimorphism while breeding, the males assuming contrasting and strikingly coloured plumage while the females are much more modest. Afterwards, however, the males lose this nuptial plumage and even their specific characters, taking on much duller eclipse plumages in which they resemble the females. The same is true of whydahs, whose males in eclipse plumage do not show the long tail plumes which are so important in their nuptial displays. Such an alternation of breeding and eclipse plumages, though by no means general, is characteristic of certain savanna birds – in contrast to those of tropical forest which retain the same plumage throughout the year, in accordance with the constancy of their environment. Other savanna birds have different moult cycles. Those species with a prolonged breeding season have only one moult, and therefore a single plumage. In those which breed almost continuously – such as the turtle-doves in particular – a given individual may moult at any time, and meanwhile ceases to nest. In others the moult begins, is interrupted by the breeding season, and is completed when the brood has been reared, as in some Red-billed Diochs *Quelea quelea*. Yet in others moult and breeding overlap (so that there is of course no difference between breeding and eclipse plumages)—a situation which arises only when the breeding season is well defined.

Like most tropical birds, savanna species lay smaller clutches than their relatives of temperate countries, but larger than those of their forest counterparts. In Africa the clutches of various savanna weaver-finches, weavers, shrikes, babblers and thrushes are larger than those of the forest species by half an egg on the average. Though such differences are not shown by African flycatchers and sunbirds, they reappear among those American birds which have been studied from this point of view (Lack & Moreau

1965). Despite this reduced number of eggs per clutch, both the laying rate (the number of eggs laid by a pair during a season, which averages 1·5) and the productivity (the number of young reared to fledging by a pair during a season, averaging 3·1) are comparable to those of counterparts in the temperate zone. The reduced number of eggs per clutch is compensated by a greater number of successive clutches during a single breeding season. Thus a pair of Red-billed Fire-finches *Lagonosticta senegala* nests four or sometimes five times during the season. However their hatching and fledging successes are among the lowest recorded for small passerines (M. Y. Morel 1969).

Migrations of savanna birds

Savannas show seasonal fluctuations in environmental conditions, to which the birds respond especially by seasonal movements. Some of these amount to no more than nomadism on a large scale, like that shown by many birds outside the breeding season, but others are so regular that they must be considered as true migrations. These seasonal movements, characteristic of a large part of the tropical savanna avifaunas, do much to enable their populations to exploit fluctuating resources, and to avoid the limiting factors of the environment at a given place.

Such movements take place across savannas with contrasting seasons in all parts of the world. However, they are poorly known in America, and ill-defined in Asia, where the intermingling of open habitats with forests of fairly constant resources allows the birds to find food without making large scale movements. This is not the case in Africa, an equatorial zone whose climate is effectively constant throughout the year, separating two regions with cycles of wet and dry seasons – opposite in phase in the two hemispheres – has profoundly affected the migratory habits of the nesting birds. As a result of these geographical circumstances, Africa is the scene of what are indisputably the most regular intertropical migrations.

Some African birds keep to one hemisphere, like the Standard-winged and Plain Nightjars mentioned above which move with the seasons from dry to moist savanna, whereas others do winter on the far side of the equator. This is true of Abdim's Stork *Sphenorhynchus abdimi*, which breeds only in a zone between Senegal and the Red Sea, nesting in the rainy season from May to September, and at the beginning of the dry season undertakes the long journey to South Africa. Other birds show the inverse cycle of movements, breeding in southern Africa and going to 'winter' in the northern hemisphere. Thus the Pennant-winged Nightjar *Cosmetornis vexillarius* nests in southern Africa from Angola and Tanzania to the Zambezi at the beginning of the rainy season (September to December). This is the most favourable season since it brings on the hatching of many insects, especially the winged phases of termites. The nightjar then moves northwards, seeming to follow the movements of the rains and the resulting abundance of insects. It passes March to August between Nigeria and Uganda, moulting there before leaving again for its southern breeding areas. Other birds of the African savannas make more limited movements, which because of their regularity must still be considered as migrations.

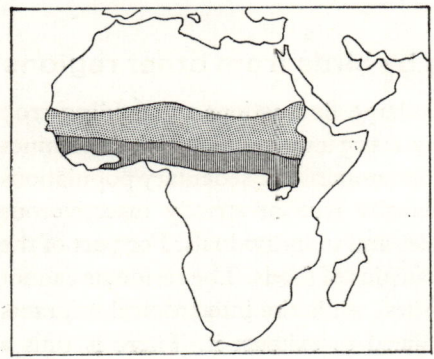

Figure 87. The migrations of the Standard-Wing Nightjar *Macrodipteryx longipennis*. Nesting area: vertical hatching, wintering area: dotted.

Thus within a single savanna area two groups of migrants can be distinguished: those which come there to breed and then leave when conditions become unfavourable, and those which arrive after breeding elsewhere.

The manner in which these intertropical migrations are controlled is fairly easy to imagine, though much harder to demonstrate. For certain birds, such as some aquatic species, environmental conditions become unfavourable at some period as a result of rising water or increasing plant cover. In Senegal, as elsewhere, the Egyptian Plover *Pluvianus aegyptius* frequents very open river beds, nesting on exposed sandbanks. When rains swell the rivers this habitat is covered by the floodwater and the plovers therefore leave for more favourable areas. The same is true of the African River Martin *Pseudochelidon eurystomina*, which nests in the river beds of the Congo and Oubangui at the time of low water in February and March. They disappear when the water rises, to shelter in the coast region of Gabon and the Congo. Other birds are greatly influenced by the density of plant cover. This is true of sand-grouse and certain nightjars, for which the plant carpet is undoubtedly too dense during the rainy season, and also of certain bustards, the hornbills and hoopoe. The movements of savanna birds are without any doubt ultimately controlled by their adaptations to food surpluses during the breeding season when they need the greatest amounts of specific foods. This allows them to exploit fluctuating resources, which show a marked minimum followed by a peak at every feeding level.

There is a very clear interaction between the movements of intertropical migrants, the cycles of sedentary or merely nomadic birds, and long-distance migrations from other regions. The intertropical migrants arrive before those from other regions return, so that to some extent the two groups draw from the food surplus not simultaneously but in a complementary way.

The exploitation of resources by birds from other regions

Savanna habitats are characterized by large fluctuations in standing-crop biomass. A very high peak is followed by a very low trough, which determines the carrying capacity of the habitat and maintains the sedentary populations at relatively low levels. This is especially true of strictly insectivorous residents, which are poor both in species and in individuals. For part of the year, however, there is a considerable surplus of foods. The residents cannot exploit this because of their low densities, while the intertropical migrants are insufficiently common and specialized to exhaust it. There is thus a place for consumers from other regions, and especially for long distance migrants with accurately controlled cycles. During the course of evolution, certain tropical savannas have in fact become the chosen wintering grounds of a large part of the migratory avifaunas from temperate regions.

This generalization is not true of the New World, since most North American migrants winter in the forests of Central America, but it is true in Asia and above all in Africa. The savannas of northern Africa, from Senegal to the Red Sea, provide the wintering grounds for most palaearctic migrants, while some of these go as far as South Africa. In the savannas which lie between the edge of the Sahara and the rainforest blocks, the standing crop biomass of plant matter and especially insects reaches its maximum towards the end of the northern summer. The wide areas of temporary flooding in the Senegal delta, the inner Niger delta and the Chad basin are filled with water and crammed with plant and animal foods, at precisely the time when palaearctic birds need these foods, which become inadequate in Europe with the onset of winter. Insectivorous passerines and water birds are the most affected by the unfavourable conditions of the European winter, and it is these which find the most advantageous conditions in tropical Africa. Migrants occupy an important place in the total avifauna of the West African savannas, and especially in the sahelian and sudanian zones. From

October to May they make up a large part of the total biomass of consumers, especially of the insectivores. In both species and individuals, the warblers form the largest fraction of palaearctic migrants, whereas native African warblers are comparatively rare. In the sahelian savannas of Senegal, there are twelve palaearctic species to two African ones (Morel).

Thus the savanna communities of tropical Africa are remarkably well regulated at the various feeding levels. The plant food resources, notably seeds, are exploited by resident populations, whether strictly vegetarian or polyphagous, with few palaearctic migrants. In contrast the insect resources largely exploited by palaearctic migrants, which can thus 'skim off' the large surplus. These birds leave their winter quarters when conditions there become unfavourable, which coincides with the return of spring in Europe. However, this is only true of the sahelian savannas, for farther south in the sudanian and guinean savannas the beginning of a rainy season results in a food surplus at that time. Here the birds' departure must be caused by other factors – internal stimuli, and perhaps the unduly dense plant cover resulting from the rains.

The same is true of aquatic birds. In the sahelo-sudanian zones, shallow basins filled by the rains provide an environment of high though temporary productivity. The resident birds cannot fully use this production because of the low carrying capacity during the dry season, when limiting factors act intensely. The migratory populations of ducks and waders come from the palaearctic to exploit this surplus, and leave for the north when conditions become unfavourable again. The numbers of palaearctic species and individuals belonging to these groups which winter in West Africa are large in comparison with those of their resident counterparts: in Senegal there are fourteen palaearctic species of the Charadriidae and Scolopacidae, as against three indigenous species. The numbers of these wintering birds are incredible, with more than a hundred thousand Garganey, Ruffs, Wood Sandpipers and godwits in Senegal and Nigeria. At least 150,000 migrant ducks of the genus *Anas* gather in the Senegal delta (Roux), as against a few thousand African ducks which belong to very different systematic groups and do not occupy the same niches. Palaearctic waders are more numerous than African ones, and do not feed in the same biotopes.

There are indications of competition between these two groups of populations. Closely related palaearctic and African species seldom meet in the winter quarters apart from a few cases: the pipit *Anthus leucophrys* with *A. campestris*, *A. cervinus* and *A. trivialis*; the swallow *Hirundo lucida* with *H. rustica*; the sand martin *Riparia paludicola* with *R. riparia*; the hoopoe *Upupa senegalensis* with *U. epops*; and the stone curlew *Burhinus capensis* with *B. oedicnemus*, with the African species given first in each case. Such mutual exclusion between birds of the same type which occupy the same niche certainly suggests competition. However, signs of competition are just as obvious among the ducks (Dorst 1962). While enormous populations of palaearctic ducks of the genus *Anas* winter in western Africa, there are no native representatives there. In eastern and southern Africa on the other hand, which are not occupied by migrants, there are many and often very common indigenous species (*A.undulata, A. erythrorhynchos. A.capensis* and *A. punctata*). None of these has penetrated into western Africa, and all keep outside the principal wintering grounds of the palaearctic species, a distribution which can be explained by competition with the migrants. In the west the food surplus can only be fully exploited by typically fluctuating populations, whereas in the east a less marked minimum in food resources allowed the native birds to adjust to the environment without extensive movements. This has allowed the development of sedentary forms confined to eastern and southern Africa, whereas western Africa, and the sahelo-sudanian belt in general, can only be occupied temporarily by migratory species, which merely exploit the excess resources of the favourable period. These enormous food resources of the western flood areas would thus remain unexploited, were it not for the populations of migrants and their remarkably precise ecological adjustments. Thus in Africa under present conditions, populations of palaearctic migrants complement rather than compete with the native avifauna.

The birds of the guinean savannas

All the foregoing applies only to the sahelian and sudanian types of savanna, whereas in wetter areas the ecological characters of savannas are modified. The rains are more abundant and more evenly distributed throughout the year. The vegetation is much richer and more varied, with more numerous trees and a dense carpet of herbs covering the soil with a continuous layer for a large part of the year. This type, called guinean savanna, thus forms a relatively closed habitat, much richer than the sahelo-sudanian savannas. It provides a transition towards the rainforests which it generally borders, since towards the boundary the trees become still denser and the canopy increasingly closed, while it is penetrated by gallery forests which actually have the characteristics of rainforest.

The avifauna of the guinean savannas itself provides a transition between those of more open savannas and those of rain forest. Species tied to very open environments are absent, but their loss is compensated by the presence of many birds of clearly forest type. The avifauna as a whole is more diverse and richer in species than those of sudanian savannas. However, though no numerical estimate has yet been made the density of individuals seems to be less. This characteristic too is transitional towards rainforest, as is the distribution of the birds by feeding groups. Graminivorous birds are rarer than in sudanian savanna, because of the reduced standing crop biomass represented by seeds. In contrast there are more frugivorous, omnivorous and insectivorous species, in accordance with the greater abundance of fruit, insects and spiders, and the less violent fluctuations in biomass. Raptors are not uncommon, and their flourishing communities are themselves intermediate between those of open habitats and of forest. These guinean birds are generally arboreal, with few terrestrial species except guinea-fowl and francolins, in contrast to the situation in both sahelo-sudanian savanna and forest. This characteristic is undoubtedly explained by the fact that the herbaceous and semi-shrubby vegetation covers the soil much less densely than in more open environments, and thus provides a less favourable habitat for ground-living birds.

The smaller fluctuations in available biomass – and especially the less pronounced minima – allow the birds much more extended breeding seasons. Certain birds such as the pigeons breed throughout the year, while the majority do so during the rainy season. The greater constancy of this environment also allows the breeding birds to be much more sedentary. Most are no more than nomadic, though a few do leave the guinean savanna during the rainy season as we have noted of the Standard-winged Nightjar. In contrast these are the 'wintering' grounds for many birds which nest in the sahelo-sudanian savannas and retire to the guinean savanna during the dry season, while a certain number of palaearctic migrants winter there. Thus from every point of view the guinean savannas provide an environment transitional between open habitats and tall forest.

L

Chapter 26

Tropical Rainforests

ACROSS part of the tropical regions, and especially astride the equator, stretch the dense evergreen rainforests.

In America they make up the huge Amazonian block (the *Hylaea amazonica* of Alexander von Humboldt) – which is the largest forest area in the world, covering more than four million km² in Brazil alone. These forests extend to the Atlantic slope of the Brazilian plateau, the Pacific coast of Colombia, the lowlands of central America and part of the West Indies. In Africa the large Congo block is separated by savannas from the smaller Guinea block, while the east coast of Madagascar is also covered in dense forest. In Asia parts of India, the Indo-Chinese peninsula and the Sunda Islands are covered in forests, and the same is true of New Guinea and of Queensland in north-eastern Australia. These isolated forests each have their own floristic characteristics, often differing widely in plant species and in the relative importance of various families in the tree flora. However, their common characters are such that rainforests everywhere are similar in physiognomy and ecology.

Climatic conditions are responsible for their incredible profusion of plants. Temperatures are high, of the order of 26°C (although the maxima are less than those recorded in other environments) and remarkably constant throughout the year: variations with season and time of day average only a few degrees. Rainfall is very high, generally exceeding 2m per annum and often reaching 4m or even much more, and is always more or less evenly distributed throughout the annual cycle. Atmospheric humidity is also always very high, usually reaching saturation. Winds are seldom violent, except in the hurricane belt. However, the tropical rainforest zones do not altogether lack seasons, since the rainfall is not quite evenly distributed throughout the year. The essential difference from the savanna regions is that even the minimum rainfall is always high, but there are seasons which are distinctly more favourable to life than others. Though the plants and animals have no need to undergo periods of dormancy as in environments with very contrasted climates, they do take advantage of periods which are definitely more advantageous than others, and which thus impose a certain rhythm upon them.

These conditions have produced a vegetation in which trees predominate, of a height unequalled in other regions. Their slim trunks carry a continuous canopy at a height of some forty metres, which seen from above gives the impression of a 'sea of foliage' – always green since the characteristic rainforest trees keep their leaves throughout the year. Wherever the rainfall regime shows a well marked dry season, rainforest gives place to deciduous monsoon forest in which leaf-fall marks a dormancy during this unfavourable season. Flowering and fruiting in rainforest also extend throughout the year.

Rainforest forms a three-dimensional world. In some primary forests the density of the canopy prevents the development of lower plant layers by excluding light, and the ground between the gigantic trunks is remarkably open. However, there is always a stratification of the plants, and some botanists such as Richards recognize no fewer than five distinct layers. More simply, the ground may be covered by a layer of low herbaceous plants; over this is a layer of dwarf and young trees, palms and tree-ferns; and over all is the layer of large trees 35 to 40m high, branching only to form the canopy, which in turn is overtopped here and there by giants. Epiphytes grow on all the trees, and lianes penetrate all the plant layers. Despite appearances, however, this forest is not uniform, and many types are distinguishable by their structure. River banks and the edges of the forest show further differences, which explain their diversity in floristic composition and physiognomy.

These forests, which are made up of trees with straight trunks supported by buttresses, of which the tallest at least have spreading crowns, are never monospecific or largely dominated by a single species as in temperate regions. Instead, the species are very numerous and intimately intermingled. It has been estimated that there are at least 3,000 tree species in Indonesia. 423 trees belonging to eighty-seven species have been counted on an area of 1·2 hectares in the Amazon forest. On an area of 175m² near Manaus in Brazil, 1,652 plants more than a metre high have been counted, belonging to 107 species and thirty-seven families. Of these, the woody species contributed 77·6 per cent of the individuals with sixty-nine species and twenty-five families, and the herbs only 6·7 per cent with eleven species and six families.

Rainforest is an environment of very high primary productivity. Although the days are relatively short all through the year, because variations in photoperiod are minimal near the equator, there is no period of dormancy as in zones of contrasting climate. The primary production is nearly constant throughout the year and reaches very high values as a result of the particularly favourable conditions of the environment, so that very flourishing food chains have been able to develop from the large plant biomass. The populations of various consumers may be large at every feeding level, so that the total rainforest biomass is very high. However, an important part of the primary production consists of plant storage substances which are not involved in any immediate transfer of energy to other levels of the ecosystem.

Thanks to constancy of the physical environment and the resulting greatly reduced annual fluctuation in standing crop biomass, productivity in both plants and animals is virtually constant throughout the year, at least at the scale of the whole ecosystem. Although the species never show the large fluctuations characteristic of more contrasting climates, each follows a definite cycle. However, these cycles are somewhat staggered in accordance with specific ecological demands, so that the total remains almost constant.

Rainforest is without any doubt the most complex environment known. Although its general characteristics – heat, humidity and constancy – are uniform, environmental conditions, both physical (especially daily variations in humidity and illumination) and biological, change within a few metres. As a result there has been unlimited specialization among the animals, and like the flora the fauna is incredibly rich in species. Rainforest is also the most enclosed environment, which involves all the animals, and especially the birds, in a series of very marked adaptations.

When primary forest is felled it regenerates as secondary forest, of different physiognomy and very distinct floristic composition. It is lower, without a continuous canopy of very tall trees, but the lower levels are much more diversified. Its productivity (both primary of plants and secondary of insects and other invertebrates) is greater and more accessible to birds than in virgin forest. As a result (though birds which are strictly confined to primary rainforest do not of course penetrate the second growth) this supports large and highly diversified avifaunas of species which are absent from the upper levels of the virgin forest.

General characters of the avifauna

Though the tropical rainforest is so luxuriant, it appears at first sight to be devoid of birds and other animals, especially in comparison with certain savanna habitats. Many factors play their parts in this. Firstly, the density of plant cover prevents one from noticing the birds. Secondly, most of the species are solitary and therefore dispersed throughout the relatively homogenous environment, whereas savanna birds tend to be more gregarious and concentrate in the most favourable spots. But most important is the remarkably low density of birds in rainforest, each species being represented by relatively small populations. This rarity is reflected in museum collections, in which certain species are represented only by single specimens. Although this is due also to the difficulties of collecting in an environment which is difficult of access, it mainly reflects the low absolute densities of the birds. This poverty in numbers contrasts with an extreme richness in species. In no other habitat have the speciation and diversification of birds reached such a degree. Thus the forest avifaunas of the Andean slopes are of unequalled richness, variations in ground relief adding to the complexity of the forest environment to produce a very great variety of ecological niches, and to bring about extreme speciation. About 200 species have been counted in 5ha of forest in western Equador, while similar evolution has occurred in Africa and Asia.

The forest community behaves as though made up of very small pieces forming a very precise and complex mechanism, whereas those of other environments (as in temperate regions) are simpler, behaving as though formed of fewer but larger pieces. These contrasts are evidence of very different ecological mechanisms.

It is not however necessary to conclude that there are no dominant species in rainforest, and that every species is represented only by a few scattered individuals. In certain states of the vegetation at least, particular species may greatly exceed all the rest in numbers. Thus in the lowest plant layer in Gabon there are almost as many Yellow-whiskered Green Bulbuls *Andropadus latirostris* as there are birds of the eighty-two other species which live in this habitat. This species' success is no doubt due to its diet, which gives it an unchallenged niche, since this is the only bird in the Gabon rainforest specialized to take fruit, berries and flowers (Brosset 1966). While certain communities characteristic of tropical rainforests are thus dominated by single species, this is not generally true and suggests that conditions locally must be unbalanced or untypical.

Rainforests have been a sanctuary for ancient faunas, which are closely tied to them by very precise ecological requirements and extreme sensitivity to the smallest environmental change. These forests, especially those of the Amazon basin, have survived since the Tertiary under constant ecological conditions. There are geological records of trogons as far from their present range as France; but they were always tied to rainforests, or dense gallery forests. Other relict forms whose survival is no doubt explicable in the same way are the Hoazin in Amazonia, the Congo Peacock (the only African representative of the pheasants, which are now a highly-diversified group in Asia) and the rock-fowl *Picathartes* in West Africa.

Rainforest faunas tend to be especially rich along the edge of the forest and in second growth, where the juxtaposition of different habitats allows components from their faunas to mingle. Forest faunas are characterized negatively by the absence of some groups represented in the neighbouring savannas, and positively by many endemic groups. The transition from one fauna to the other is generally abrupt, paralleling the change in habitat since the forest often ends clearly in a defined face. Despite their mobility even the birds change in parallel with the rest of the fauna. Although certain forest species do occupy the guinean savannas these do not include any whose ecological needs restrict them to the dense forest.

The avifaunas of tropical rainforest are complex. Some of the characteristic groups are cosmopolitan, but almost all are represented by different species in each sector of their world ranges. The most characteristic of these groups are the parrots, pigeons and trogons; and the woodpeckers, barbets and other piciforms which are absent from east of Wallace's Line and from Madagascar. The New World rainforest faunas include tinamous, humming-birds, some endemic piciform groups such as the jacamars (Galbulidae) and puffbirds (Bucconidae), motmots (Momotidae), wood-hewers, ant-birds, manakins, cotingas and cocks-of-the-rock, tanagers and honey-creepers. In Africa the dominant groups are the turacos, forest guineafowl, hornbills and especially the passerines – above all some highly specialized thrushes, weavers, sunbirds and cuckoo-shrikes. Those in Asia are the hornbills, babblers and pheasants (many characteristic species of which inhabit the lowland rainforests). In New Guinea or northern Australia or both one finds cassowaries – the only ratites adapted to closed environments – birds of paradise, and many endemic groups of small passerines. There are raptors in these forests represented by remarkable and sometimes gigantic forms. Examples of the tendency towards large size are the Monkey-eating Eagle of the Philippines, the Crowned Eagle of Africa and the Harpy Eagle of South America, while there are smaller species. However, raptors are generally rarer than in open environments such as savanna, since closed ones are less favourable to birds of prey. These find difficulties in hunting there, where their prey can hide more easily and where it is more difficult to pursue them among the trees and branches. It is not surprising that, apart from a few species which are highly evolved and narrowly specialized for this environment, raptors are relatively rare and poorly diversified in rainforest.

Birds in the hot, wet environment

As we have seen, tropical rainforests are characterized by high though not excessive temperatures associated with very high humidities – an air temperature of 30°C with a relative humidity of 95 per cent are not exceptional – and such conditions persist throughout the year. Birds have thus had to adapt physiologically to an environment in which effective heat loss is difficult. This has apparently not necessitated any special adaptations, since the thermoregulatory mechanisms common to all birds are broadly sufficient. Metabolic rates are not slowed to cope with these conditions, despite the tendency for body heat produced by combustion to build up. Instead the birds counteract this tendency mainly by their behaviour, seeking out the most favourable microclimates. The birds in the tree tops are not exposed to such high humidities, since their habitat is dried out during the day by the sun; but on the other hand they have to protect themselves against the direct effects of this insolation. The birds of the lower layers are screened from sunlight, but have to contend with a relative humidity which is always very high. Forest birds in general seek out the shadows, so as to avoid direct solar heating, and (like all birds though still more obviously) are especially active at daybreak and again at nightfall, remaining inactive during the heat of the day. They bathe frequently in puddles, thus increasing their heat losses. Thus on the whole neither the physiology nor the behaviour of rainforest birds is very unusual. Normal function, with only quantitative adjustments, seems adequate to ensure an equilibrium with the environment.

The coloration of forest birds

Rainforest avifaunas are remarkably varied, with some species which are extraordinary in form or ornamentation: brilliant birds such as the humming-birds, the most richly-coloured such as the manakins and tanagers, and the most strangely ornamented such as the pheasants and birds of paradise.

However, this generalization needs to be qualified by reference to the ecological niches occupied by the various species, and in particular to the plant layers which they occupy. Canopy birds, living at the tops of the tall trees, are generally very brightly coloured with metallic glosses or vivid hues. In contrast the birds of the lower layers, and especially the undergrowth, are much duller, their coloration dominated by melanic pigments which give them brown, black or olive-green plumages. They are usually lacking in contrast, with patterns which are either vaguely defined or made up of bars or streaks, like many of the Old World thrushes and babblers and the antbirds and wood-hewers of the New World rainforests. Thus there can be striking contrast, even within a single family, between the birds of the lower layers and of the canopy. Among the hummingbirds, the species of *Phaeotornis* live in the low bushes of the undergrowth and have dull plumages dominated by brownish colours; whereas the more typical species which frequent the treetops or the riversides show the flashing coloration for which the family is famous. Similarly among the African sunbirds, *Cyanomitra olivacea* which keeps to the lower layers is dull olive-grey; whereas the remaining species – all of which belong to the canopy or other exposed habitats – are brilliantly coloured.

Gloger's rule, that colours (and especially those due to melanic pigments) are more intense in wetter environments, applies best to the birds of the lower layers. There are remarkable exceptions to this rule, and some forest birds such as Rothschild's Mynah *Leucopsar rothschildi* of Bali and several South American cotingas (*Procnias*) are wholly white. This also raises the question of cryptic coloration. Treetop birds probably have no need of camouflage and can carry their gorgeous plumage without disadvantage; but their seemingly conspicuous coloration may in fact be cryptic. It is possible that in the play of colours of the leaves, the bright colours of the flowers, and the brilliant light, these birds are actually less conspicuous to their enemies than if they were dark in colour. In contrast the blackish, brownish or greenish colours of the undergrowth birds blend admirably into the shadows and dense vegetation. Birds which keep to the tree-trunks are often coloured so as to blend with the bark, like some brown or black woodpeckers, (in which, even where there are patches of bright colour, these are neither vivid nor glossy), and especially the South American wood-hewers, whose plumages are mottled with brown, whitish and blackish. The pootoos, with their marbled plumage, blend perfectly with the branches on which they sit motionless, providing one of the best examples of cryptic coloration.

It may be suggested as a hypothesis that, finding themselves in an environment rich in foods and physically favourable, they are able to use part of their energy resources in 'luxurious' ornamentation. However, the high density of species in rainforest must favour the development of characters which assist in discrimination between the species and the sexes. The diversification of coloured markings and ornaments in these forests is certainly correlated with the unrivalled degree of speciation and specialization.

Distribution among the strata of vegetation

Both temperate and rain forests form three-dimensional worlds, in which birds are located not merely by their ground position but more significantly by their height above the ground. It has been shown that several distinct communities can exist one above the other with a minimum of inter-penetration.

The species are distributed in this way primarily according to their food preferences. Optimum illumination must also play a part in habitat selection since the amount of light varies so much between the different layers. Thermal conditions too, and especially the daily temperature fluctuations, differ considerably and this must have a great influence on the birds. In combination these factors certainly control the regular stratification of the avifauna, as a result of which the birds compete less intensely and can exploit the whole standing crop biomass throughout the thickness of the forest mantle.

The lowest layer is that of the ground itself, on which dead leaves and other plant debris forms a thick carpet rich in foods, of both plant and animal origin. This layer, which is often very open in primary forest, is occupied by a series of terrestrial birds, often with weak powers of flight, which glean their food from the ground litter. Examples are the pheasants, partridges and forest guineafowl of the Old World and the tinamous and curassows of the New World. Most of these birds even nest on the ground, only the more arboreal curassows nesting in trees. Certain forest passerines are of equally terrestrial habits, as is shown by their strong legs and long tarsi. This is especially true of the Amazonian forest antbirds (Formicariidae) which hunt insects (particularly ants) by scratching the ground vigorously with their feet. Certain Old World thrushes and babblers share this way of life.

Most rainforest birds, are arboreal. Some, such as the Old World pittas which readily descend to search for insects and small molluscs on the ground, keep to the low or intermediate layers. Certain Old World flycatchers, especially the blue flycatchers *Hypothymis* of the Far East and the African broadbills *Smithornis* and certain American tyrant-flycatchers and hummingbirds (including *Phaetornis*), similarly live in the low bushes and dwarf trees.

A little higher above the ground in the Old World live white-eyes *Zosterops*, whistlers *Pachycephala*, and certain far-eastern kingfishers. This level is occupied in the New World by different hummingbirds such as

Mellisuga, many wintering wood-warblers Parulidae, certain antbirds, tyrant-flycatchers, manakins and a few tanagers. Woodpeckers, woodhewers and other birds of the trunks and large branches are found mainly at this level.

The crowns of the tall trees which form the canopy are also densely populated. Here are concentrated the fruit pigeons, parrots, toucans, hornbills, trogons, and the birds of paradise and many smaller passerines. Sunbirds, tanagers and many finches are abundant, attracted by the fruits and flowers. Particular species of flycatchers, warblers and tyrant-flycatchers frequent this level where they catch many insects, while this is also the habitat of most South American hummingbirds. Finally, it is from this canopy that raptors (especially the monkey-eating eagles, harpies and others which feed mainly on primates) fly out over the forest to seek their prey.

The distribution of bird species among the plant layers in Malaya is as follows (Harrison 1962):

Zone	No. of species	Diets: Vegetarian	Mixed	Insectivorous	Carnivorous
Air	17 (6%)	0	0	12 (4%)	5 (2%)
Canopy	79 (26%)	38 (13%)	26 (8%)	15 (5%)	0
Middle-zone	163 (54%)	0	30 (10%)	121 (40%)	12 (4%)
Ground	47 (15%)	4 (1%)	17 (6%)	26 (8%)	0
Total	306 (100%)	42 (14%)	73 (24%)	174 (57%)	17 (6%)

This shows that vegetarian birds are dominant in the treetops, though insectivores may be almost as well represented there and clearly dominate the intermediate layers. This distribution corresponds to the productivities of the layers at different trophic levels. Similar results have been obtained in Australia, while the mammals also show this correspondence between vertical distribution and diet.

Adaptations to arboreal life

The most important requirement for survival in a forest environment is adaptation to arboreal life. Apart from a small number of terrestrial birds, all forest birds are adapted to living in the trees – so closely that mainly because of their short tarsi most of them are unable to settle on the ground. These birds have feet like pincers with which to grip the branches on which they perch. In the passerines only the first digit is turned backwards, but although short it is strong, and sometimes broadened so as to oppose the middle and other front toes. Especially in certain kingfishers and bee-eaters and even some passerines, the third and fourth digits may be fused together by their basal phalanges so as to ensure a firmer grip. This configuration is carried much further in those birds which early systematists grouped as Scansores or climbers. In the parrots, cuckoos, woodpeckers and other piciforms the fourth digit is turned backwards, so that the foot is transformed into a real pincer with two forward-pointing toes opposed to two pointing backwards. The reduction of the first digit seen in certain woodpeckers hardly weakens the grip, since the opposed anterior and posterior elements remain of equal strength. In the trogons it is the second digit which is turned backwards to join the first, while the parrots climb using their bills as well as their modified feet. However, the abundance of such birds in tropical rain-forest (though explicable also by their diets) is indisputably related to their anatomical peculiarities, which have to be considered in terms of their environment.

Woodpeckers and woodhewers have strong feet whose toes end in recurved claws, giving them a firm grip of the bark. Their tails are specialized, with stiffened quills to the rectrices so that they can be braced against the trunk, and so balance the force which tends to topple the bird backwards. Thanks to this peculiarity these birds are remarkable climbers, able not only to run about on trunks with great ease, but even to work there digging in the wood and gathering insects. Pygmy parrots *Micropsitta* of the New Guinea region also have such rectrices, and use them in their similar way of life on the tree-trunks.

In contrast, the flying abilities of most rainforest birds are poorly developed. Some fly weakly and only over short distances. A few, such as the flycatchers and tyrant-flycatchers which hawk in clearings and along the forest edge, can fly fast but only do so in chase. The remainder seek their food among the branches and foliage, and thus have no need to travel at high speeds or over long distances. The wings of these birds are generally rounded, a characteristic shown by all rainforest groups from passerines to parrots and trogons, and associated with relatively slow flight. Together with the large wing area this very rounded shape gives high lift, allowing slow agile flight through spaces closely packed with many obstacles, giving these birds superior manoeuvrability. The same evolutionary trend is shown by the raptors, allowing them to weave dextrously among the trees.

As a final arboreal adaptation, the wings of young Hoatzins *Opisthocomus hoazin* carry free clawed digits, with which they cling securely to branches and move about among them from an early age. This is especially important for these birds, since they nest in trees overhanging rivers, into which they would fall if they lost their grip. Thus these structures, though they undeniably represent the survival of an ancient organization, are of positive value as an adaptation to the forest environment.

513

Forest water birds

Forest regions are not lacking in aquatic habitats. At the height of the rains the great rivers flood the low-lying parts of the forest, to produce an environment which shares the characteristic of aquatic and wooded habitats. This is especially true of the Amazon and parts of the Congo basins, where huge areas are permanently or temporarily flooded.

Part of the forest avifauna, (especially ibises, wood-ibises and other storks, herons and egrets), finds its food in these watery habitats. These birds find within the forest an environment remarkably well suited to their ecological needs, since they can feed beside the rivers and in the flooded areas, while the bordering trees offer sites for their nesting colonies. They are therefore widely spread along forest watercourses and many species are even confined to the forest, including the tiger herons *Tigrisoma*, found from New Guinea to central Africa and tropical America whose plumage is remarkably cryptic. These birds migrate regularly in accordance with the height of the water, low water being most favourable to them since it leaves uncovered higher areas rich in food. Though these migrations do not cover great distances they are none the less regular, especially in South America and Africa.

Ducks too are represented by specialized forms in this environment, including species of the genus *Anus* and whistling ducks *Dendrocygna*, as well as species of *Cairina* distributed across the forests of tropical America (*C. moschata*, whose domesticated form is known as the Muscovy Duck), Africa (*C. hartlaubi*) and South-East Asia (*C. scutellata*). Of nocturnal habits, these birds roost in tall trees during the day, and at twilight begin to feed in flooded and swampy areas, while they usually nest in holes in trees. They thus show a double adaptation, to life both in the forests and in the water.

Diet

Rainforest is very productive, especially at the primary level. Complex food chains develop from the considerable plant biomass, and very large energy transfers take place along them at variable rates.

Productivity at every level is remarkably constant throughout the year, thanks to the stability of the physical environment and especially the climatic conditions. These conditions are thus in marked contrast to those in savannas, where vital functions are forced into cycles by the annual climatic fluctuations.

The constancy of the environment also allows the birds of tropical rainforest to remain in limited areas throughout their cycles, so that they are exceptionally sedentary with only a small proportion of migrants. Sedentary habits also explain the morphology and structure of the locomotor apparatus characteristic of these birds, and especially their wings which are ill-suited to long flights. On the other hand, the frugivorous species especially may be more or less nomadic. Significant, though often very small differences in rainfall from season to season and from place to place result in variations in plant cycles, especially in the ripening of fruit. As a result frugivorous birds move about within limited areas in order to exploit this varying food supply. Parrots, hornbills, toucans, and many frugivorous passerines, as well as nectar feeders, thus travel on a local scale in response to fruiting and flowering seasons, but these small though regular migrations do not take any of the birds outside the forest.

Further, almost no migrant from temperate regions winters in the forest, at least in the Old World and Amazonia where only swallows regularly penetrate the equatorial forest blocks. While the forest environment is much too closed for the birds of temperate regions, a special ecological situation is certainly largely responsible for this absence, since the rainforest – enjoying constant conditions throughout the year and providing correspondingly regular food supplies – does not supply the seasonal surplus to support a varying population. All the available ecological niches are thus firmly occupied by precisely adapted sedentary species. However, wintering migrants do immediately occupy closed areas, which resemble their breeding habitats in physiognomy (Brosset 1968).

Among the very wide range of foods available to rainforest birds, a few classes are especially important because their productivity is so great. Certain levels of the food chains are especially favoured, whereas others are neglected. The only bird known to use foliage for food is the Hoazin of the Amazon forest, which feeds mainly on the leaves of the water plants (*Monstera*) choking the beds of the forest rivers.

The nectar secreted by the flowers of many trees is much more highly esteemed, and hummingbirds, honey-creepers, sunbirds, flowerpeckers and honeyeaters all make it an important component of their diets. Fleshy fruits provide the most important plant resource, playing a principal part in forest communities in contrast to the situation in savannas. As a result fruit-eating birds are highly diversified in the forest, where they form large populations. Finches, tanagers, cotingas, hornbills, toucans, turacos, trogons and green pigeons are all primarily frugivorous, although some of them also take a sizeable proportion of insects and other small prey. Most parrots too are fruit eaters, and each species takes a diet to which the development of its buccal apparatus is adapted. Even a few raptors very willingly take fruit, like the Palm-Nut Vulture *Gypohierax angolensis* which feeds largely on the fruit of the Oil Palm. It is in rainforest that birds play the most important parts in disseminating the seeds of the various plant species whose seeds they take, as in the important interaction between parrots and the oil palm. In contrast grass seeds are rare in tropical rainforest. As a result there are few gramini-vorous birds there, and no truly specialized species.

Insects are certainly the most important food resource for carnivorous birds, since they are the main consumers of the plant cover and their biomass and productivity are enormous. Apart from innumerable tree-living and wood-boring forms, the populations of termites and ants are especially large, and include plant-eating species of considerable importance in the transfer of energy from plants to animals. Molluscs, worms, spiders and millipedes also play large parts in this, and many of them are eaten by birds. Insectivorous birds, highly diversified in the rainforest, include a very long list of species belonging to the most varied families and exploiting all available food resources. Many, such as bulbuls and some thrushes and tyrant-flycatchers, hunt insects among the foliage, searching for them on every leaf, while others including the most typical flycatchers and tyrants pursue them in flight. Wood-boring insects are sought by woodpeckers and woodhewers and certain birds are specialized for taking caterpillars. The many ant-eating birds deserve special mention. Many ants, notably the soldier-ants *Eciton*, pass through the forest in long columns whose move-ments are followed by flocks of birds. This is especially true of the ant-birds of the Amazonian forest whose diet is based largely on these insects, but they are far from being the only ones since bulbuls and babblers take the same diet in the Old World. Others again, such as another South American antbird the White-backed Fire-Eye *Pyriglena leuconota*, specialize in the capture of insects which are thrown into turmoil by the intrusion of the ants.

Other carnivorous rainforest birds feed on vertebrates. Toucans, hornbills and certain raptors hunt arboreal reptiles, while other predators specialize in the capture of mammals – especially the Harpy and Monkey-eating Eagles and the African Crowned Eagle, which hunt arboreal mammals (monkeys and squirrels) and sometimes parrots and other canopy birds. Finally, in the great African forests the Bat-Hawk *Machaeramphus alcinus* is specialized in the capture of bats, its crepuscular habits and aerial agility enabling it to follow them in flight.

Sociability

Birds typical of rainforests are mainly solitary, or live in pairs rather than gregariously, large concentrations appearing only where a flowering or fruiting tree attracts appropriate species. This characteristic solitariness is especially obvious when the members of a single systematic group occupying different habitats are compared. Thus among weavers (Ploceidae) gregarious species are common in the savannas, where they form large flocks during the breeding season, whereas the characteristic forest species live much more retiring and solitary lives; and the same is true of guinea-fowl. However, there are notable exceptions to this rule, especially among parrots, most of which are notoriously gregarious. The birds of the lowest plant layers are generally the least gregarious of all, and those of the canopy somewhat more sociable.

Forest birds use various means of communication. Calls are especially important and many species have very resonant voices, so that one hears far more birds in the forest than one sees. The bellbirds *Procnias* among Amazonian cotingas, and certain African birds such as the Giant Turaco *Corythaeola cristata*, are famous for their resounding calls. Visual means of communication are equally important, at close quarters at least, especially to canopy birds. Visual stimuli play an important part alongside auditory ones, in the very elaborate nuptial displays of manakins and cocks-of-the-rock. In contrast aerial displays are less widespread, though those of hummingbirds and of the African Lyre-tailed Honey-Guide *Melichneutes robustus* (whose significance is still unknown) are notable.

Whereas forest birds never form large concentrations of a single species, they do freely associate in multispecific flocks. This peculiarity is almost exclusive to rainforest, being much rarer in other habitats. These flocks – which may include up to seventeen species, each represented by only a few birds – have real social structure, and are not as was long believed mere concentrations of individuals. Certain species are dominant and lead the flock, which forms round these 'nuclear species'. Although such bird 'patrols' are not permanent, but form every morning and break up during the afternoon, they maintain relatively uniform composition within a particular avifauna. This composition allows a better utilization of the environmental food resources, each bird benefiting from the activities of the others as it seeks its own prey. The 'leader' of the flock (which is essentially a hunting party) reveals itself by special agitated behaviour, showing common traits whether it is a tanager, hummingbird, tyrant-flycatcher or cuckoo. These behaviour-patterns have a social role in maintaining the cohesion of the flock, and serve to confuse potential prey to the advantage of all members of the party. There is generally no competition, since the diets of the various species are more or less complementary, and though all will exploit an abundant food this involves no disadvantage to the flock as a whole. Finally, this interspecific association provides better defence against predators (Brosset 1969).

Reproduction

The forest environment being relatively constant throughout the year, birds can breed in it without such obvious periodicity as is seen in regions with contrasting climates. In fact, nesting usually extends over a large part of the year, and often there is no more than increased activity during one part of the annual cycle and reduction at another. The breeding of different species within most groups is desynchronized, mainly as a function of diet, so that in most tropical rain forests nesting goes on throughout the year. The succession of rainy and of less wet, or even dry, seasons has impressed its rhythm on the annual cycles shown by the plants and most of the animals. It influences the birds, which seek the best dietary breeding conditions, and thus mould their rhythms to those of the environment.

Thus, rainforest birds do not breed indiscriminately throughout the year. Though their nesting seasons are much longer than those of savanna birds, they are still restricted. There are usually one or two peaks of breeding, and a trough or even a complete stop during the driest season.

In Guyana birds (notably the pigeons, hummingbirds and tanagers) have breeding seasons which are prolonged or even continuous – though often with a well-marked peak in March to April at the end of the minor dry season,

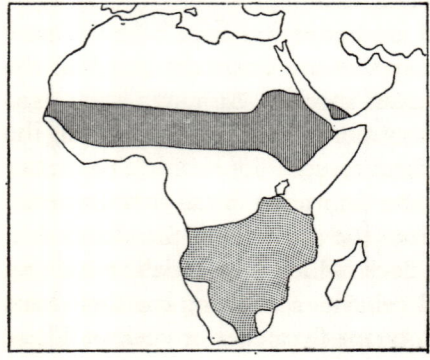

Figure 88. The migrations of the White-bellied Stork *Sphenorhynchus abdimi*. Nesting area: vertical lines, wintering area: dotted.

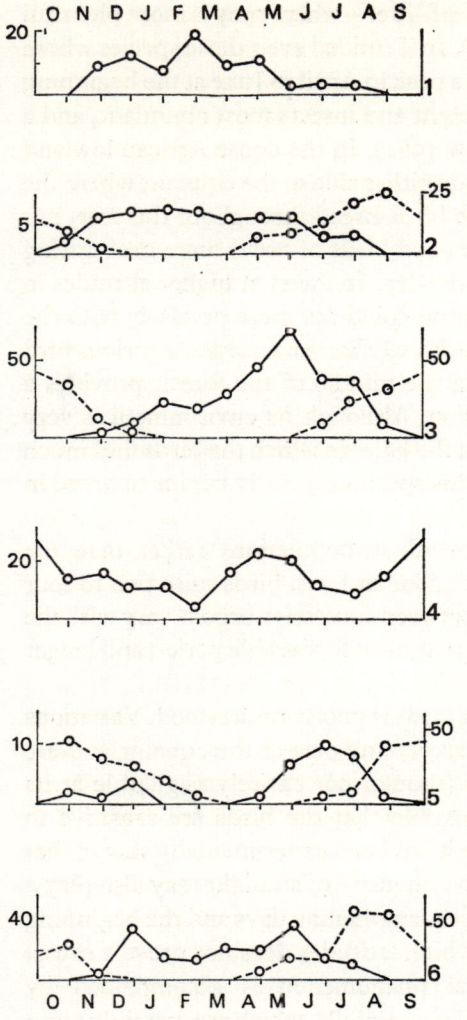

Figure 89. The breeding and moulting seasons of various tropical birds in Trinidad. 1. the Oil-Bird. 2. humming-birds. 3. a manakin. 4. a wren. 5. a honeycreeper. 6. a tanager. Continuous line, number of nests (left-hand scale). Broken line, percentage of the birds caught which were in moult (right-hand scale).

and another less distinct peak in September at the start of the major one. In Surinam few birds breed during the long dry season from August to November and many begin to nest in January, though some (especially the ducks) do nest during the dry season (Haverschmidt, Davis 1953). In Costa Rica there is a well-defined breeding season with its peak during the rainy season from April to June, after which breeding slows down considerably during the last months of the year to increase again from January. Hummingbirds and honey-creepers nest mainly from November to February, during the dry season when most plants flower. Graminivorous birds also nest rather late, as do those which find their food in the leaf-litter – where prey is more plentiful after thorough soaking (Skutch 1950). In Trinidad even those species whose breeding seasons are prolonged show a peak in April to June at the beginning of the rains, when fruiting is at its height and insects most abundant, and a less marked peak in November (Snow 1964). In the dense African lowland forests extending up to 3° of latitude on either side of the equator, where the dry season is scarcely discernible, the birds breed throughout the year, but (apart from the pigeons, woodpeckers and birds of prey) more nest during the wettest season and fewer during the dry. In forest at higher altitudes in East Africa (Kenya) the breeding season coincides more precisely with the wettest period. The White-necked Rockfowl *Picathartes oreas*, a curious bird which nests in rock-holes opening in the middle of the forest, provides a remarkable example of synchronization. Although its environment is very constant, and the microclimate within the holes in which the birds find much of their insect food is even more so, this species regularly begins to breed in March (Brosset).

The foregoing statements concern whole populations rather than the breeding cycles of individuals. In Trinidad at least, birds raise two to four broods each season and the intervals between successive broods vary with the species, while they are shortest during the most favourable period and longer outside it (Snow).

Control of breeding cycles in forest birds is poorly understood. Variations in the length of daylight can play scarcely any part at the equator and are minimal throughout the forest zone (though not entirely negligible at its highest latitudes). However, it is possible that the birds are sensitive to small variations in photoperiod, since it has been experimentally shown that some can react to such variations. The intensity of sunlight may also play a part. However, the coincidence between lengthening days and the beginning of breeding in forested areas at fairly high latitudes does not prove a causal relationship, since the astronomical phenomenon is accompanied by changes in the environment – especially in rainfall – which are certainly more important. Control of breeding in this environment is in fact to be sought much more in the rhythm of the rains, which produces fluctuations in the biomass available to each feeding class. In places where photoperiod is constant but the rainfall regime markedly seasonal, the breeding seasons are very distinct and co-ordinated with the latter. The lack of synchronization between the breeding peaks of the various feeding classes is another argument for this relationship. An internal rhythm controlled by the endocrine system may be postulated, in which a pituitary cycle uninfluenced by external factors could determine the phases of breeding and at least initiate it. This mechanism is especially evident in certain birds which begin to nest on the same fixed date year after year, however late the rains may be. Thus rainforest birds universally show annual rhythms, though with much reduced amplitudes varying according to local conditions. Rainforest birds show undeniable signs of increased sensitivity to environmental factors.

The moult shows a rhythm parallel to that of breeding, to which it is of course linked. In the forest birds have only one adult plumage and moult once a year. This moult takes place at a season which is usually clearly defined, at the same time from year to year in the same place. A young bird's first moult, by which it passes into the adult condition, is probably very important in synchronizing its cycle with those of the rest of the population.

The reproductive rates of most forest birds, like their population structures, are poorly understood. From what is known of the White-bearded Manakin *Manacus manacus* in Trinidad (Snow 1962), its reproductive rate must be low since the clutches consist of only two eggs. Nest mortality is high, only 40 per cent of the eggs hatching and 19 per cent yielding fledged young. Almost all the losses (86 per cent) are due to predators, tree-snakes taking an especially heavy toll. An average of only 0.33 young fledge from each clutch, but a female lays several clutches during the year, each following three to six weeks after the preceding young have fledged. The number of clutches per year varies from two to five, but is most often three. This means that a female raises an average of one young per year, which is decidedly less than the passerines of temperate regions achieve. This low reproductive rate, which appears to be general among rainforest birds, is surprising. Most American birds characteristic of this habitat lay clutches of not more than two eggs (Skutch). In Africa rainforest birds have smaller clutches (by an average of half an egg per clutch among the waxbills, shrikes, weavers, babblers and thrushes) than closely related forms which nest in savannas, while both are less prolific than their representatives in temperate regions (Lack & Moreau). On first principles these birds should be able to raise more young because of the richness of their environment, even where (as among the manakins) the female alone is occupied in raising them. Although it takes longer to hunt the insects on which the young are fed than to gather the fruit on which the adults feed themselves, it does not seem that the limiting factor can be the food available: the White-bearded Manakin devotes only 10 per cent of its day to gathering food for its young. It has been suggested that the low reproductive rate is related to the heavy predation, since a number of young would demand more frequent visits by the adults to the nest and correspondingly greater danger of location by predators. Rainforest birds do seem to show a more extended rhythm in their visits to and absences from the nest, as in the Red-crowned and Sooty Ant Tanagers *Habia rubica* and *H. gutturalis* (Willis).

Balancing the low rates of reproduction and productivity, the mortality rates of adults are also low. The annual mortality rate among adult White-bearded Manakins does not exceed 11 per cent. Populations are thus renewed only slowly in marked contrast to the situation in other habitats.

The bird populations of rainforest also seem not to be in harmony with the amount of food available, since there must surely be a surplus of food which the birds do not exploit. However, this needs to be confirmed by quantitative study of the available foods, and it may be that certain specific foods – especially animal proteins – are in short supply. Psychological factors may also play a part in restricting populations, especially through the delimitation of territories.

Finally, it is noteworthy that few birds of the great equatorial forests lay down periodic fatty reserves. Birds from regions of more contrasting climate, fatten at the end of the favourable period so as to lay down reserves ready for the dry season, the variety and quantity of foods which are more or less constantly available in rainforest make such cyclical changes unnecessary.

The bird communities of tropical rainforests

We have repeatedly stressed the complexity of the avian communities of the great equatorial forests, paralleling the complexity which characterizes all their plant and animal groups and correlated with the luxuriance and thickness of the plant cover and the structure of a biocenose very rich in ecological niches. Here speciation has reached its maximum rates, so that many birds have evolved. Some of these are strictly confined to this habitat by their ecological demands whereas others have colonized other habitats, which have thus acquired species moulded by novel ecological imperatives.

Birds have established themselves and diversified so as to exploit all the food resources of the rainforest environment. Chief among these resources are fruit (but not seeds) and insects – especially termites and ants. The majority takes a mixed diet so that it can exploit the greater part of the environmental resources and few are confined to narrow diets.

The avifaunas of all tropical rainforest blocks include a common element, largely of an ancient type maintaining itself since the Tertiary in an environment which has remained homogeneous and stable throughout the ages. Such an element is provided by the trogons, which are found in America, Africa (though poorly represented there) and tropical Asia. However, the early separation of the continental masses, and the encirclement of rainforests by wide zones of very different ecological conditions, have caused most other groups to evolve independently in each of the resulting widely separated areas within the belt of equatorial forest. Most families are represented by different genera or even subfamilies in each sector, whose avifaunas are even more distinct than this suggests, since certain families have evolved in single regions to which they are endemic. This is especially true in Amazonia, whose avifauna is clearly distinct from those of Africa and Asia, whereas these show much clearer faunistic affinities with each other.

Although distinct, the evolution in each continent has been along parallel lines. Thus equivalent ecological niches in the rainforests of each continent are occupied by birds which are not at all closely related, but which have evolved in parallel to the point of showing the same morphological and biological characters. The following table summarizes several of these cases of parallel evolution and ecological replacement from one geographical area to another.

Niche				
Layer	Diet	America	Africa	Asia
Ground	fruit	tinamous, curassows	guineafowl, Congo Peacock	pheasants, partridges
	invertebrates	antpittas, antpipits	pittas	pittas
	ants	antbirds	thrushes	babblers, thrushes
Dense undergrowth	insects	ovenbirds, tyrant-flycatchers	thrushes, flycatchers	babblers, flycatchers
Treetrunks	insects	woodpeckers, woodhewers	woodpeckers	woodpeckers, nuthatches
Middle layer	insects taken in flight	puffbirds, jacamars,	bee-eaters,	bee-eaters, broadbills,
		tyrant-flycatchers	shrikes, drongos	drongos
	caterpillars	cuckoos	cuckoo-shrikes	cuckoo-shrikes
Treetops	fruit	tanagers, parrots,	weavers, parrots	parrots, green pigeons
		toucans	green pigeons, turacos hornbills, orioles	hornbills, orioles
	nectar and insects	hummingbirds, honey-creepers	sunbirds, white-eyes	sunbirds, flowerpeckers, white-eyes, lories
	insects	tyrant-flycatchers	flycatchers	flycatchers

Though these comparisons may appear coarse at first sight they deserve detailed analysis, which shows many precise convergences between two families which have evolved in parallel.

Chapter 27

High Mountains

Altitudinal zonation

Living things are distributed on the slopes of high mountain masses in a precise altitudinal zonation, which recurs with variations in every region of the earth. Thus a series of more or less clearly defined zones may be distinguished by their climatic and biological characteristics. In European mountains – especially the Alps – a zone of hills reaches an average height of 600m or more, beyond which begins a highland zone which extends up to about 1,500m and is characterized by deciduous forests. Beyond this up to 1,700m (or sometimes to 2,300m under special circumstances) is the subalpine zone covered by conifer forests. Beyond this again, in the alpine zone which reaches the limits of permanent ice, trees are absent and the vegetation consists only of low plants forming scrubs and meadows, with the plant cover becoming ever thinner with altitude. This zonation is not rigid, since the limits of the zones can vary widely even within Europe, depending on the exposure and the local climate. It varies still more between tropical mountains, on which the alpine zone is much higher. In the Colombian Andes the zonation consists successively of a tropical zone up to 1,500-2,000m, a subtropical zone up to 3,000-3,200m, and then a temperate zone up to 3,700m. Here the alpine zone does not begin until about 4,000m. High altitude compensates for low latitude.

The distribution of the birds as a whole shows a broad altitudinal variation, from sea-level to the very limits of life at the feet of the glaciers, an upper altitudinal limit which may reach almost 6,000m. Indeed, birds can fly over the highest peaks, as they have been repeatedly seen to do in the Himalayas and Andes. Life in the highland and subalpine zones present scarcely any problems to birds. Their adaptations are of the same order as in similar environments at other latitudes, as can be seen especially in the deciduous and conifer forest of the Alps. The avifaunas of these zones are broadly similar. The alpine zone, in which throughout the world the birds have to face problems of adaptation to altitude and rigorous climatic conditions, is a different matter, and it is only this zone which we shall consider here.

The alpine climate

The alpine zone as a whole is characterized by extremely severe climatic conditions. This is partly due to the direct effect of altitude, which involves rarefaction of the atmosphere with very marked reduction in the partial pressure of oxygen. At 4,000m this pressure is only two-thirds of the sea-level value, while at 5,500m it is no more than half. This causes a reduction, directly proportional to the altitude, in the total amount of oxygen available and produces symptoms of hypoxia. The reduced pressure also affects the absolute humidity of the air – the quantity of water vapour which it contains – which diminishes still more rapidly with altitude. Thus at 4,000m, while the air pressure is two-thirds of that at sea-level, the absolute humidity is no more than a quarter (Rivolier).

Altitude also profoundly modifies the climate. Although the climates of alpine regions are far from uniform, a number of common characteristics can be recognized. Air temperatures tend to fall with increasing altitude, with a theoretical decrease of 0·6° C for 100m difference in level, though the actual values vary considerably with latitude and locality. In the French Alps in summer the 15° C isotherm is at 1,000-1,200m, the 10° C isotherm at 2,000m and the 0° C isotherm at 3,500m, while in winter the − 5° C isotherm is at 1,500-1,700m (Martonne). In the Swiss Alps the mean temperature at 3,000m is − 11·5° C in January and − 4·5° C in July, the annual mean being − 4·3° C (Corti 1955). The temperature drop is especially marked in the tropics. In the Peruvian Andes, at Vincocaya which is at an altitude of 4,320m, the mean temperatures are as follows (Weberbauer):

Warmest month (November)	4.8°
Coldest month (July)	− 2.3°
Annual	1.9°
Mean annual range	7.1°

Other localities at high altitudes in the tropics show correspondingly low temperatures.

Fluctuations in temperature with time of day are still more striking. Insolation may be intense during the day since solar radiation at ground level increases with altitude – by 2 to 4 per cent per 100m up to 2,000m and 1 per cent per 100m beyond that – because the rarefied air intercepts it less. Thus while the air is colder the soil receives more heat energy from the sun, rising from 0·8 calories/cm²/min at 100m, to 1·6 at 4,000m, so that it and objects exposed to the radiation are considerably heated. In shade and during the night, on the other hand, the rarefied air does not act as an effective insulator, and the soil loses heat by radiation, so that night cooling is intense. The fluctuations reach their maxima during dry periods, when no screen of clouds filters the daytime radiation and there is no precipitation to temper the nocturnal minima. High in the Peruvian Andes, in a tropical region, a diurnal maximum shade temperature of 20°C contrasts with a nocturnal minimum of −15°C, while surfaces exposed to the sun suffer still greater increases in temperature during the day. Of such a temperature regime it has been said that on a high mountain it is winter every night and summer every day (Hedberg). These rapid variations, which occur even during the daytime when clouds blow up to cut off the solar rays, must present greater problems to the animals even than the reduction in mean temperatures.

The direct effects of radiation on animals is still poorly known. At these altitudes sunlight includes a much greater proportion of ultraviolet wavelengths: four times as much at Briancon at 1,330m in the French Alps as at sea level. Besides the heating produced by insolation, living things are thus subject to radiations from whose highly specific action they need to protect themselves.

The lowered atmospheric pressure combined with the strong diurnal radiation results in drying of the air. The humidity, already absolutely low, falls sharply in relative value on insolation. According to observations made on Mount Kenya (Coe 1967) the relative humidity, 90 per cent at dawn, falls below 20 per cent within ninety minutes after the sun has climbed above the mists. Desiccation, especially marked in mountains which are either permanently dry or endure prolonged dry seasons, is accentuated by the violent and regular winds which are another feature of the alpine climate. They blow incessantly through certain corridors hemmed in by mountain masses, and over high plateaux such as those of the Andes. They also considerably increase the cooling power of the atmosphere, even when they are not charged with snow or ice, and their influence on life in high mountains is crucial.

The alpine zone of certain mountains is remarkably dry, as in some massifs in the Alps, and in the Mediterranean region and a large part of the Peruvian Andes. In the latter, rainfall during the southern summer is considerable (up to 1,100m) but at other times there is severe drought. Other mountains in contrast enjoy heavy rainfall spread over a longer period: the outer ranges of the Alps receive up to 3,000mm a year. Summer rain and a thick layer of snow in winter are most favourable to vegetation on temperate mountains, and also produce glaciers which further damp down climatic fluctuations. In the tropics certain massifs enjoy constant humidity throughout the year, as is especially true of the alpine zones of the Andes in Colombia and Northern

Equador and the mountains of East Africa, where rains and fogs are constant throughout the year. This humidity significantly affects the climate making it much more regular, without the fluctuations in temperature seen elsewhere on high mountains. The temperatures at 4,191m on Mount Kenya are as follows (Coe 1967):

Annual mean ...	...	...	...	...	...	+1.8°C
Mean maximum	...	...	...	...	...	+5.2°
Mean minimum	...	...	...	...	...	−3.5°
Maximum temperature recorded	...	...	...	...	+11.0°
Minimum temperature recorded	...	...	...	...	−6.7°

However, the constant high humidity itself presents problems to the birds, which have to resist the resultant cooling.

Although the climate in the heights of certain mountains is remarkably constant, as a general rule quite the reverse is true. At altitude in temperate mountains the summer is very short, just as in arctic regions. In Europe the growing period shortens by six to seven days for every 100m gain in altitude: in Switzerland it is only seventy-five days long at 2,100m, and scarcely fifty days long at 2,400m (Gensler). This period is also the only one favourable to the birds, which thus have to adjust their annual cycles and in particular their breeding seasons to this rhythm.

Winter snow is a major obstacle to birds in the mountains of the temperate zones, making their food inaccessible to most except the ptarmigans, and preventing them from moving about easily. Species of high levels in the tropical zone do not meet the same difficulties, since the snowline scarcely varies during the year. The frozen zone remains permanently uninhabitable, while below this the ground is free of snow which melts soon after falling.

These generalizations are subject to innumerable variations, since no other habitat is affected by so many local modifications producing such a wide range of microclimates. Exposure plays a major part, as is shown by the altitudinal differences in vegetation zones between the northern and southern slopes of a single valley. Exposure to the wind also has a dominant influence. Thus the alpine zone breaks up into a mosaic of environments, more or less favourable depending on their physical characteristics. As a result the distributions of species in high mountains are characteristically discontinuous, which has affected the course of their evolution.

Together these factors have produced a very specialized alpine vegetation, whose most obvious characteristic is the almost complete absence of treelike plants because of the short summer and the thermal conditions. Most of the vegetation is close to the ground, belonging to the herbaceous and shrub layers. Its physiognomy is often very characteristic, with a preponderance of cushion-shaped plants. The mostly perennial alpine plants can pass through their annual cycle very quickly: since they are ready to bloom from the melting of the snows, this is followed by a floral explosion. The alpine flora as a whole is poor, the number of plant species and their density decreasing rapidly with altitude (at levels which depend upon latitude and exposure) so that the vegetation soon becomes frankly desert-like.

Biotic factors are thus scarcely favourable to birds, since the vegetation offers them only meagre resources in food and shelter. The invertebrates which make up their potential prey are scarce, so that the standing crop biomass is relatively small.

Montane avifaunas

The severity of this environment has rigorously weeded out stocks originating in the lowlands. This faunal impoverishment is particularly striking in the tropics, in contrast to the very diverse lowland avifauna. The lower Amazonian slopes of the Andes form one of the richest regions in numbers of species, whereas the high Andean plateaux are singularly poor.

Physical and biotic conditions have excluded all species dependent upon closed habitats and all tropical birds which need high temperatures. Nevertheless, the alpine zones of certain mountains harbour some unexpected types of bird. Hummingbirds occupy the high mountains of South America up to 5,000m. Penetration by several members of this family, which is so diversified in the warm tropics, is explained by the very marked adaptations of these high Andean representatives, very different from their representatives in the Amazonian forests. The presence of a sunbird (*Nectarinia johnstoni*) in the alpine stage of high African mountains is more easily understood, since it occupies relatively moist and richly vegetated habitats – though the conditions are still very harsh for a bird which usually shows narrow thermic tolerances.

The avifaunas of the alpine zone in the various parts of the earth still show affinities with their lowland relatives, forming with them a single faunal entity.

The diverse origins of birds which occupy corresponding ecological niches in montane habitats in different regions of the earth make it all the more interesting that they show remarkable convergences. Many examples might be cited including the ptarmigan *Lagopus* – grouse of the snow zone of the Alps and parts of the Arctic – which closely resemble certain tinamous of steppes on the high Andean plateaux, where the seed-snipe

Thinocorus also show certain biological parallels. Thus similar environments, acting on birds of very different origins and systematic affinities, have moulded them into similarity.

In Europe the subalpine zone of conifer forests is occupied by characteristic birds: nutcrackers, crossbills, Hazel Grouse (though these are equally widespread in broad-leaved forests) and Capercaillie. Beyond this the true alpine zone shelters a small number of endemic birds, such as the Alpine Chough *Pyrrhocorax graculus*, Chough *P. pyrrhocorax* (which is also found locally on steep coasts), Snow Finch *Montifringilla nivalis*, Ring Ouzel *Turdus torquatus* (which also occupies Scandinavia and Scotland), Alpine Accentor *Prunella modularis*, Wall Creeper *Tichodroma muraria*, Water Pipit *Anthus spinoletta* (though curiously enough a specialized race, Rock Pipit *A. s. petrosus*, occurs only along the sea shore), Rock Partridge *Alectoris graeca* (which also has a Mediterranean range, seeking dry and sunny areas) and Ptarmigan *Lagopus mutus* (which does not avoid cold northern slopes).

In central Asia the montane avifauna is much richer, including (besides several species also found in Europe) game-birds such as the Tibetan Partridge *Perdix hodgsoniae*, Snow Partridge *Lerwa lerwa*, the snowcocks *Tetraogallus tibetanus*, *T. himalayensis*, *T. caspius* and *T. caucasicus*, some waders such as the Ibisbill *Ibidorhynchus struthersi*, and highly specialized passerines such as larks, corvids (Hume's Ground Chough *Pseudopodoces humilis*), finches (several rose finches *Carpodacus*) and thrushes (the White-capped Redstart *Chaimarrornis leucocephalus*). The alpine avifauna of the high mountains in the northern part of North America is related to that of similar areas in the Old World. However, it is not sharply distinct since the true alpine zone is very restricted, whereas the forested zones which reach very high altitudes carry a fauna closely related to that of the lowlands. The few species confined to the alpine zone – a ptarmigan, the Water Pipit and finches of the genus *Leucosticte* – have Asiatic affinities.

In South America a great mountain chain stretches north and south across the tropics. This spread of the Andes, bearing large areas of diverse habitats, has resulted in the differentiation of an alpine avifauna which is very diverse, though naturally much poorer than those of the lowlands. The dominant groups are tinamous, some raptors (of which condors are the best known), seed-snipe, ovenbirds, earth-creepers and miners (*Upucerthia* and *Geositta*), terrestrial tyrant-flycatchers (*Agriornis* and *Muscisaxicola*) and finches (*Phrygilus*), besides hummingbirds, several genera of which are confined to the high Andean zone. These almost continuous chains, within relatively easy reach of the mountains of North America across ecologically similar blocks in Central America, have been colonized by several northern elements, including certain water birds (coots, stifftails and moorhens) and passerines such as the Grass Wren *Cistothorus platensis* and pipits (*Anthus*). In contrast the birds of Patagonia, whose fauna evolved very long ago in the southernmost part of South America, have been able to return northwards thanks to the altitude and the similarity of habitats in Patagonia and in the high Andes from Peru to Columbia. Examples are Darwin's Rhea *Pterocnemia pennata,* the Andean Goose *Chloephaga melanoptera,* and various finches, ovenbirds (*Upucerthia* and *Cinclodes*) and tyrant-flycatchers. As a result of the unique situation and extent of the Andes, the mountain avifauna is distinctive.

In contrast the high mountains of Africa support only a very much impoverished alpine avifauna. The reduced area of the alpine zone and the wide gaps between alpine habitats, which appear as tiny islets set within expanses of very different ecology, explain the absence of a highly special-ized avifauna. The upper zone of Mount Kenya holds only sixteen species, while the same biotope on the other mountains is still poorer. Apart from species which are also found in the lower highland zones – especially raptors, swifts, grass warblers and serins – only three are characteristic of the African alpine environment: the Grey-Wing *Francolinus afer psilolae-mus,* the Scarlet-tufted Malachite, Sunbird *Nectarinia johnstoni* and the Hill Chat *Pinarochroa sordida* (which however is found in Ethiopia far below the alpine zone).

The alpine avifauna comprises several components, not all of which are truly montane.

Many other birds are found in the high mountains because this environ-ment offers them the crags which they need, so that they are not so much alpine as rupicolous. This is true of the Chough, which occurs also on coastal cliffs. Others are found only in the mountains because human interference has eliminated them from the lowlands. Examples are the large raptors, and especially the Golden Eagle *Aquila chrysaetos.*

Finally, some birds of the alpine zone are ubiquitous species, whose great ecological flexibility allows them to occupy very diverse habitats from the plains upwards. The Black Redstart *Phoenicurus ochruros* is found up to the snowline, while the Wheatear *Oenanthe oenanthe* is as well established on dry hills, moors and tundras as on rock-scattered alpine meadows. In South America the Olivaceous Cormorant *Phalacrocorax olivaceus* occurs from the seashore to the high Andean plateaux at almost 4,500m, near lakes which assure it ample supplies of fish. These aquatic environments, very well represented in the high Andes, have attracted communities of water-birds – ducks, coots and even a gull *Larus serranus* – despite the inhospitability of the Peruvian coastal deserts and Amazon basin to such an avifauna.

The populations, distribution and evolution of alpine birds

Alpine avifaunas in general are thus poor in species. The populations are also small, since primary production is low. However, the density of populations varies very much from place to place. Alpine communities are established in the most favourable places leaving others unoccupied. This can be well seen among the birds whose distribution is therefore character-istically discontinuous, with fairly densely populated 'islands' separated by 'deserts'. Within the populated areas, the density of nesting pairs often involves the reduction in area of their true territories (the hunting and feeding grounds sometimes being used in common by all the pairs establi-shed nearby). There is also a definite relationship between the areas of the territories and their richness in food resources. Thus the density of the Scarlet-tufted Malachite Sunbird *Nectarinia johnstoni* on Mount Kenya is directly proportional to the richness of the habitat, the populations being limited by the carrying capacity.

Fragmentation of populations has produced isolates which have pro-gressively diverged. As a result speciation in the montane zones has often been intense, especially in the mountains of central Asia and in the Andes. Hummingbirds of the genus *Oreotrochilus* are differentiated in the highest, and usually the most arid, zones of the Andes into a series of species which occur from Chile to Ecuador. The same thing is found among birds of the lower Andean zones. Thus finches of the genus *Atlapetes*, characteristic of the subtropical zone from Mexico to north-western Argentina, include very narrowly localized species alongside others of wide range: of the twenty-five species no fewer than thirteen are each confined to a very limited area in Ecuador or Peru. Thus the most favourable conditions for the differentia-tion of new forms occur together in the montane zones through the truly 'insular' isolation of small populations.

In other cases the course of evolution can be explained by the lack of competition brought about by faunal impoverishment. A founding stock has thus been able to differentiate into a series of sympatric species, each occupying a distinct ecological niche which under other circumstances would have been occupied by birds of different origins. Thus in the Andes

531

no fewer than five finch species of the genus *Phrygilus* (*P. gayi, P. fruticeti, P. unicolor, P. plebejus* and *P. alaudinus*) occupy the high plateaux of southern Peru, each showing morphological differences in accordance with its choice of habitat and diet.

It is noteworthy that this type of evolution is virtually confined to the Andes, probably because only this range has many large areas at high altitudes: most other alpine zones are too small to allow such extensive differentiation.

Evolutionary diversification among alpine birds has not significantly affected their plumage, since bright colours are rare among them. Most are more or less uniformly coloured, depending on the environments in which they live. Their brownish or greyish tints blend with those of the rocks, in contrast to birds of the forested zones many of which are distinctly more brightly coloured.

Some birds of the alpine zone take on white coloration for the winter. Thus the Ptarmigan is brown in summer, with a very complicated pattern, but turns white in winter so that it merges with its snow-covered surroundings.

Physiological adaptations of mountain birds

High altitude results in reduced atmospheric pressure, and in particular reduced partial pressure of oxygen. Like all montane animals, the birds have to overcome this as one of the most important adverse factors of their environment, through physiological modifications of their gas metabolism. In mountain species of mammals the number of red corpuscles is notably higher and the gas transport mechanism is modified. In particular the oxygen affinity of the haemoglobin is greater, so that sufficient gas is found despite its lower pressure.

Birds probably show comparable adaptations, though nothing is directly known about this. It is known that the hearts of mountain birds are more highly developed than those of their lowland representatives. According to studies made on a dozen North American species, the heart is always heavier (by at least 10 per cent) in individuals from high altitudes than in those from the plains. The pulmonary part of the heart is especially well developed (Norris & Williamson 1955) thus increasing the flow of blood through the lungs where gaseous exchanges take place.

Modifications of the egg-shell structure of mountain birds also seem to be related to the low oxygen tension. The canaliculi which traverse the shell are more numerous and often wider, so as to avoid asphyxia of the embryo which needs relatively intense gas exchanges during development. It is known that pigeon eggs do not hatch when the oxygen pressure is less than two-thirds of its value at sea level, corresponding to an altitude of about 3,300m, although various birds nest above 5,000m in the Himalayas and Andes.

In a quite different way reduced atmospheric pressure must make flying more difficult since the lifting power of the air is lower. This demands greater effort by the birds, especially during take-off and landing. Despite this, most mountain birds are excellent flyers, with long and often pointed wings. However, these anatomical modifications cannot really be considered as responses to the rarefied atmosphere, since other factors – especially the need to fly far and fast in search of food – have produced selection in this direction.

Physiological adaptations to altitude take place quickly. Strong-flighted lowland birds can fly at very great altitudes during their migratory flights. Ducks and geese fly at more than 8,000m over the Himalayas, during migrations from the Siberian tundras to the plains of India: the height record is held by geese flying above Dehra Dun in India at almost 9,000m. Such performances are astonishing in view of the low temperature at such altitudes, the considerably reduced oxygen pressure, and the muscular effort needed for flight in the rarefied atmosphere. However, they are limited in duration, and the problems of animals which have to live permanently at great heights are different. Apart from an immediate response, reflected in hyperglobuly, long-term adaptation involves much more profound physiological changes, which are more difficult to bring about and do not always take place.

Adaptations to cold

Despite the importance of these physiological modifications, the main adaptations to the alpine environment are ecological, altitude acting not so much directly as through other environmental factors which have much more effect on animals. Altitude as such does not set the upper distributional limit to any bird or other animal, since the availability of food – depending on the climatic factors which control primary production – is much more important. Thus birds regularly occur up to the limits of plant life.

The first adverse environmental factor with which mountain birds have to contend is the cold, to which they show some morphological and physiological adaptations. In accordance with Bergman's rule, the montane forms of any bird species are larger than their lowland relatives. Along the eastern slope of the Bolivian Cordillera, ten of the twelve species with altitudinal races are represented by larger forms at high altitudes. In addition, many mountain birds such as Ptarmigan have remarkably thick plumage, which efficiently protects them from the cold.

However, these morphological adaptations are only minor, and cannot explain the survival of birds in such a rigorous climate. Ecological adaptations are of much greater importance, and the most obvious of these is the search for the most favourable sites. We have already repeatedly stressed the very uneven distribution of bird populations within the alpine zone, depending on the considerable differences between local microclimates. Population densities vary considerably with the direction of slope, and with shelter from the wind – an especially adverse factor because of its high cooling power. As a result alpine birds seek out not merely the sunniest places, but more especially those best protected from wind. Rocks, including scree, piles of boulders and cliffs, are especially favoured. If properly orientated they provide very effective shelter, and contribute further towards a much more favourable microclimate by their high thermal absorption, warming up much more quickly and intensely in the sun than soil does. The following measurements of temperature in C°, made in the Peruvian Andes at Pasto Bueno (Pallasca Province) at 3,950m, show this eloquently:

Time:	0600	0800	1200	1400	1600	1800
Air	4·6°	6·9°	11·0°	10·3°	7·3°	6·0°
Soil	6·0°	7·2°	16·5°	17·0°	14·4°	12·6°
Rock	5·3°	8·3°	21·0°	23·6°	16·4°	14·2°

Thus the rocks act as a heat reservoir. The microclimate they produce favours the growth of various plants, which cluster in denser and taller associations than in less-sheltered sites. The microfauna, especially the insects, is also richer, and it is not at all surprising that the birds too reach greater densities.

Such variations in density between habitats are especially clear in the high Andean plateaux. The steppe-like pampas, covered with grass (*puna*) is occupied by a few highly specialized species. In contrast the sheltered valleys – especially the foot of any appropriately orientated rocky cliff – and gorges are occupied by many more birds belonging to more diverse species. Many alpine birds are found especially near rocks, scree and boulder-strewn meadows. In Europe the Snow Finch keeps to glacial moraines and escarpments, while the Alpine Accentor, Black Redstart, Rock Thrush,

18 Guillemots *Uria aalge* in their cliff colony. The adult in the foreground is of the 'bridled' morph, so called because of the white line which runs back from the eye.

19 Oystercatchers *Haematopus ostralegus*. They feed largely on bivalves (cockles and mussels), but despite their name do not eat oysters. A single bird can take in a day from 300 to 400 two and three year old cockles.

20 The mutual nuptial display of two Adelie Penguins *Pygoscelis adeliae* on the antarctic continent.

21 Emperor Penguins *Aptenodytes forsteri* of the antarctic continent. In the foreground young chicks gather in a 'huddle', an example of social thermoregulation which enables the individuals to resist low temperatures.

22 A female Spotted Sandgrouse *Pterocles senegallus* on its eggs. North Africa and the Middle East.

23 A pair of Whiskered Terns at their nest on a lake in the middle of France. The clutch consists of three speckled eggs.

24 Nuptial display of a male Desert Lark *Alaemon alaudipes* (Alaudidae) from the Sahara and the Middle East.

25 A Cream-coloured Cursor *Cursorius cursor* with one of its two eggs. It breeds between March and June, and does not build a nest.

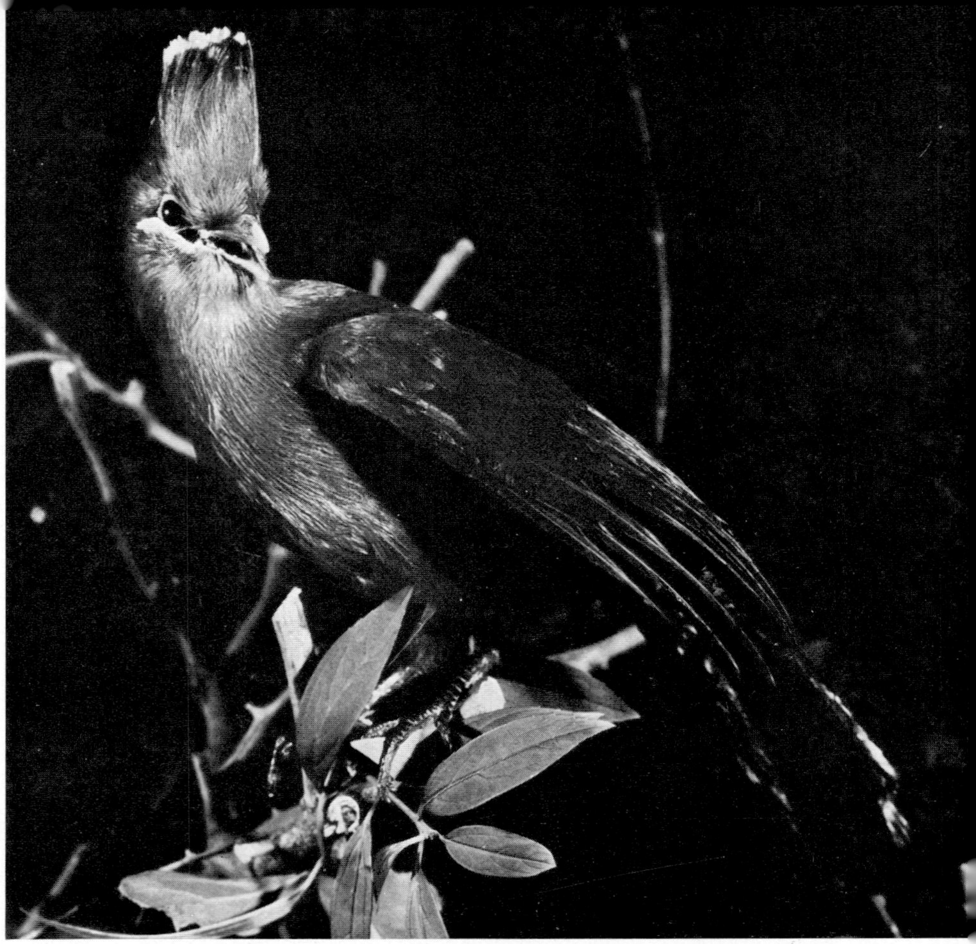

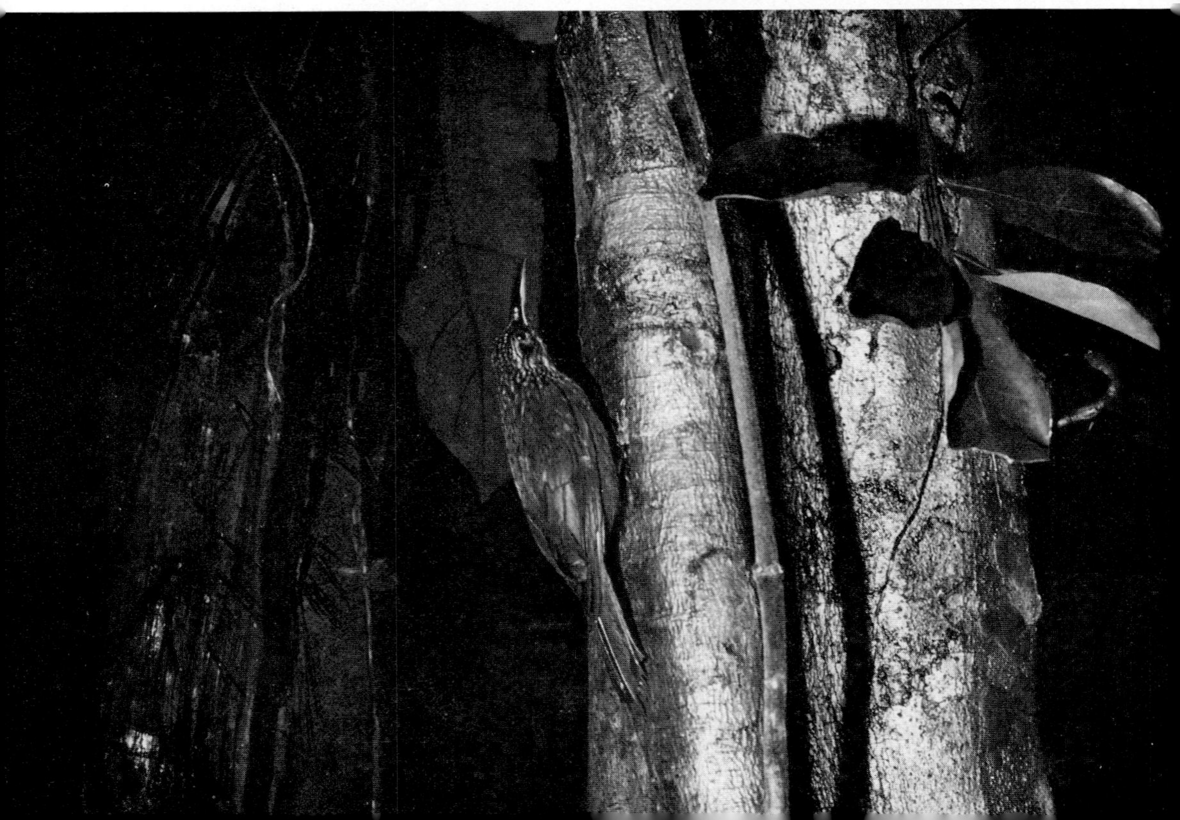

28 A Blue-tailed Pitta *Pitta guajana* from Sumatra, Borneo and Java. In Borneo it lives in the forests of mountain spurs, between 600 and 1200 m.

26 *Above left :* The Knysna Turaco *Tauraco corythaix* of South Africa. The red colour of the wing is due to turacin and the green of the remainder of the plumage to turacoverdin, pigments peculiar to the turacos.

27 *Left :* A Buff-throated Woodcreeper *Xiphorhynchus guttatus*. A climbing passerine from the forests of Central and South America, including Trinidad and Tobago.

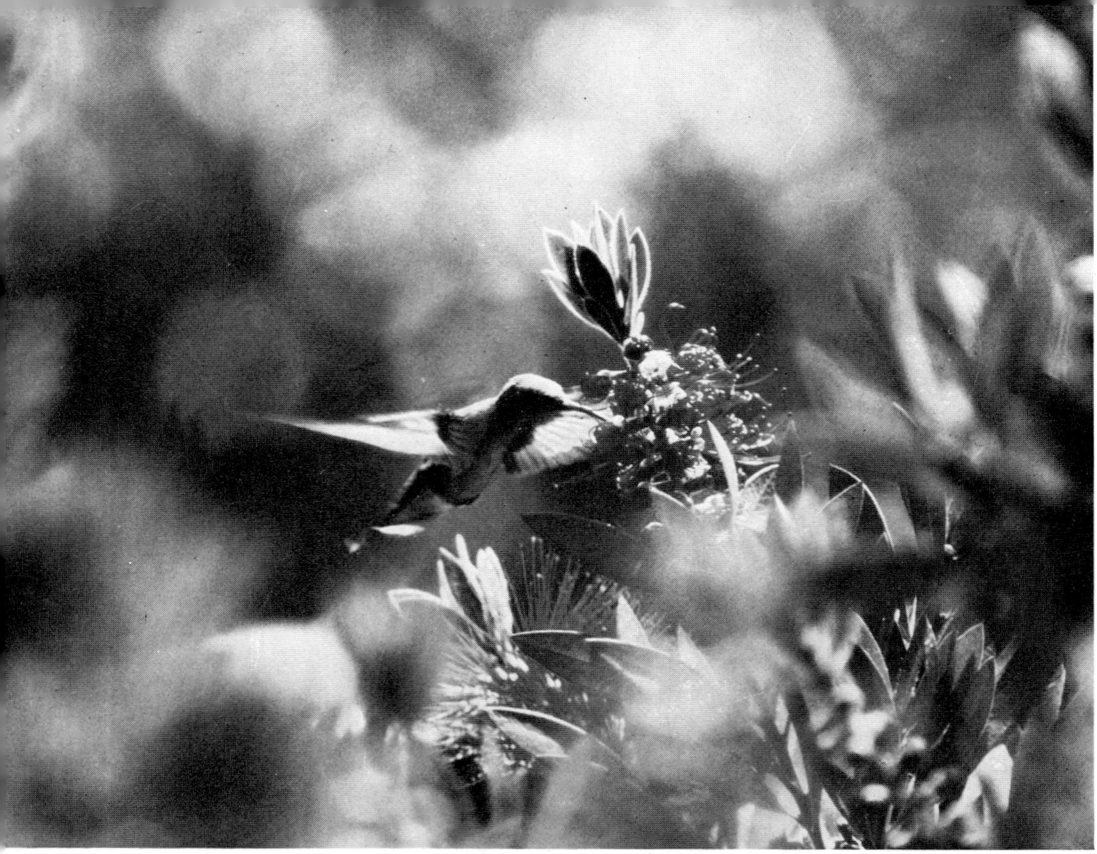

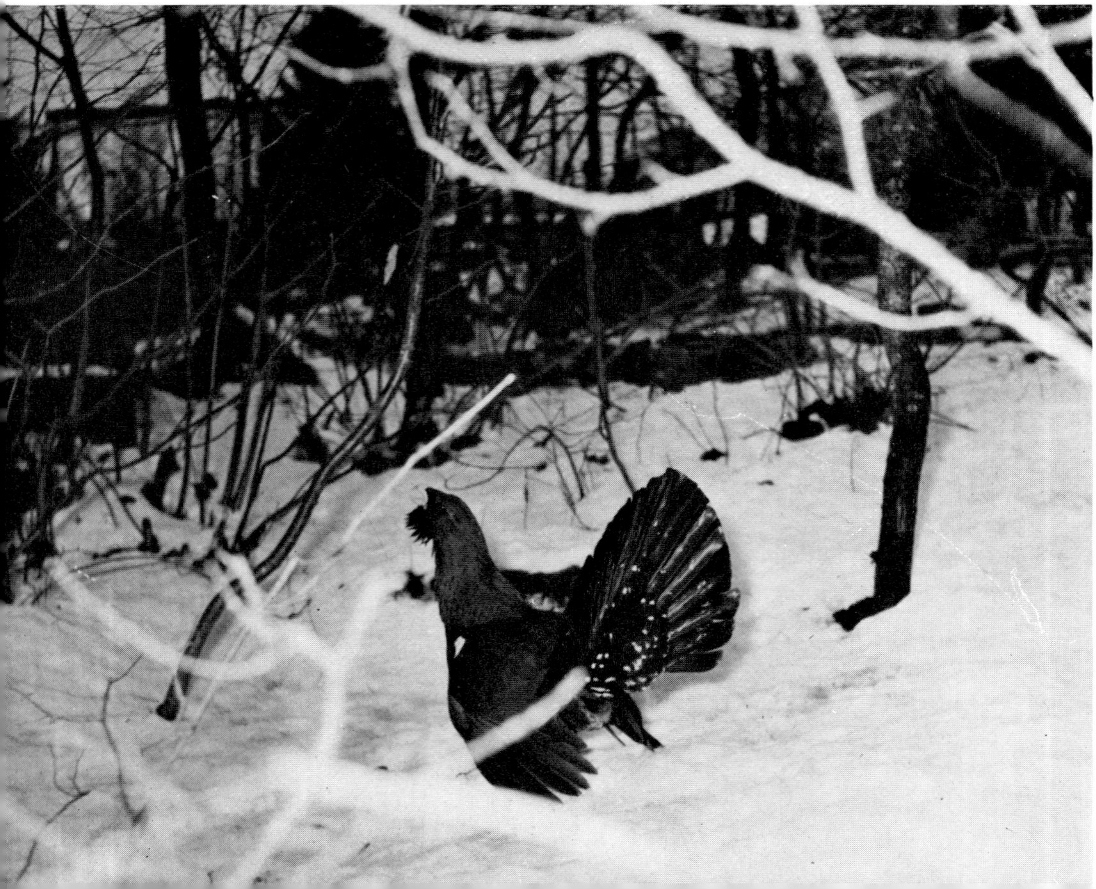

31 The flightless Cormorant *Nannopterum harrisi* of the Galapagos Islands, an example of a bird whose wings are so much reduced that it is incapable of flight. Other members of its family (the Phalacrocoracidae) fly normally.

29 *Above left* : A sunbird (Nectariniidae), member of an Old World family whose diet – the insects and nectar deep in the corollas of flowers – approaches that of the American hummingbirds (Trochilidae).

30 *Left* : A Capercaillie *Tetrao urogallus*.

32 Canada Geese *Branta canadensis* on migration. They nest in northern
North America, and winter in the southern states and on the east and west
coasts of the USA.

Wheatear, Ptarmigan and Rock Partridge prefer bare rocks and scattered or piled boulders. These tendencies are still more developed in certain birds which are confined to rocky walls, like the swifts and swallows which hunt along cliffs and nest on their flanks, thus maintaining their ancestral habits in the mountain environment. In the Alps the House Martin *Delichon urbica*, Crag Martin *Ptyonoprogne rupestris* and Alpine Swift *Apus melba*, are found well above 2,000m and nest in large colonies under the ledges of high vertical walls, as do other species, notably in the high Andes (e.g. *Oreochelidon murina* and *Petrochelidon andecola*).

The bird best adapted to rocks is the Wall Creeper of European and Asian mountains. Its long toes armed with sharp claws allow it to hop and walk about on vertical walls while seeking food (small insects, their eggs and larvae, and spiders) in the cracks. It interrupts its search with irregular flights, during which its large rounded wings beating unrhythmically make it look like a butterfly. In the Andes hummingbirds have also adapted to rocks. Armed with longer nails than the lowland species, they, like the Wall Creeper, pursue the insects which form the basis of their diet along rocky walls. Cliffs criss-crossed by narrow ledges shelter a restricted flora, in which develop insects which swarm in the sunshine. Thus this environment is so favourable that some birds actually build their nests there, another adaptation to rocks and so to the montane climate.

Adaptation to climate is shown also in the synchronization of daily cycles of activity with the variations in temperature. Many mountain birds are especially active during the warmest part of the day, retiring to shelter when it becomes cooler in the evening. However, this is not true even of all the most fragile in appearance such as the high Andean hummingbirds, which are active early in the day when the temperature is still very low and the ground is covered with hoar-frost. These birds need a lot of food in order to meet their very high expenditures of energy, and since the relatively short days at low latitudes leave them just enough time to feed they must put all the daylight hours to use in gathering food.

The search for night-time shelter is equally important to birds which are exposed to a large drop in ambient temperature. Here again piles of rock play a major part and many alpine birds, notably the Wall Creeper and Wheatear, shelter during the night in crevices. In the Andes many passerines – especially finches (*Zonotrichia capensis*, *Phrygilus gayi*, *P. plebejus* and *P. unicolor*) and the Bar-winged Cinclodes *Cinclodes fuscus* – come to such sites for the night, using them as communal roosts. Twenty or more birds pack in beside each other under a single rock. More than 200 finches *Diuca speculigera* have been found roosting in a glacial cleft at

5,300m on the Chacaltaya in Bolivia (Niethammer). Such collective defence against the cold is effective because each member of the huddle loses far less heat than if it were alone.

As a final adaptation, hummingbirds of the Andean alpine zone are partly poikilothermous. Those of the genus *Oreotrochilus* are really torpid at night, and so economize considerably in energy during that part of their cycle when they cannot feed (see Chapter 5).

Diet

The diets of alpine birds cannot easily be interpreted in terms of their environment. Their food resources are diverse, although the standing crop biomass may be small, and the birds therefore belong to several feeding groups. Nevertheless, there are few vegetarian birds. The Ptarmigan is outstanding among these few since its diet consists of the leaves and buds of alpine plants, especially willows (forming up to nine-tenths of its stomach contents), bilberries, saxifrages and various composites and Rosaceae, completed by berries of *Empetrum* and *Vaccinium* and a few insects. The Blackcock, which scarcely crosses the forest edge, feeds mainly on berries during the summer, supplementing them with leaves and buds which become its main winter food. The Rock Partridge feeds on buds, sprouts and seeds, but in summer insects and spiders form an increased part of its diet.

Most alpine birds are insectivorous or at most omnivorous, also taking berries and various seeds. Grasshoppers, beetles, butterflies and caterpillars, and flies (bibionids, tipulids, etc.) are their basic prey, together with spiders which are very important in the higher areas. Among European birds the Water Pipit, Alpine Accentor, Wheatear and Snow Finch seek insects on the ground, while the Black Redstart flies out from a perch and takes them in flight. The Chough and Jackdaw feed on insects, worms and all sorts of rubbish.

Other alpine birds are carnivorous. The Golden Eagle feeds in the Swiss mountains mainly on marmots surprised at the entrances to their burrows – it is their principal predator, and they make up 44 per cent of its prey – and also on Black Grouse and Rock Partridges (15 per cent), young chamois, ibexes, red and roe deer (14 per cent), Varying Hares (12 per cent), and various other prey from snow voles to foxes and some birds. A representative in the high mountains of East Africa, the Verreaux' Eagle *Aquila verreauxi*, feeds mainly on rock dassies *Procavia* which it takes at their colonies where they live something like marmots; they also attack small antelopes. The Eagle Owl, whose last refuge is mainly in the mountains, feeds on very varied prey from hares, marmots and voles to Ptarmigan, Blackcock and small birds.

Birds belong to the same feeding groups in other zones of the alpine biome. Though this is true in the high Andes, supplies of seeds are much more abundant on the plateaux and especially on the grass steppe (*puna*) which extend over much of them. These graminivorous birds – especially finches which are one of the dominant groups – have been able to establish themselves in large numbers. The high Andes cannot sustain strictly nectarivorous birds since nectar-bearing flowers are rare and have limited flowering seasons, so that in penetrating the alpine zone hummingbirds have become almost entirely insectivorous. While their warm-country representatives feed largely on nectar, they almost exclusively hunt insects, especially those which have found shelter on cliffs and among piles of rocks. So, in a parallel African situation, does the Scarlet-tufted Malachite Sunbird *Nectarinia johnstoni* of the alpine zone on Mount Kenya (Coe 1967). Though belonging to a largely nectarivorous group, this species feeds almost exclusively on insects, and more particularly on bibionid flies which come to feed among the flowers of *Lobelia*. These account for 90 per cent of its food, the remainder consisting of chironomid midges which breed in the little pools at the bases of the lobelia rosettes. A similar insectivorous diet recurs among the hummingbirds which frequent the *Espeletia* paramos of the Columbian Andes, where conditions are so like those of the alpine zone, with its lobelias and senicio, of the high African mountains.

Nesting adaptations

Adaptations to the alpine environment appear again in modes of nesting. Many mountain birds frequent rocks, and conceal their nests within the rocky piles. In the Alps the Water Pipit hides its nest under a stone or a thick bush. The Alpine Accentor uses a fissure in a wall or a cavity under a stone. The Wheatear nests only in cavities under boulders or in fissures, sometimes 50 cm from the entrance, as do the Black Redstart (which also puts walls and old buildings to use), the Wall Creeper and the Snow Finch. In such sites the overlying blocks of rock function as thermal flywheels to the nest, and the young birds benefit from the very favourable microclimate in an enclosure well protected from wind, bad weather and cold nights as well as from predators.

Other birds, such as the swallows and swifts of the Alps, hang their nests from the sides of rocky cliffs. This tendency recurs among Andean hummingbirds: unlike other species which make their nests among vegetation, the best adapted of these (especially the species of *Oreotrochilus*) build against cliffs, taking advantage of crevices and overhangs. Some even nest in dark caves or old mine shafts, in a markedly more favourable microclimate. About 75 per cent of the nests of *Oreotrochilus estella* of southern Peru are under great slabs of rocks split off from the sandstone cliffs. To obtain optimal conditions for their nests these birds even place them so as to catch the rising sun, so that they are appreciably warmed during the first hours of daylight when sites not exposed to the sun experience minimum temperatures. Though favourable then, direct heating by the sun becomes dangerous later in the day; and the nests are in fact placed so that from 9 am onwards they are protected by some over-hanging rock, which shelters the nestlings and brooding female from direct rays which have become too intense. Very significantly the great majority of nests (70 per cent) are in cliffs facing the east, and all in positions where they are shaded from the sun during the heat of the day.

Such nests fixed directly to the rocky wall are very large: though varying considerably so as to take advantage of the facilities for support and anchorage offered by different sites, they are always much larger than those of hummingbirds which nest in warm climates, and are built of finer materials. The tendency towards large nests, shown too by certain hummingbirds in the high mountains of Mexico and by Andean finches, is another response to the cold at these altitudes. The warm enclosure thus formed gives the eggs and later the nestlings effective protection from the cold and maintains their warmth despite the repeated absence of the brooding bird, which are the more frequent since in this family the female alone broods.

Other mountain birds nest underground in large burrows driven through soft soil. This tendency is especially marked among the birds of the high Andes. By making burrows the birds shield their broods from the effects of bad weather and ensure them favourable temperatures. While the ambient temperature always fluctuates widely the daily variation within the burrows does not exceed 2-3°C, with the mean steady around 8-10°C. This remarkable constancy is a great advantage to the nestlings, allowing them a considerable economy of energy. Andean birds which have adopted this mode of nesting belong to very diverse systematic groups, including ducks (e.g. *Anas flavirostris*), doves (*Metriopelia melanoptera*), raptors (*Falco sparverius*), woodpeckers (*Colaptes rupicola*) and many passerines, especially ovenbirds (*Upucerthia, Geositta*) and tyrant-flycatchers (*Muscisaxicola*). The burrows usually consist of long galleries, opening on to talus slopes or in the cliffs of soft soil resulting from erosion by rivers. They may be more than a metre long before opening out into an incubation chamber about 30cm in diameter. Some birds such as the Andean Flicker *Colaptes andicola* lay their eggs on the bare earth, while others such as *Upucerthia* lay down a litter of plant matter as additional protection for the young. The birds mostly dig these burrows themselves with their bills and feet, repairing them so that they can be used from year to year, and sheltering in them at night outside the breeding season. However, birds sometimes use the burrows of other animals such as rodents. Thus in Tibet the burrows of Pikas *Ochotona rufescens* are occupied by snow finches (*Montifringilla ruficollis* and *M. taczanowskii*, especially the latter) which are permanently associated with the lagomorph colonies. Similarly the Little Owl *Athene noctua* lives in the burrows of marmots.

The breeding seasons vary considerably in length, depending on the area concerned, and as always they occur at the most favourable times of year. They may be very short in mountains at high latitudes, as in Europe, where only two or three months are favourable, though many passerines of the alpine zone (such as the Water Pipit, Alpine Accentor and probably the Snow Finch) manage to raise two broods. In those tropical regions where the seasons are well marked, breeding takes place only during the well defined period when the rains produce an increase in the standing crop biomass. In contrast some mountain ranges are characterized by almost constant climates, and on Mount Kenya – where high humidity, fog and rain persist throughout the year – the Scarlet-tufted Malachite Sunbird *Nectarinia johnstoni* nests without a distinct breeding season.

Since the number of eggs per clutch is adjusted to the number of young a pair can raise with the food available, clutch size tends to be reduced in the birds of the high Andes. *Zonotrichia capensis*, a finch with a very wide distribution across South America, lays only two eggs there, whereas the normal clutch in the lowlands consists of four or five eggs. The sunbird *Nectarinia johnstoni* only lays one egg, while related lowland species usually lay two or three. However, this tendency is not shown by the birds of the Alps, since (according to the available information) the Snow Finch, Alpine Accentor, Water Pipit and Wall Creeper lay essentially the same number of eggs per clutch as related lowland species.

Slowed development of the young is also often observed among mountain birds, expressed both in longer incubation and in slower post-embryonic growth. This is most clearly shown by the hummingbirds, as is brought out by this comparison of the high Andean *Oreotrochilus estella* with several lowland species (Dorst). The timings are given in days:

	Archilochus colubris	*Selasphorus rufus*	*Colibri thalassinus*	*Oreotrochilus estella*
Range	eastern N. America	western N. America	Mexican mountain forests	High Peruvian plateaux
Incubation	11–14	12–24	16–17	22–23
Eyelids open	about 7	?	11	16
Feather buds appear	?	?	5	12–14
Fledging	6–18 in the north 14–28 in the south	about 20	19–28	30–40

Thus the whole development of the mountain species is considerably retarded. A slight retardation is shown by several passerines of the high Alps. Thus the duration of incubation is fourteen to sixteen days and of a fledging fifteen days for the Water Pipit, and twelve to twenty-four and ten to fourteen days respectively for the Tree Pipit; thirteen to fourteen and eighteen to twenty-one days for the Snow Finch, and eleven to twelve and thirteen to sixteen days for the House Sparrow. The retardation is still more marked in the Wall Creeper, for which incubation probably lasts eighteen to nineteen days and fledging twenty-one to twenty-three days or more, as against fifteen and sixteen to seventeen days in the Tree Creeper. This extension of the breeding period no doubt partly explains why the Wall Creeper raises only one brood a year.

Altitudinal migrations

Where the climate does not fluctuate unduly, alpine birds may be more or less sedentary even in the depths of winter. This is the case for most species of the high Andes. Elsewhere the climatic fluctuations are too severe for most birds to remain at altitude throughout the year. This is especially true of mountains in the temperate zone, where the winter is very rigorous and a thick layer of snow makes the food inaccessible. Nevertheless some birds are sedentary even here. In the high mountains of central Asia grouse do not leave the alpine zone, but remain in small shelters or tunnel for food like their arctic equivalents. The dense feathering of their toes, which increases their bearing surface and gives them a better grip of the soft substrate helps them to move over the snow (though they still sink into fresh snow since their feet do not act as snowshoes). The Snow Finch too scarcely leaves the highest zones of the alps, where even in winter it occurs up to 3,000m and rarely below 1,000m.

Though other alpine birds are forced to migrate they, unlike those which nest in the lowlands, may find conditions very different in valley bottoms and on the plains. They can thus escape from a climate which has become decidedly unfavourable by undertaking small-scale movements, since they need only change altitude within the same mountain range to pass from one zone to another. Though these altitudinal migrations therefore take place over relatively short distances, they are none the less regular and as precisely regulated as the long-range migrations of other birds.

For some especially demanding mountain birds descent into the valleys is by no means sufficient. In Europe the warblers, thrushes, swallows and Alpine Swifts entirely leave the mountains and travel like other migrants. Many, however simply take refuge in the valleys, leaving the highest zones which are their summer habitat. In the Alps this is true of the Wall Creeper, Alpine Accentor and Water Pipit, which descend at the beginning of winter into the low-lying valleys and even disperse over the plains. In Asia the River Chat *Chaimarrornis leucocephalus*, which nests up to 5,000m in the Himalayas, spreads in winter to the hills and plains of India. In North America such movements are made by the Pine Grosbeak *Pinicola enucleator*, the Brown-capped Rosy Finch *Leucosticte australis* and the Gray-headed Junco *Junco caniceps*.

There is a fundamental difference in patterns of migration between alpine birds and arctic ones, which are also forced to leave their breeding grounds during the winter. Thus of eighty-two species present in Swedish Lapland during the summer, fifty-four or 65 per cent leave the region entirely in winter; whereas of a similar number of species nesting in the Alps of Savoy only 30 per cent are long-range migrants. Of the rest some are sedentary, remaining in the high zones, while most of the others are content to settle in the valley bottoms for the winter (Barruel). Some migrate at fixed dates, while others are controlled by the onset of winter.

Daily movements also take place, certain birds descending every day during the winter to feed in the valleys, and then ascending again every evening to pass the night among the higher crags. This behaviour is very characteristic of the Alpine Chough (Barruel). The reverse movements take place on tropical mountains. Thus in California hummingbirds ascend during the day to feed on nectar from the flowers of the alpine zone, and descend again to pass the night in lower and warmer regions.

As a result of temperature inversions, higher zones sometimes enjoy milder winter weather than lower ones. During a calm, cold air accumulates in the valley bottoms which remain misty, while the warmer air above keeps the higher zone warmer and mist free, and allows solar heating. At such times the valley bottoms appear deserted, the birds having reascended to take advantage of the more favourable conditions. When the weather changes the birds revert to their normal altitudinal distribution, taking refuge again in the valley bottoms.

Chapter 29

Island Avifaunas

ISLANDS occur throughout the world, in every climatic and biogeographical zone. Most natural habitats are to be found on one or other of them, and the birds which have colonized them show the very diverse adaptations mentioned in earlier chapters. However, islands also show special characteristics as a result of their isolation, the manner in which they have been populated, and the evolution of those animals which have become established there.

Several categories of island may be distinguished. Some are merely parts of the continent, recently separated from it and remaining on the continental shelf. Their faunas are very similar to the neighbouring continental ones, since the species have not had time to evolve profoundly. Other, *oceanic*, islands have in contrast never been in any way connected with a continental mass. Of coralline or volcanic origin, they have risen from the ocean bed and have always been separated. They have therefore been populated at random, animals having arrived on them either by air or by water – sometimes on the gigantic floating rafts of vegetation uprooted by tropical rivers, which are carried great distances by marine currents at speeds of two knots or more (at least 50 miles a day). Darwin and Wallace were the first to emphasize the interest of these facts, which explain the characteristics of the island populations. Only oceanic islands are considered here, since the others are little different from the continents, of which they are usually mere fragments.

General characteristics of oceanic islands

Oceanic islands always carry impoverished faunas. Very few animals have been able to reach them and even birds prove disinclined to cross straits within their capabilities. Many birds, especially those restricted to closed habitats, do not as a general rule leave them and never venture over the sea, so that they have little chance of being carried away by the wind. Of the 265 species of birds known from that part of New Guinea facing New Britain, only eighty or less than 30 per cent are represented on that island,

543

a strait seventy km wide having stopped the majority. Much narrower stretches of water often act as equally effective barriers (Mayr 1939). Among those birds which do attempt the crossing, whether voluntarily or not, very few succeed.

Insular communities of animals, and especially of birds, are thus descended from a very small number of founding stocks. They include a high proportion of endemic forms, evolved in a closed system and biologically replacing others – which may or may not be their geographical representatives – in continental faunas. Often however the representative species occupy markedly different ecological niches, so that the 'ecological spectrum' is quite different from that on the continent. A sort of substitution is especially noteworthy, in which the gap left by a missing species is filled by a very different one which, if it exists at all in the related continental fauna, there occupies a narrower ecological niche.

Islands are furthermore the sites for many examples of *adaptive radiation*. From a limited founding stock divergent lines develop so as to occupy very varied ecological niches, which in continental communities are occupied by unrelated forms. Communities confined to oceanic islands (and to some extreme environments with harsh physical conditions) thus show evolutionary short cuts which allow the processes at work to be better understood. They are of prime importance ecologically since they show that (in a particular habitat offering a given number of ecological niches) there is room for a certain number of animal forms which always evolve convergently, whatever their origins and modes of evolution. This ecological convergence, directed by the factors of similar habitats, is a remarkable example of adaptation and the evolution of living things in response to their physical and biotic environments.

Oceanic islands are ideal localities for studying the genetic aspects of evolution and speciation, and many works have been devoted to these instructive natural experiments. While few studies have been devoted to the evolution of entire biocenoses, they can illuminate many problems by showing how an ecosystem is constituted and how the species divide up available ecological niches. Given the opportunities of access to founding stocks, the faunistic richness of an island depends in general on the variety of habitats which it offers. The total area of the island has an important effect on the number of animals which succeed in colonizing it as well as on the differentiation of new endemic forms. There is thus a relation, which various biologists have formulated mathematically, between the number of faunistic elements on the one hand and the inhabitable area and the variety of habitats on the other, which explains the faunistic differences between various islands in terms of their geographical and physical conditions.

Island ecosystems, comprising only a reduced number of elements, are thus less complex, with simpler food chains, and are therefore much less robust. They are the more affected by external factors as the component species are archaic on the one hand or overspecialized on the other, as a result of differentiation within a closed system sheltered from the intense competition and predation experienced on continents. It is not surprising that such communities should have suffered more from human interference than those of other and wider environments. The proportion of extinct and threatened species is higher on islands than anywhere else on earth.

The Canary Islands

The Canary Islands belong to the Palaearctic region as their whole fauna shows. Their birds, partly derived from Europe, have established themselves according to their ecological needs in habitats like those of the Mediterranean region. However a clearly Saharan influence is shown by the presence of such birds as the Cream-coloured Courser *Cursorius cursor*, the Hubara Bustard *Chlamydotis undulata* and the Trumpeter Finch *Rhodopechys githaginea*, whose presence is explained by the desert conditions which are especially marked in the eastern part of the archipelago (Fuerte Ventura). The fauna is markedly impoverished, lacking a certain number of species characteristic of the European avifauna despite the presence of their preferred habitats.

The Canaries harbour several endemic species, including Meade-Waldo's Chat *Saxicola dacotiae* which is intermediate between the Stonechat and Whinchat, 'Berthelot's Pipit' *Anthus bertheloti* (also on the Salvages and Madeira), and two pigeons confined to the laurel woods: *Columba trocas* (also on Madeira) and *C. junoniae*. Many widespread species are represented on the Canaries by local races, and sometimes by several, each of which is confined to particular islands.

Because certain species are absent the ecological distribution is different from that in Europe, even when the nature of the habitat is taken into account. Thus, the Willow Warbler being absent from the Canaries, its place is taken by the Chiffchaff, which is here found from the low bushes which are the former's characteristic habitat to the higher layers of vegetation which are its own. Similarly the Coal Tit, characteristic of conifer forests in Europe, is absent from the Canaries where it is replaced in this habitat by the Blue Tit, which on the continent is confined to deciduous trees. Thus a widening of ecological distributions is produced by the absence of a number of species, which on the great continental masses are in competition and are therefore forced into well-defined habitats and narrow ecological niches. Further, in the absence of related competing species, the songs of Canary Island birds have evolved independently: the song of the local Chiff-chaff is very markedly different from that of the closely related European race (Hüe & Etchécopar).

Among the chaffinches this situation is reversed. Over a long time, and probably as a result of two successive invasions, the genus *Fringilla* has evolved two distinct forms in the islands. The first (represented by three endemic subspecies) is the common Chaffinch *F. coelebs* which is so widely distributed in Europe, while the second is a distinct species endemic to Teneriffe and Gran Canaria, the Blue Chaffinch *F. teydea*. Whereas the common Chaffinch is found in Europe in all types of woodland including conifers, it is restricted in the Canaries to broadleaved woods and is replaced among conifers (notably the pine *Pinus canariensis*) by the endemic species. Thus competition between the two species of the Canaries has led to ecological specialization, which has not taken place in Europe with its single species. These few examples show how birds may redeploy themselves ecologically in accordance with competition and avifaunal composition.

The Galapagos Islands

The Galapagos, situated in mid-Pacific at about 1,000km from the coast of South America and close beneath the equator, are unquestionably of prime importance to biology. The group consists of thirteen islands (apart from many islets) of which five are relatively large, with a total area of 7,500km². It represents the peaks of a cluster of submarine volcanoes which are certainly very old and wholly volcanic in origin, so that they have never been in any way attached to the American continent. However, at certain times during the Tertiary an outpost of Central America extended in their direction, reducing the width of sea separating them from the continent. Across this animals have arrived at intervals almost randomly, which explains the extreme poverty of the fauna, and especially the complete absence of certain groups. Birds themselves are poorly represented by some eighty species, the great majority of which are seabirds.

The exceptional oceanographic situation of the Galapagos, bathed by the cold waters of the Humboldt Current, is equally noteworthy. This has led to modifications of the marine fauna, paradoxical for such equatorial latitudes, and explains especially the presence of a penguin (*Spheniscus mendiculus*) closely related to species on the coasts of Chile and Peru. The low temperature of these waters has also had profound effects on the climate, which is markedly dry, producing deserts (except in higher regions where mists condense and rains fall in a wet season though this is rather irregular).

The Galapagos are a locality of the highest interest for the differentiation of birds. A special case is the Flightless Cormorant *Nannopterum harrisi*, strictly endemic to these islands (Snow 1966). This large heavy cormorant is indeed incapable of flight, its vestigial wings being quite functionless, and therefore lives somewhat like a penguin, swimming with the greatest ease and landing on the low banks where it nests. Finding food near the coast throughout the year, this bird has in the course of evolution lost the ability to fly, probably because there are no mammalian predators in the Galapagos.

The most remarkable group among the birds are Darwin's finches, the Geospizinae (Bowman), while other land birds include only an endemic dove *Nesopelia galapagoensis*, a cuckoo, and a few passerines. The mocking-birds (*Nesomimus*) represent a stock which arrived long ago and has had time to evolve distinct species and subspecies on the various islands. Three tyrant flycatchers (*Pyrocephalus* and *Myarchus*) and the Yellow Warbler *Dendroica petechia* are more recent arrivals, which like all the Galapagos birds came from the American continent, showing no affinity with the Polynesian fauna.

Darwin's finches form a group of fourteen species, endemic to the Galapagos apart from *Pinaroloxias inornata* which is confined to Cocos Island, and forming a specialized subfamily of the Fringillidae. They came a long time ago from the continent (perhaps from Central America since there are certain avifaunal affinities between the Galapagos and the Caribbean) where they now have no close relative. It is almost certain that all the Geospizinae are descended from a common ancestor, which presumably shared the characters of the least specialized existing species. It was probably a graminivorous passerine with a heavy but short bill, unless it belonged to a group of still undifferentiated passerines from which various American families, from the Fringillidae and Thraupidae to the Coerebidae and Parulidae, were later derived. When this ancestor arrived the group of islands must have been occupied by few if any land birds, so that there was neither predator to limit its populations nor competitor to oust it or keep it within ecological bounds. The original stock would therefore have begun to proliferate and occupy all the available terrain. Being originally graminivorous these birds would at first have confined themselves to exploiting seeds, but as their numbers increased would have expanded into other ecological niches. Thanks to a series of mutations and very intense genetic drift, the original type was gradually modified and gave rise to several lines.

Without any doubt it is the bill which has evolved most spectacularly in the Geospizinae, differing widely between species and genera, so that it

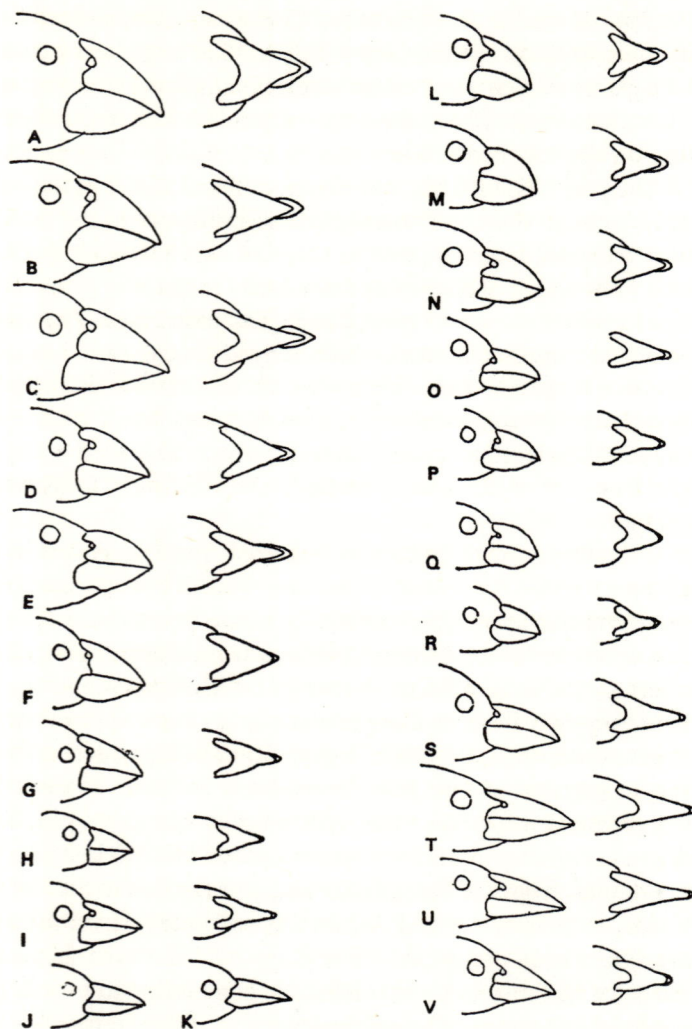

Figure 90. Different shapes of bill among Darwin's Finches (Geospizinae), passerines endemic to the Galapagos Islands.

provides useful systematic characters. In *Geospiza* (six species) the bill is thick and forms a strong pincer capable of exerting strong pressure basally. While still of modest size in some species, it is much stronger and more swollen in the most highly evolved, resembling that of a grosbeak and serving to crush hard seeds which the bird grasps in the base of its bill. In *Camarhynchus* (three species) it is also very strong, but is shaped to exert stronger pressure at its tip, and can be used to split hard wood in extracting xylophagous larvae. In *Cactospiza* (two species) it is still strong basally but elongated, and can grip strongly by the tip when used to explore cavities in the wood. In *Platyspiza* (one species) the bill is short and swollen, somewhat like that of a parrot with a very convex culmen which allows it to exert strong pressure throughout its length. In *Certhidea* (one species) on the other hand it is very fine, so that the bird can explore vegetation and delicately pick prey from branches and leaves. In *Pinaroloxias* (one species on Cocos Island) the bill is incurved, so as to seize delicate prey, slice flowers and pulp fruit. The bill musculature, especially the adductor muscles of the mandible, is developed in parallel to the bill, being especially powerful in those species (*Geospiza magnirostris* and *G. fortis*) which use their strong bills to crush seeds. The alimentary tract has also evolved in parallel in response to the bird's diet, the intestine being longer in the vegetarian forms.

In comparison to the bill, the plumage is only slightly different from group to group. However, it does show differences between genera in pattern and coloration, in accordance with the layers in which the birds live. *Geospiza*, which seek their food on the earth – which is usually formed from black lava – have black or predominantly black plumage, usually with marked sexual dimorphism (the adult males being entirely black while the females are streaked with black on a pale ground). *Camarhynchus*, which only occasionally come to the ground and primarily explore the vegetation, have plumage which is predominantly grey-brown, the males sometimes having a black hood. *Cactospiza* and *Platispiza*, which are strictly arboreal, are grey or olive green with little contrast, blending well with their surroundings.

Thanks to their different specializations, Darwin's finches have been able to occupy very diverse ecological niches. They are distributed across the archipelago, specializing in the different habitats: notably the lowland desert zone covered with cacti, the dry transition forests at intermediate altitudes, and the few wet upland zones covered in dense forests. They also habitually occupy clearly defined strata, some being primarily terrestrial and others arboreal, and specialize particularly in their diet. Those

549

Food habits of the GENUS	Strata inhabited	Bill shape	Bill capabilities	Food habits of the species	Comparable birds
GEOSPIZA Seeds, some insects	Ground, low vegetation	Culmen convex Gonys straight Bill strong and conical, small to huge	Strong tip Crushing at base	G. magnirostris Hard seeds	Coccothraustes coccothraustes (Fringillidae)
				G. fortis moderately hard seeds	Melanospiza richardsoni (Fringillidae)
				G. fuliginosa Small and soft seeds	small Fringillidae
				G. scandens Opuntia fruit and nectar, soft seeds	Tangavius aeneus (Icteridae)
				G. conirostris Opuntia fruit and flesh, seeds	Saltator albicollis (Fringillidae)
CAMA- RHYNCHUS Insects extracted from wood by digging; occasionally seeds	Vegetation, more rarely the ground	Culmen and egonys convex Bill short and laterally compressed	Very strong tip Crushing at base	C. psittacula Large insects, some seeds	Psittiparus (Paradoxornithidae)
				C. parvulus Small insects, some seeds]	Parus (Paridae)
CACTOSPIZA Insects extracted from wood by searching cracks	Vegetation, more rarely the ground	Culmen and gonys slightly convex Bill relatively weak	Strong tip Crack-searching Tool-using (spine or twig)	C. pallida Insects and larvae, some fruit Land habitats	Tachyphorus coronatus Thraupidae
				C. heliobates Insects Mangroves	
PLATYSPIZA Buds, flowers, fleshy fruits, some seeds	Vegetation and ground	Culmen strongly convex Gonys narrow Bill short, deep and wide	Strong tip Crushing throughout length	P. crassirostris as the genus	
CERTHIDEA Small insects from leaves and branches	Strictly arboreal Very active	Culmen and gonys straight, slightly re- curved basally Bill fine and weak	Grasping tip Surface- searching	C. olivacea as the genus	various tits and warblers

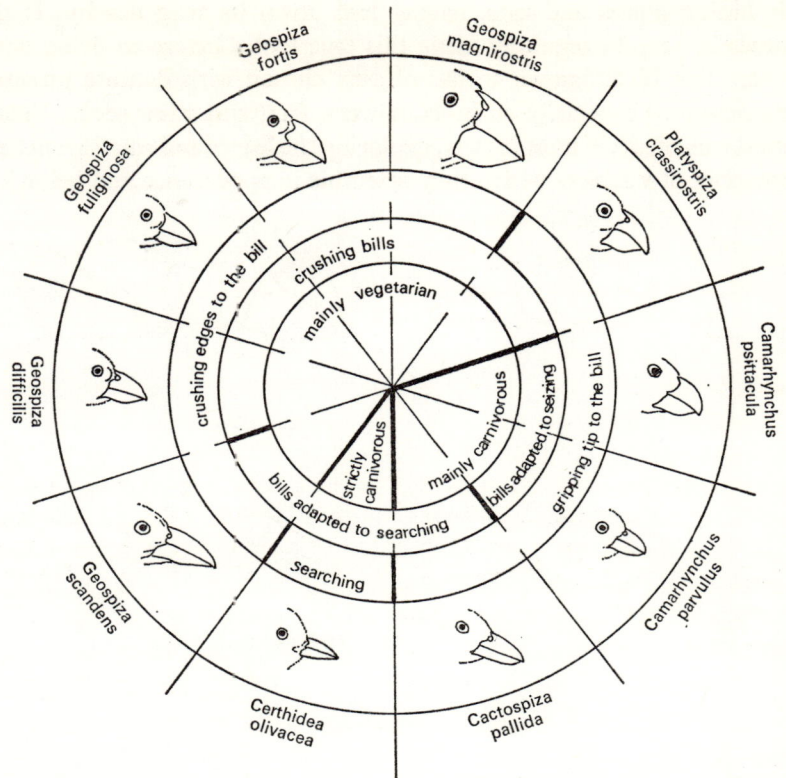

Figure 91. Diagram illustrating the relation between bill-form and diet, in the ten geospizine species on Indefatigable Island in the Galapagos.

Geospiza with relatively weak bills (*G. fuliginosa*) feed on small soft seeds while those with powerful bills (*G. fortis* and *G. magnirostris*) can crush large hard-shelled seeds and thus live on a wide range of different plants. *Geospiza scandens* shows marked preferences for the pulpy fruit of cacti and the nectar from their flowers, together with a few seeds. *Camarhynchus* feeds especially on insects extracted from wood. *Cactospiza* hunts insects hidden in the cracks of bark, in which both *C. pallida* and *C. heliobates* show quite remarkable behaviour. These birds in fact occupy a niche rather like that of woodpeckers, since they explore the trunks in order to find the insects which are their exclusive food. Though their bills are shorter than those of woodpeckers and their tongues are by no means as highly modified, they compensate for this deficiency by the use of a twig or a cactus spine which they trim to the proper length. Holding this in the bill a *Cactospiza* uses it to explore holes dug in the wood by larvae, which it extracts from

o

their hiding places and eats, having laid down its twig nearby. It then resumes its search, regularly using this true tool, *Cactospiza* being one of the very few birds known to use objects chosen with definite intention. *Platyspiza* feed especially on buds, flowers, fruit and a few seeds. Finally, *Certhidea* exclusively hunt insects, exploring the leaves and small branches in the manner of warblers which they resemble in appearance and behaviour.

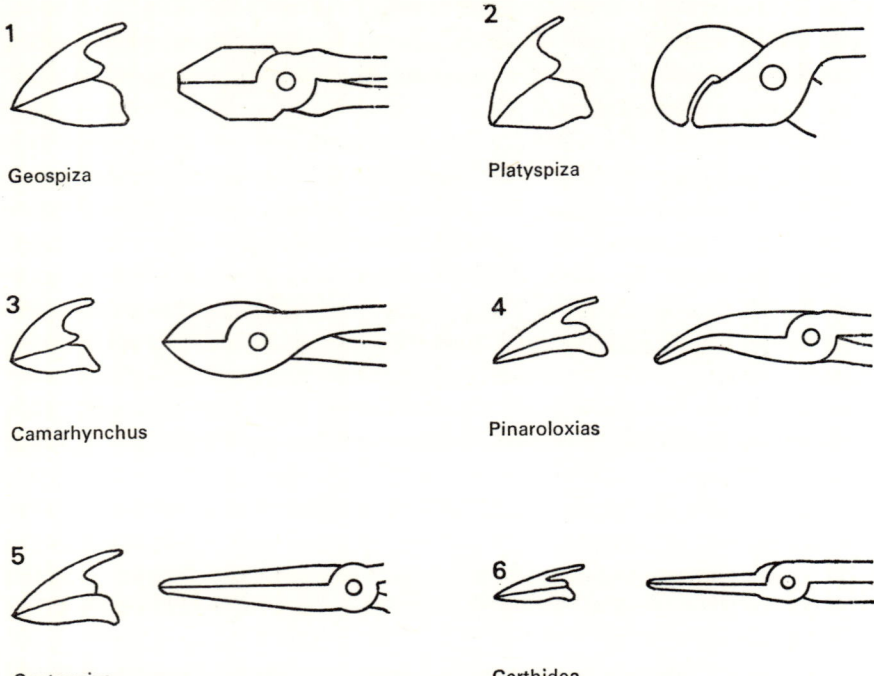

1 Geospiza   2 Platyspiza

3 Camarhynchus   4 Pinaroloxias

5 Cactospiza   6 Certhidea

Figure 92. The bills of six geospizine genera compared to six types of pincer.

The Geospizine are of the greatest evolutionary importance, since this group presents an impressive summary of passerine evolution. None the less some niches remain unoccupied. The stock has developed neither 'larks' nor 'shrikes', apparently because it lacks these potentialities. On the other hand, while no true nectarivore has developed on the Galapagos, this is because the local habitats offer no opportunities to such a bird, whereas on the richer Cocos Island the endemic geospizine *Pinaroloxias inornata* is primarily nectarivorous like the honey-creepers which it resembles.

The importance of this evolutionary phenomenon, whose discovery was decisive for Charles Darwin in establishing his theories, may be imagined. In a large continental mass, the attempts of a bird 'trying' gradually to become a grosbeak or a warbler would quickly have been stifled by competition with similar birds, already specialized and better fitted for such competition. The complexity of the community is a stabilizing factor, since it keeps each group evolving in a particular direction, by eliminating any divergent individuals. On the other hand, according to Darwinian evolutionary theory, it is intra-group competition itself which has encouraged the divergence of the lines. Under the pressure of environmental conditions, especially the quality and quantity of available food, their specialization proceeds so as to reduce competition, allowing a selection of birds to spread out across the ecological spectrum.

The evolution of Darwin's finches is the more interesting since it is still in its earliest stages. It is often difficult to classify certain individuals in one of these species rather than another, since if a large collection is arranged according to the shape and development of the bill it forms an almost continuous series. The variation within a species, and often within a single population, is considerable. Further, the species are not uniformly distributed across the islands, none of which is simultaneously occupied by all thirteen. Each population has evolved differently, depending on the composition of the local avifauna. Thus the same bird may vary somewhat from island to island, in ecological niche as well as in behaviour (including vocalizations).

This example of adaptive radiation is thus of the greatest interest. It may be interpreted in Darwinian terms but, even if new observations as a whole are coming to invalidate this theory of evolution, Darwin's finches remain ideal material in studying the adaptations of an avifauna to its environment.

The other Galapagos land birds have arrived more recently in the islands, and differ only slightly from the continental forms to which they are so clearly related. This is especially true of the buzzard *Buteo galapagoensis* and the two owls *Asio galapagoensis* and *Tyto alba*. Nevertheless these raptors show very remarkable dietary characteristics, giving evidence of great adaptive flexibility. Depending on circumstances they feed on sea birds (especially petrels), land birds (being with the endemic snake *Dromicus dorsalis* the principal predators of Darwin's finches), young iguanas, and authochthonous and introduced rodents.

The Hawaiian Islands

The Hawaiian Islands are in somewhat the same situation as the Galapagos, since they are volcanic and have never been linked to any continent. It is

553

even probable that they have never been connected as a single land mass. Having no doubt emerged only since the Pliocene, they have been colonized by a reduced number of stocks – those which succeeded in the difficult crossings over the vast stretches of sea which separate this archipelago from the great continental masses. Despite a very favourable warm humid climate, the fauna is therefore poor and groups including the insects and molluscs show characteristic adaptive radiations. Hawaiian birds are especially remarkable, since they show a mixture of American and Polynesian forms which arrived at various times – as is proved by their different degrees of differentiation – from which no less than fourteen successive invasions can be distinguished. The oldest stage of endemism is unquestionably shown by the Hawaiian honeycreepers, Drepanididae, a family endemic to the islands and broadly dominant among the passerines, which divides into two subfamilies, nine genera and about twenty-two species. These birds differ one from another to such an extent that their earliest systematists placed them in different families, notably among the Fringillidae, Dicaeidae and Meliphagidae. Yet anatomical studies have proved that they belong to a single type, and that they are all derived from a group of which the Coerebidae or honeycreepers form the present nucleus. Arriving a very long time ago in Hawaii, which must then have been devoid of land birds or at any rate much impoverished, these birds have specialized so as to colonize the vacant ecological niches and to realize their whole genetic potential. This evolution must be ancient, since the generic level of differentiation has been reached (Amadon 1950).

Some drepanidids have remained relatively primitive, close to their ancestral stock as derived from the group centred on the Coerebidae. The Apapane *Himatione sanguinea* has a short fine bill in accordance with its primarily nectarivorous diet supplemented with a few insects. This diet is shared by the Iiwi *Vestiaria coccinea*, with a longer recurved bill, and the Mamo *Drepanis pacifica* whose bill is still longer. In contrast the Ula-ai-hawane *Ciridops annae* has a stronger short bill like that of a finch, and its diet consists mainly of fruit and seeds.

At the base of another lineage (the Psittirostrinae) the Amakihi *Loxops virens* also has a short recurved bill, as with variations have the other species of this genus. These birds are insectivorous, though nectar still plays a significant part in their diets. Species of *Hemignathus* have very fine strongly recurved bills with the lower mandible distinctly shorter than the upper, so that the bill ends in a fine curved needle. The most primitive of these species too are partly nectarivorous, although insects form a major part of their diets. The most specialized, especially the Akiapolaou

H. wilsoni, are wholly insectivorous, hunting insects over trunks and branches, under bark or in rotten wood. Opening its bill widely one will hammer the wood strongly with its lower mandible, producing a wood-pecker-like drumming audible at great distances, and then extract the insect or lava using both mandibles. Species of *Psittirostra* show considerable development of the bill, which in the most specialized species has become swollen and very strong like that of a grosbeak or small parrot. These birds feed almost exclusively on fruit and seeds, whose husks they break open with their strong bills after removing them from their pods (Amadon 1950).

The Drepanididae thus provide striking example of adaptive radiation, a single stock having occupied very diverse ecological niches as a result of the absence of competitors. The profound modifications involved result from the early establishment of the founding stock and the variety of habitats offered by the Hawaiian islands. Much wetter than the Galapagos and partly covered by tropical forest, these islands provide a much more favourable site for evolution than the latter, whose plant resources and variety of habitats are incomparably poorer.

However, in contrast to the Galapagos, Hawaii carries a larger proportion of elements other than the nucleus. Rails, honeyeaters, a thrush, a flycatcher and a goose are all sufficiently specialized to be recognized as distinct genera; a crow, a buzzard and a duck have reached the stage of specific differentiation; while an owl, a moorhen, a stilt and a coot are no doubt much more recent immigrants, only sub-specifically distinct. Some of these birds show Polynesian affinities, whereas others have clearly come from the American continent. The honeyeaters (*Chaetoptila* and *Moho*) are rather distantly related to Samoan forms, while the flycatcher *Chasiempis sandwichensis* is related to the Melanesian *Monarcha* and Polynesian *Pomarea*. The others are from America – most clearly the thrushes of the genus *Phaiornis* which is related to *Myadestes* – although long isolation has allowed them to diverge widely. The Hawaiian Goose or Ne-ne *Nesochen sandvicensis* is a well-marked form, within a widespread holarctic group.

Madagascar

Madagascar presents a somewhat different example of insular colonization. This large island is really a small continent, with strong relief and a great diversity of natural habitats. Though it was never connected to any continent, its fauna has been very profoundly influenced by the proximity of Africa. Thus its faunistic affinities are clearly Ethiopian, though with a few Asiatic elements whose presence biogeographers have explained in very diverse ways. The presence in Madagascar of clearly Indian elements is probably due to their having originally had much wider ranges, and being eliminated from the intervening areas.

The Madagascan fauna is strongly characterized by many relict species, by the high proportion of endemics (some representing entire endemic families), and by the absence of many typically African animals. This exclusion has been very severe among the birds, which including the wintering migrants number only 238 species.

Many stages of differentiation may be distinguished, resulting from earlier or later invasion by the founding stocks and different speeds of differentiation from group to group, so that the degree of endemism varies from the familial to the subspecific level.

Madagascar harbours several survivors of an archaic fauna, some of which are real ornithological enigmas like the Mesitornithidae, a family of three species (*Mesitornis variegata*, *M. unicolor* and *Monias benschi*) belonging to the Ralliformes. Though in their ground-living habits these resemble the gamebirds, this is certainly due to convergence resulting from their way of life, since these aberrant and very primitive birds are closer to the sunbitterns (Eurypygidae) of tropical American forests than to any others. Both groups are probably the remnants of an archaic assemblage, once world-wide in distribution but now represented only by widely dispersed relict groups. The situation of the asitys (Philepittidae), primitive passerines formerly placed near the sunbirds, and the Cuckoo-Roller *Leptosomus discolor*, intermediate between cuckoos and rollers, is similar. Apart from these highly endemic forms the avifauna is derived mainly from Africa, but has a small Asian component (such as the rail *Amaurornis olivieri*, the owl *Ninox superciliaris*, the swiftlet *Collocalia francica* and the thrush *Copsychus albospecularis*). Though this is a relatively diverse avifauna, it has wide gaps and empty ecological niches which have allowed certain stocks to undergo adaptive radiation like that of the Hawaiian honeycreepers. This is seen to some extent among the cuckoos, a family which (apart from a more recent immigrant the Madagascar Cuckoo *Cuculus poliocephalus*, closely related to an Indian form) is represented in Madagascar by the endemic genus *Coua*. The ten distinct species show clear ecological segregation, which must be ancient since they differ so much in accordance with their habitats. Some are climbers and keep to the trees, flying and leaping from branch to branch, while others are runners, using their wings at most to help them move over the ground. Several very distinct niches are thus occupied by the species of *Coua*, which are broadly specialized in ecology (Milon).

A similar but much more striking adaptive radiation is shown by the vangas (Vangidae), a family endemic to Madagascar apart from one species in the Comoros. Like all such radiating groups they diverge widely in external appearance and biology, and in the morphological peculiarities related to their ways of life, but the unity of the group is clearly shown by fundamental anatomical structures which are sheltered from disruptive selection.

Some vangas (species of *Leptopterus*, and *Tylas eduardi*) live in the treetops where they actively hunt insects among the branches. Others (*Vanga curvirostris* and *Euryceros prevosti*) larger and with powerful bills ending in strong hooks, supplement this diet with small vertebrates, especially chameleons and batrachians, behaving like real little raptors. Others in contrast (*Oriolia* and *Xenopirostris*) hunt like flycatchers, sitting still on a perch and swooping on insects which come within range. Species of *Calicalicus* search for insects along branches like tits, which they replace biologically in Madagascar. Similarly *Falculea*, with long sickle-shaped bills occupy the place of woodpeckers, hunting wood-boring insects and those which hide in the crevices of bark. Finally, *Hypositta* replace nuthatches, which they so much resemble that they are still often included among the Sittidae.

Thus the Vangidae provide a further example of adaptive radiation. They differ from Darwin's finches in their more advanced stage of evolution, with ten genera for their fourteen species, and from this point of view are more like the Drepanididae. However, their direction of evolution has resulted from the carnivorous specialization of the stock from which they originated. Whereas the primitively vegetarian and insectivorous Drepanididae have evolved several lines which differ in diet, the Vangidae have all remained carnivorous but have specialized in hunting methods. Thus while this radiation has gone far in certain directions, it has been subject to certain limits. Even originally the Vangidae were not alone in Madagascar, and the competitive pressure of other groups confined them to a part of the ecological spectrum. It is also possible that the original stock was already too specialized to exploit further opportunities. Nevertheless the Vangidae offer a remarkable example of adaptation, showing once more how a fauna is moulded by the environmental influences to which it is subjected, so as to occupy all the available space and the whole spectrum of ecological exploitation.

Pacific Islands

East of New Guinea and Australia is a multitude of islands, scattered over thousands of square kilometres and grouped in more or less well-defined archipelagos. These are oceanic and often volcanic in origin, being situated beyond the limits of the continental shelf, and having never (at least since the beginning of the Tertiary) had any connection with a continent. They are either low flat atolls with biotopes which necessarily show little variety; or volcanic structures, often with steep gradients, and with much more diverse habitats.

Completely isolated, at vast distances from each other and much further still from the areas of faunistic evolution – faunas are still very impoverished. All their animals come from the west, showing affinities with those of New Guinea and Australia. There is a progressive but rapid faunal impoverishment eastwards from the edge of the Australo-Papuan continental shelf. Thus New Guinea has 520 species of land birds, the Solomon Islands 127, Fiji 54, Samoa 33, the Society Islands 17, the Marquesas 11 and Henderson Islands 4. However, while the number of species decreases rapidly, the proportion of endemics increases. Thus the islands nearest the Asiatic continent have 520 species of birds, of which 12·7 per cent are endemic, while those of the western and central Pacific have only 225 species but with 27 per cent endemic, and those of the eastern Pacific only 42 species of which 78·6 per cent are endemic. It is just as though 'waves' of colonization had rolled in and spent their force, many of them failing to reach the eastern limits of the Polynesian archipelagos; while the rate of progress of each 'wave' has allowed time for forms to evolve.

There are flourishing communities of seabirds. The most productive marine zones abound with boobies, tropic birds, frigate birds and terns and with petrels, many of which nest high in the mountains. While there are fewer land birds the proportion of non-passerines is unusually high. One hundred and eighty-five of the 469 Polynesian species and subspecies and as many as 59 of the 104 genera, are non-passerine. The avifauna is distinguished by conspicuous gaps, the cockatoos, bee-eaters, rollers, hornbills, pittas, drongos, sunbirds and dicaeids being absent while other groups – especially the birds of prey and ardeids – are poorly represented.

In contrast Polynesia is noted for its pigeons, parrots and kingfishers, though even among these the number of species is less than in Papuasia. Parrots and pigeons, being frugivorous and with highly developed faculties for dispersal are well known to search out those islands which offer them shelter from predators, even nesting by choice on islets off the main islands because of the security which they find there. This is especially true of the fruit pigeons (*Ducula*) and the vividly coloured fruit doves (*Ptilinopus*). The parrots are represented by species belonging to several groups. While many of these are frugivorous, the brilliantly coloured lories are nectarivorous with brushed tipped tongues adapted to their diets.

Brush tongues are especially well adapted to the cup-shaped flowers which are commonest in the Polynesian vegetation, especially among the myrtles and eucalypts. The same adaptation recurs in the honeyeaters, a group of nectarivorous passerines typical of Australo-Papua and Polynesia whose tongues also are divided into fine strips. Insectivorous birds are represented by a variety of passerines, among which the dominant types are campephagids or cuckoo-shrikes (*Lalage* and *Coracina*), flycatchers (*Rhipidura*, *Monarcha* and *Petroica*) and whistlers (*Pachycephala*). It is noteworthy that many of these birds are strong flyers, easily capable of crossing wide water gaps.

Speciation has reached a very advanced stage, especially among the passerines and pigeons. White-eyes (*Zosterops*), whistlers (*Pachycephala*) and fan-tailed flycatchers (*Rhipidura*) especially are represented by many species, each endemic to a group or even to a single island, while subspecific differentiation is correspondingly marked. Thus a single type of bird may be represented across Polynesia by many local forms which replace each other geographically.

Bird communities have remained much impoverished with only a limited number of elements. Many ecological niches remain empty or are very imperfectly filled, and the birds have not radiated adaptively as in other parts of the world. As a result the ecosystems show no special differentiation. Quite possibly most Polynesian islands are too small to give rise to such evolution, which are known to be produced only under precise geographical conditions.

Islands and loss of flying ability

A high proportion of island birds are wingless, or have lost the power of flight by the secondary regression of their wings. Others, though their wings are normal, scarcely fly and prefer to escape on foot, sneaking off through the vegetation.

This is not a strictly insular tendency, since one does meet such birds on continents, but is very conspicuous on islands where the best present examples live. It is probably related to the absence of mammalian predators, and the introduction of dogs, cats, rodents and even monkeys to many islands has been followed by the reduction or extinction of most of the birds.

Moreover the large running birds, apart from the ostriches and rheas which live in specialized habitats, are or were all confined to islands. The most spectacular of these were *Aepyornis* of Madagascar and especially the moas (*Dinornis* and related genera) of New Zealand. The latter, whose fossil and subfossil remains have been found in these islands and also in parts of eastern Australia, formed a diverse though homogeneous group within which two families, seven genera and at least twenty-eight species can be recognized. These gigantic birds, some of which reached a height of 3·5m, had entirely lost their fore limbs, even their scapular girdles being vestigial. They formed a true community, no doubt with definite ecological specializations, since some lived in grassy savannas and others in swamps. Without natural enemies, terrestrial mammals being absent from New Zealand, these birds were probably the main vertebrate users of the plant cover.

Now totally extinct, the moas were probably victims of natural catastrophes resulting from floods, as is shown by the numbers of bodies engulfed in swamps and peat bogs. However, man also contributed to their extinction, for the moas were contemporary with the Maoris who hunted them and used their hides. The remains of at least six species have been found in kitchen middens, and carbon C14 analyses have shown that these date back 700 years or more. The practice of firing pastures probably accelerated their disappearance. New Zealand is also the exclusive territory of kiwis, very primitive birds whose wings are reduced to stumps. They live in dense wet forests where they seek the worms, insects, larvae and fruit which form their food by smell, a sense otherwise undeveloped among birds. A clear tendency towards flightlessness in New Zealand is also shown by the Takahe *Notornis mantelli*, a species related to the Purple Gallinule which was believed extinct until it was rediscovered in 1948 in a few valleys of South Island, and the Kakapo *Strigops habroptilus*, a monospecific genus of parrot confined to New Zealand. The latter, owl-like in appearance and of exclusively nocturnal habits, spends the best part of its time on the ground and only glides down from trees up which it has climbed. New Zealand thus harbours a remarkable series of birds, belonging to the most diverse systematic groups, whose wings are absent, atrophied or functionless. This convergent evolution results from New Zealand's total isolation, without a single native predator.

A similar tendency can be seen in the Mascarene islands, also faunally much impoverished because of their isolation, with only twenty-eight species of birds and without native mammals. These islands were formerly occupied by gigantic terrestrial pigeons, with wings reduced to mere stumps quite incapable of supporting the birds in flight: the Reunion Solitaire *Raphus solitarius*, the Dodo *R. cucullatus* of Mauritius, and the Rodriguez Solitaire *Pezophaps solitarius*. Weighing a score of kg these massive birds led a terrestrial life and fed on plants. They too were over-evolved, in equilibrium within a community without predators but unable to resist for more than two centuries the impact of man.

Many rails on oceanic islands have also lost the power of flight. This ancient world-wide family shows a distinct tendency towards flightlessness with regression of the fore limbs. This development is especially marked on islands, where rails have developed endemic species and even genera. They are represented on Pacific islands by a series of strongly differentiated forms, the rails of Wake Island *Rallus wakensis*, Tahiti *R. pacificus*, Lord Howe Island *Tricholimnas sylvestris*, New Caledonia *T. lafresnayanus*, Fiji *Rallina poeciloptera*, Kusaie *Aphanolimnas monasa*, Iwo Jima *Poliolimnas cinereus*, and Samoa *Pareudiastes pacificus*. To this may be added the Kagu *Rhinochetus jubatus* of New Caledonia, a monospecific family within the Ralliformes. These flightless birds could have evolved only in the absence of any terrestrial predator. The importance of this factor is shown by a similar development among the pigeons, since the Tooth Billed Pigeon *Didunculus strigirostris* endemic to Samoa is largely terrestrial and flies only for very short distances, and even among the passerines, as is shown especially by the New Zealand wrens *Xenicus* which are confined to off shore islands. The rails show parallel developments on islands elsewhere, notably on the Atlantic islands Gough and Tristan de Cunha (*Atlantisea rogersi*).

Finally we may recall the Flightless Cormorant on the Galapagos. The tendency towards flightlessness is thus very general on islands – so much so that it is quite characteristic of evolution in these ecologically very specialized habitats.

Chapter 28

The Place of Birds in the Living World

THE previous chapters have demonstrated birds' high degree of adaptation to different habitats, reflected in a great number of morphological and anatomical characters, ethological and especially ecological traits. The class as a whole is very adaptable as we have repeatedly seen, since its potentialities are the greatest among all terrestrial vertebrates. Birds have thus had the opportunity, which they have successfully taken throughout their evolution, to colonize ecological niches closed to other vertebrates.

It is important to examine the place which birds occupy in various ecosystems, and especially the part they play in the transfers of energy. The consecutive elements of an ecosystem are not interchangable, each occupying a clearly defined place determined by the species' adaptations, and this qualitative aspect must not be neglected. A mammal cannot replace a bird or vice versa, even if both live in the same habitat and get their energy from the same food source. We know that, even where birds play only a modest part in the energy transfers in comparison with other animals, their very special adaptations make them the only ones which can ensure a transfer to the place which they occupy in the ecosystem. Their functional significance is thus greater than the part that they represent in the total biomass of the habitat would suggest.

Some principles of quantitative synecology

It is above all important to recall some fundamental principles of quantitative ecology (Odum; Bourlière Lamotte 1962). Birds represent only a small fraction of their *biocenoses* or biotic communities, a term for all the plants and animals in a natural environment, dependent on the inorganic substrate which together with these living things make up the *ecosystem*. The totality of living communities is the *biosphere*. The whole problem of 'metabolism', or energy exchanges within the communities which form biological systems, must be considered qualitatively as well as quantitatively.

A biocenose consists essentially of two trophic levels: the *producers* or auto-
trophs – effectively the green plants – capable of using the energy of solar
radiation and transforming it into chemical energy in the form of organic
matter (which they synthesize from mineral salts, water and carbon
dioxide); and the *consumers* or heterotrophs – effectively the animals –
which feed either on the plants or on other animals. The *decomposers*,
which are mostly microscopic (bacteria and fungi), live on all these
organisms, degrading their dead organic matter and 're-cycling' it in
simpler form so that it can be re-used by other living things.

Consumers occupy many trophic levels. Primary consumers are vegetarians which feed directly on producers. Secondary consumers are carnivores, predators on the first, and may in their turn be the prey of super-
predators (tertiary consumers), which are themselves sometimes hunted by
animals of a still higher level. They thus make up a *food chain*, each link
of which is at a different trophic level. However, the situation is rarely so
simple since the chains are interlinked, the same animal acting as a link in
several chains of a single biocenose – sometimes at different levels, as with

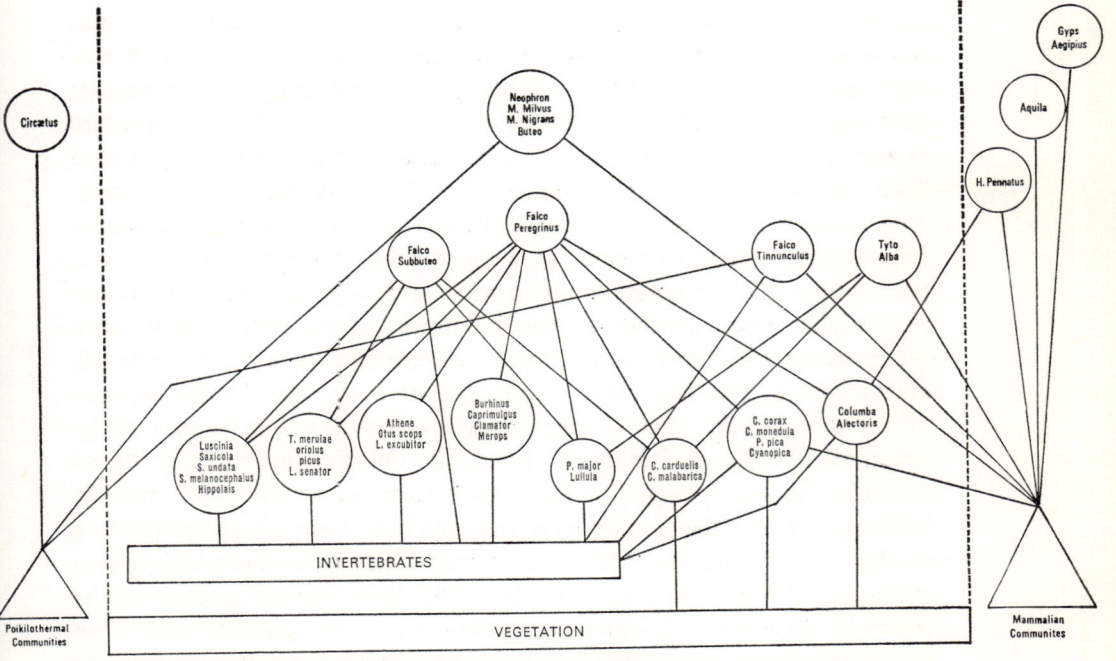

Figure 93. A simplified food-chain for the birds of the Coto-Donana in southern
Spain. The names are shown of the super-predators (raptors and owls), and some
of the primary consumers.

polyphagous species which are simultaneously primary and higher-level consumers. Thus all the food chains of the biocenose unite to form a network or food web.

The position of a species within such a web depends upon many factors, since its diet, feeding methods, habitat, adaptation and behaviour can greatly alter its place and function in the ecosystem. Its total characteristics define its ecological niche. As Odum has said, an organism's habitat is its address while its niche is its profession. The number of ecological niches varies greatly. In simple ecosystems such as arctic habitats the numbers of niches and food chains are reducd, whereas they are greatly increased in such extraordinarily complex ones as tropical rainforests. In principle every niche is occupied by one species, which has evolved in response to all the factors of the environment. It is only in very complex habitats that several species appear to occupy the same niche, and even this may be an illusion which will be dispelled by greater knowledge. Within different biocenoses on the other hand, the same niche may be occupied by very different and entirely unrelated species, and a single species may occupy different niches in the various environments which it has colonized.

This qualitative appraisal of biological systems needs to be supplemented by quantitative studies. However, the number of individuals which makes up a community or one of its levels is of little significance. It is the *biomass* or standing crop, the total mass of living matter (whether of an entire community or of the population of a particular species), which provides a fundamental datum for evaluating the place of each species in the biocenose. The biomass is calculated by multiplying the number of individuals by their average weight to obtain the living weight, which may be expressed per unit area. The biomass of any trophic level is greater than that of the next higher level, so that all the biomasses of an environment, each representing a category of consumers, are arranged as an erect pyramid, apex upwards. This is because the metabolism of each level obeys the laws of thermodynamics, the biomass pyramid corresponding to an energy pyramid. The life of a biological community depends on the ability of the plants to fix solar energy, at very low efficiencies up to 1 per cent. In feeding on plants the primary consumers in turn waste a large proportion of the energy in the food they eat. Their efficiency is often only about 10 to 12 per cent and sometimes less. Different types of animals are of different efficiencies, the homoiothermous vertebrates notably lower than the cold-blooded ones, because of the extra energy needed to maintain their temperature at a constant high level. Again, since energy requirements are inversely proportional to body size, the smaller the animal the more intense the metabolism, so that the biomass of a population at a particular trophic level is less the smaller the individual organism. This is allowed for in calculating the 'consuming biomass' (see below).

Thus to reconstruct the energy pyramid it would be necessary to estimate the *energy budgets* at different trophic levels. For this the amount of organic matter formed in a given time would have to be measured, and the energy dissipated in all forms to be estimated. This leads to the estimation of *productivity* or rate of production, a parameter more important even than the biomass since it reflects the dynamic aspect of the ecosystem. Unfortunately few data are available, and certain ecosystems are still virtually unknown.

The various ecosystems differ greatly. The most productive environments at the plant level are tropical rainforests (which in Java produce 5,400 to 6,900g dry weight per m² per annum) and conifer forests (which in England produce 3,180). In the sea the cold zones are far more productive than the warm ones. At other trophic levels the rate of secondary productivity falls as one ascends the food pyramid, since nothing further is produced beyond the plant level. The organic substances are ceaselessly rearranged and energy circulated from one trophic level to another, with a loss at each transfer due to a generally rather low rate of assimilation.

One often speaks of the 'richness' of a habitat, and the richest communities are generally those with the most species and the highest biomass. However, the true 'richness' of a habitat is much better measured by its productivity.

Bird densities and biomass

Densities depend especially on the amount of specific food available in a given habitat, and thus on its productivity at various trophic levels, but other ecological and ethological factors also play important parts. From all these factors the carrying capacity of a habitat may be defined. This is the maximal or optimal threshold which populations forming an avian community (whether as a whole or species by species) do not exceed, in a biome which is in equilibrium without large fluctuations. Equilibrium is of course only rarely attained, even without human interference, and any fluctuations in the numbers of birds follow environmental fluctuations.

The numerical densities of all animals including birds also depends of course on the trophic groups to which they belong.

Censuses undertaken during the past few years, though still unfortunately insufficient in number, allow us to specify the densities and numbers of

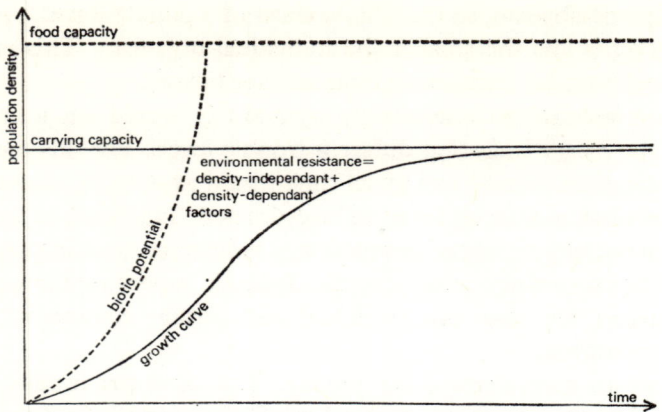

Figure 94. The influence of various limiting factors on the growth-curve of a population.

birds in their principal habitats. These data allow us to calculate the *biomass* – the weight of living matter – per unit area of one or many species or even of a whole community. This parameter is fundamental in establishing the energy balance-sheet and productivity of an ecosystem, though since quantitative ecology is in its infancy, little numerical information of this kind is yet available. A few examples indicate the biomass of the birds established in certain habitats.

a Conifer forests A population of birds in a Czechoslovakian forest of spruce *Picea excelsa* consisted of sixty-three species, with 1,286 individuals and a biomass of 48,338g per 100 hectares, or roughly thirteen birds and 483g per hectare. Bird numbers and individual weight contribute differently to the biomasses of various species, some of which excel in abundance and others in body size. The data below (simplified from Turcek 1956) allow comparison between the dominant species.

BIOMASS PER 100 HECTARES OF SOME BIRDS IN A SPRUCE FOREST

Species	Number of individuals	Average weight (g)	Biomass (g)
Coal Tit *Parus ater*	347	12	4164
Chaffinch *Fringilla coelebs*	234	24	5616
Song Thrush *Turdus philomelos*	70	70	4900
Goldcrest *Regulus regulus*	56	5	280
Chiffchaff *Phylloscopus collybita*	43	10	430
Crested Tit *Parus cristatus*	40	13	520
Robin *Erithacus rubecula*	40	15	600
Jay *Garrulus glandarius*	14	160	2240
Ring ouzel *Turdus torquatus*	14	110	1540
Blackbird *Turdus merula*	13	80	1040

Species	Number of individuals	Average weight (g)	Biomass (g)
Great Tit *Parus major*	13	19	247
Mistle Thrush *Turdus viscivorus*	10	120	1200
Nutcracker *Nucifraga caryocatactes*	8	150	1200
Hazel Hen *Tetrastes bonasia*	8	400	3200
Wood Pigeon *Columba palumbus*	7	480	3360
Crow *Corvus corone*	6	550	3300
Woodcock *Scolopax rusticola*	4	360	1440

Thus in these forests the Coal Tit and Chaffinch are *numerically* the dominant species. They also contribute importantly to the total biomass, each accounting for about 10 per cent despite their low individual weights. Others such as the Hazel Hen, Wood Pigeon and Crow make similar contributions, since their high weights compensate for their low densities. Ecologically the species with the greatest biomasses obviously play the most important part in the energy flow, since they are the major links in the principal food chains. These are not necessarily the numerically commonest species (though a correction for individual size must be made in estimating the *consuming biomass*, a point dealt with below). Thus although the order of numerical densities is *Parus ater*, *Fringilla coelebs* and *Turdus philomelos*, that of ecological dominance[1] is *Fringilla coelebs*, *Turdus philomelos* and *Parus ater*, followed by *Columba palumbus*, *Corvus corone* and *Tetrastes bonasia*. However, the Wood Pigeon and Crow feed mainly outside the forest, and thus scarcely take part in the economy of this ecosystem.

The birds of this biocenose may also be divided into feeding categories. The 1,286 individuals in 100ha, belonging to sixty-three species, divide as follows:

Diet	Number of birds	Percentage of the avifauna	Biomass (g)	Percentage of the biomass
Vegetarian	116	9	13,029	27
Polyphagous	911	71	26,437	54
Insectivorous	249	19	5,018	10
Carnivorous	10	1	3,854	9

This shows that polyphagous birds are markedly dominant in numbers and in biomass, while carnivores, represented by a few raptors and owls, are predictably rare.

b Deciduous forests Biomasses in deciduous forests are usually distinctly higher. Thus the birds in a Slovakian forest of beeches, maples, firs and

---

[1] Some prefer the term 'importance' to 'dominance', so as to avoid confusion with other meanings (Blondel). It is noteworthy that the three 'dominant' species in this example form together less than 30% of the crude total biomass of the habitat, though elsewhere certain species may provide a larger proportion of the total biomass.

spruces provide a biomass of 1,168g/ha, and in an oak-hornbeam forest 1,447g/ha (Turcek). In a riverside forest in the Marne, eastern France, eighty-four pairs of insectivorous birds corresponding to sixty-four birds at fledging, have been recorded per eight hectares, the biomass varying from 200g/ha in winter to 1,630g/ha in summer (Erard).

The biomasses of forest birds in the USA have been estimated as follows (from Wing):

	per 10 hectares:	
Association	Number of birds	Biomass (g)
Oak–hickory	37	1088
Oak–maple	55	1147
Beech–maple	53	1735
Beech–maple (climax)	85	1,455

c *Mediterranean environments* These are poorer than the last, with much smaller biomasses of birds. The density on a bushland (*garrigue*) in the south of France has been estimated as 21·2 pairs per 10 ha, made up of 3·6 pairs of Nightingale *Luscinia megarhynchos*, five of Sardinian Warbler *Sylvia melanocephala*, 12·1 of Subalpine Warbler *S. cantillans*, 0·4 of Dartford Warbler *S. undata*, and 0·1 of Magpie *Pica pica*. The density in a most favourable year was 24·9 pairs per 10 ha, and the average may be taken as 23 pairs (Blondel 1969), corresponding to a biomass of about 57g/ha. This may be lower still during the summer, because of the un-favourable climatic conditions, while it rises in winter with the arrival of wintering migrants which exploit the local seasonal maximum in available food.

Semi-desert and desert habitats are much poorer still. In the Ksours Mountains of the northern Sahara (Blondel 1969) counts on one km² of clay and stone steppe showed three Black-tailed Sand-Larks *Ammomanes cinctura*, eighteen Short-crested Larks *Galerida malabarica*, five Desert Wheatears *Oenanthe deserti* and two Houbara Bustards *Chlamydotis undulata*. The large bustards bring the biomass up to about 52g per hectare. However, among fallen rocks on undulating and rocky ground, counts on 1km² showed on the average fifty Sand-Larks *Ammomanes deserti*, 0·6 Blue Rock-Thrushes *Monticola solitarius*, 0·6 Black Wheatears *Oenanthe leucura*, a few Trumpeter-Bullfinches *Bucanetes githaginea*, and one Little Owl *Athene noctura*, with a biomass of only about 15g/ha.

d *Intertidal environments* According to estimates made on the Bay of Aiguillon in the Vendée (Spitz 1964), the total biomass of carnivorous waders wintering on the mud-flats reaches about 8,880kg. In mass per unit area, taking account only of the area uncovered at high water which deter-

mines the limiting capacity of the habitat, the maximal biomass of carnivorous waders varies from 570g/ha in the summer (June) to 6,590g/ha
during wintering and passage (November), remaining above 4,000g/ha
throughout the winter. In addition the omnivorous ducks provide a similar
biomass, which reaches 6,107g/ha though it does not remain at this level
for so long. In the Gulf of Morbihan in Brittany (Chaucheprat in Spitz),
wintering Brent Geese *Branta bernicla* and ducks – especially Wigeon *Anas
penelope* – form concentrations with a total biomass of about 42,700kg,
which at high water corresponds to 21,350g/ha. This very high value is due
to the fact that these birds are primary consumers.

e *Tropical environments* Almost the only data available are those of Morel
(1968), obtained in the sahelian savannas of Senegal. These shrub savannas
are dominated by various acacias such as the Gum Tree *Acacia senegal*,
under which grasses form a discontinuous and relatively thin carpet. The
following table gives a selection of results from censuses undertaken in an
area of 25ha during 1961, with the values for different feeding groups in
May – the unfavourable second half of the dry season – and in October –
the favourable second half of the rainy season:

Feeding Group		No. of species May	No. of species October	No. of individuals May	No. of individuals October	Biomass (g) May	Biomass (g) October
Vegetarian:	Sedentary	6	12	22	79	1,815	3,185
	Migrant	0	0	0	0	0	0
Polyphagous:	Sedentary	2	10	4	94	58	7,851
	Migrant	0	4	0	12	0	193
Insectivorous:	Sedentary	5	7	15	32	727	1,106
	Migrant	0	5	0	40	0	1,264
Totals:	Sedentary	13	29	41	205	2,600	12,142
	Migrant	0	9	0	52	0	1,457
Grand Totals		13	38	41	257	2,600	13,599

These figures reflect the great seasonal fluctuations in tropical savannas
which we have already noted (Chapter 25). The birds are forced to follow
the fluctuations in available food by travelling, which explains the seasonal
variations in their own biomasses. The total biomass is further increased
during the winter by the arrival of Palaearctic migrants, which are almost all
insectivorous or polyphagous. Reducing to biomasses per unit area, it
seems that at the most unfavourable season the carrying capacity of this
type of savanna is about 1 to 1·5 birds a hectare, with a biomass of about
100g; while at the most favourable season the density is about eight
residents plus two migrants a hectare, with biomasses of 485 and a little less
than 60g. Since the maximum densities of residents and migrants coincide,

the mean biomass of these savannas towards the end of the rains is of the order of 540g/ha. It is noteworthy that the vegetarian and especially the polyphagous birds have the highest biomasses, while the insectivorous birds are less numerous though they form the great majority of the migrants. In terms of biomass the raptors are negligible. Though vultures are commoner than hunting raptors the commonest of all, the White-necked Vulture *Pseudogyps africanus*, is represented by only one individual in every 420ha, which at a body weight of 5,500g contributes some 13g/ha to the biomass; while there is only one Martial Eagle *Polemaetus bellicosus* in every 5,000ha.

The Senegal gallery forests of *Acacia scorpioides* are the richest in birds, because of their greater variety of ecological niches and higher productivity. In contrast to the situation in savanna the maximum population occur during the dry season and the minimum during the rains, so that these forests act as compensators, while the highest densities of migrants do not coincide with those of residents. The carrying capacity of this environment is about ten times that of savanna. During the rainy season there are from 36 to 41 birds, with a biomass of the order of 3,200g, per hectare, while during the dry season up to 101 birds have been counted per hectare, with an estimated biomass of 8,845g/ha. It is unfortunate that we still lack information about other types of savanna and (largely because of the difficulties of census in such closed habitats) about tropical rainforest.

Tropical *aquatic environments* too are very productive, so that their bird biomasses are correspondingly large. The following are some values for an area of 12,000ha south of the Senegal River, the sites chosen for census including the river banks, the shores of Lake Guier, a semi-permanent swamp, and paddies which occupied over half the prospected area.

Feeding group	No. of species	No. of individuals	Biomass (kg)
Residents			
Piscivores	16	12,292	11,676
Insectivores[1]	19	26,662	1,990
Vegetarians[2]	16	1,204,487	25,740
Predators		50	80
Total		1,243,491	39,492
Migrants			
Piscivores	9	900	2,100
Insectivores[1]	26	83,700	2,535
Vegetarians	7	108,525	18,215
Predators		2,800	905
Total		195,925	23,755

[1] Passerines (warblers and wagtails), pratincoles, sandpipers, plovers, herons etc.
[2] Including the Red-billed Dioch *Quelea quelea*, with an estimated population of more than a million individuals and a biomass of 15,000kg.

These figures indicate average biomasses of 3,200g/ha for the residents and 1,900g/ha for the migrants. The importance of the latter deserves emphasis in a habitat where even the autochthonous forms are highly nomadic. The very nature of the environment, subject to huge fluctuations during the seasonal cycle, necessitates and regulates these movements, and as a result the biomass at a given place varies considerably.

f The seas Although detailed estimates of seabird populations are available, it is difficult to relate these to meaningful areas of distribution. Many of these birds disperse over huge areas outside the breeding season, though it would be misleading to include these in estimating biomass per unit area. Even while nesting they may exploit large fishing grounds. Thus Manx Shearwaters *Puffinus puffinus* from Skokholm Island off South Wales go to fish deep in the Gulf of Gascony. However, it is impossible to take account of such distributions in relating biomass to unit area since one would have to know precisely the area searched by the members of a particular colony. Information on this is very vague, so that it is not certain that the results have the same meaning as in terrestrial habitats.

An attempt has been made to estimate the biomass per unit area of the Emperor and Adelie Penguins which, with petrels and a few seals, form the greater part of the animal biomass established on the coast of Adelie Land. Their total biomass is estimated at 3,000kg for 100km of coastline (Prévost). Since during migration and feeding journeys these birds travel about 1,000km from the coasts, their feeding area may be taken as 1,000 × 100km, giving an estimated biomass of 300g/hectare. However, the biomass of birds in antarctic seas as a whole must be lower than this, since favourable local circumstances raise the values for Adelie Land above the average (Holdgate).

The available data on biomasses are obviously still inadequate, and apply to only a fraction of the world's biomes. Nevertheless the comparison of biomasses in various habitats does reveal large variations. In general they vary with the total productivity, as long as the production is suitable for the birds' specific dietary needs. Thus deserts and wastelands are very poor, while deciduous forests are much richer. Because of their high productivity aquatic environments, and especially the intertidal zone, support the greatest biomasses.

When the biomasses of birds are compared with those of other animals, especially invertebrates, they appear much lower. Whereas the mammals often make up a much greater fraction, the contribution of birds to the

GROSS BIOMASSES OF THE BIRDS IN SOME COMMUNITIES

Habitat	Locality	Biomass kg/ha	Authority
Heathlands	NW Germany	0·005	Schuhmann
Dry Forest	NW Germany	0·080	Schuhmann
Conifer forest	Finland	0·220	Palmgren
Spruce forest	Czechoslovakia	0·483	Turcek
Mixed forest	Finland	0·580	Palmgren
Beech–maple–conifer forest	Czechoslovakia	1.168	Turcek
Oak–hornbeam forest	Czechoslovakia	1·447	Turcek
Riparian forest	France	0·200–1·630	Erard
Oak–hickory forest	USA	0·109	Wing
Oak–maple forest	USA	0·115	Wing
Beech–maple forest	USA	0·173	Wing
Climax beech–maple forest	USA	0·145	Wing
Pine forest	USA	0·024	Salt
Pine–fir–spruce forest	USA	0·079	Salt
Fir–spruce forest	USA	0·113	Salt
Aspen association	USA	0·719	Salt
Halophile vegetation	Vendée, France	0·444	Thiollay
Polders	Vendée, France	0·434	Thiollay
Water–meadows	Vendée, France	0·853	Thiollay
Meadows	Vendée, France	0·268	Thiollay
Shore dunes	Vendée, France	0·555	Thiollay
Prairie with bushes	USA	0·272	Salt
Swamp with sedges and willows	USA	0·220	Salt
Mediterranean garrigue	France	0·057	Blondel
Desert steppe	Algeria	0·052	Blondel
Desert scree	Algeria	0·015	Blondel
Sahelian savanna:	Senegal		
dry season		0·052–0·106	Morel
rainy season		0·543	Morel
Gallery forest:	Senegal		
dry season		8·845	Morel
rainy season		3·276	Morel
Ponds and river banks	NW Germany	1·300	Schuhmann
Intertidal mudflats:	France		
carnivorous waders		0·570–6·590	Spitz
omnivorous ducks		6·107	Spitz
geese and ducks		21·350	Chaucheprat
Aquatic habitats (river banks, marigots, paddies):	Senegal		
residents		3·200	Morel
migrants		1·900	

GROSS BIOMASSES OF SOME NATURAL COMMUNITIES

Habitat and zoological group	Locality	Biomass kg/ha	Authority
MARINE COMMUNITIES (benthos)	Mediterranean	100	Demel & Mulicki
	Gulf of Guinea	117	Longhurst
	Baltic	330	Demel & Mulicki
	Bering Strait	1,650	Beliaev & Ushakov
	North Sea	3,460	Demel & Mulicki
	Antarctic	13,470	Beliaev & Ushakov
FISH			
Trout lake	USA	62·5	Carlander
River	USA	187	Carlander
Sea	The Channel (Plymouth)	162	Harvey
Atoll lagoon	Eniwetok	446	Odum
INVERTEBRATES			
Grasslands with Cyperaceae	Guinea	7–25	Lamotte
Savanna with Andropogonæ	Guinea	250	Lamotte
Mountain grasslands, Mt Nimba:			
all invertebrates	Guinea	75·5–175·3	Lamotte,
oligochaetes only	Guinea	53·5–131·9	Lamotte, Aguesse & Roy
Sansouire, Camargue	France	55	Bigot
INSECTS			
Sahelian savannas	Senegal	2·4	Morel
Guinean savannas	Lamto, Ivory Coast	10	Gillon
REPTILES			
Guinean savannas: lizards	Lamto, Ivory Coast	0·1–0·3	Barbault
MAMMALS			
Arctic tundra	Canada	8·0	Bourlière
Beech–maple–conifer forest	Czechoslovakia	6·6	Turcek
Prairie	North America	35	Bourlière
Steppe	Southern USSR	3·5	Bourlière
Saharan ergs	Mauritania	0·05–0·2	Bourlière
Sahelian steppe	Chad	0·8	Bourlière
Masai steppe	Kenya	132	Bourlière
Savanna	Kivu, Congo	235	Bourlière & Verschuren
Tree savanna	Rhodesia	44	Bourlière
Tropical rainforest	Ghana	0·75	Bourlière
Voles (Microtus)	Vendée, France	1·0	Spitz
Field-mice (Apodemus)	Czechoslovakia	0·3	Turcek
Guinean savannas: rodents	Lamto, Ivory Coast	0·04–1·64	Bellier
insectivores	Lamto, Ivory Coast	0·01–0·28	Bellier

total biomass of a habitat is relatively small[1]. This may be explained both by the relatively high trophic levels which vertebrates occupy in ecosystems, and by the high specialization of birds which play very specific functional roles in biocenoses.

Qualitative place of birds as consumers within ecosystems

As we have seen, birds have adopted every possible diet and are established in all available niches (Chapter 5). We now have to return to this question in considering the division of birds into trophic groups.

a   Primary consumers Apart from a few ducks, none of the birds confined to marine environments is a plant feeder. In terrestrial habitats the proportion of the whole avifauna is still small, but varies greatly between environments. Very few birds feed on leaves and buds because of the low food content of the vegetative system, and those which do adopt this diet are relatively large and sometimes show regression of their flying abilities. Most plant-eating birds are graminivorous or frugivorous. The former, characteristic of temperate habitats and tropical savannas, make use of foods of high nutritive value, available almost throughout the year at a single locality. Strictly frugivorous birds are virtually restricted to the tropics and some of them even to rainforest.

Occupying low levels in the food chains, plant-eating birds may be very numerous and gregarious, as is shown by the ducks, geese, pigeons, weavers and finches.

b   Secondary consumers Many birds are secondary consumers, at trophic levels which are often difficult to distinguish. Some are at the first carnivorous level, feeding largely on vegetarian animals. This is true of birds which feed on phytophagous insects, such as storks and starlings on grasshoppers and cuckoos and tits on caterpillars; on rodents, such as buzzards feeding on voles; and on phytophagous fish which are common and often the easiest to catch. These birds live off a relatively large biomass and therefore they can still be fairly numerous and sometimes even gregarious. However, the biomass they live off is subject to large fluctuations and the predatory birds suffer the consequences, especially in relatively simple environments such as those of the arctic and temperate regions.

---

[1] Indeed, the terrestrial vertebrates as a whole make only a very modest contribution to the global biomasses. According to recent calculations, animals often represent less than 1 per cent of the plant biomass, while the invertebrates make up 95 to 99·5 per cent even of this animal biomass.

Other birds are higher in the food chains, feeding on predatory insects, such as honey-buzzards on bees and bee-eaters on various hymenoptera; on spiders and centipedes which passerines hunt both on the ground and on tree-trunks; and on insectivorous mammals, such as buzzards and owls on shrews. Some raptors (such as the Sparrow Hawk) feed on birds which are at least partly insectivorous, others on insectivorous reptiles and even on carnivorous mammals. Snake-eagles mainly hunt snakes which are themselves predatory, and Ospreys show a certain partiality for carnivorous pike. These super-predators can only be very rare, because of their levels in the food chains, while few are closely tied to particular types of prey – an excessive specialization which would endanger their survival.

Finally certain birds are carrion feeders. Small vultures and American vultures are scavengers and feed on rubbish of all sorts, while large ones live especially on the carcases of hoofed mammals. For example the Lammergeyer mainly eats their bones and is thus at the top of a defined food chain, which makes it particularly vulnerable to any disturbance of the environment.

c *Omnivorous birds* While some birds can thus be located at particular trophic levels, it is difficult to place many others. Although a few species are stenophagous, most are of catholic tastes and distinctly polyphagous. Almost all raptors hunt prey belonging to various trophic levels. Other birds – including woodpeckers, hornbills and toucans, and many passerines from the thrushes, warblers and wood-warblers to the tanagers, tits and bulbuls – are both vegetarian and insectivorous.

Vegetable foods play as important a part throughout the year in the diets of many of these birds as those of animal origin. For example some turdids feed at the same time on berries and on any small animals they find upon the ground, while crows and gulls eat anything they can.

Other birds' diets vary with the time of year. Many are predominantly carnivorous during the summer but become vegetarian in winter, while others remain carnivorous but hunt other prey. These seasonal changes in sedentary species may take place on the spot, whereas in migrants they may be related to habitat changes. Thus on the arctic tundras near fresh water waders feed primarily on insects; but becoming distinctly marine while wintering, they then feed almost exclusively on molluscs and crustaceans. Similar changes in diet and hence in trophic level occur with age. Many birds such as weavers and various finches, which are graminivorous as adults, are fed on insects and other animal prey when young. There are even sexual differences. In certain raptors, notably Goshawks and Sparrow Hawks, the females are perceptibly larger than the males and do not feed on precisely the same prey. In some woodpeckers such as the Hairy Woodpecker *Dendrocopos villosus*, the female lives exclusively on insects living on the bark and the male on xylophagous larvae dug out from deeper in the wood (Kilham 1965). Thus polyphagy and variation in diet are very general among birds, a tendency much less widespread among mammals, few mammals being truly omnivorous.

Birds have principally turned to types of food which the mammals have neglected. Many of them exploit the large biomass provided by insects without incurring any real competition. In aquatic habitats fish form another large part of the standing crop biomass, on which a sizeable fraction of sea and freshwater birds feed. In contrast the mammals are broadly dominant among the vegetarian animals. Attempts to return to life on the ground – the only habit compatible with this diet in terrestrial environments – have almost all been doomed to extinction during evolution. Birds do compete with other vertebrates in many sectors. Thus there is competition between certain insectivorous birds and reptiles (shrikes and lizards), and between owls and nocturnal carnivores (mustelids). However this can be of scarcely any importance at the level of the biocenose as a whole, if only because, apart from their shared food, the ecological niches of the competitors differ profoundly.

It may be noted that birds very largely live and are dependent on members of other communities. Apart from a few raptors which hunt other birds, all carnivorous birds live on insects and on vertebrates of other classes.

Birds are specialized for precise niches. Being small they cannot take foods which are too heavy, awkward or out of their reach. In contrast their liveliness, the speed of their reflexes, and their bills precisely adapted to take the most minute prey fit them all to catch insects. Birds are also the only vertebrates except bats able to fly and thus to move about with little effort in a three-dimensional world. Without leaving the ground entirely, they have mainly occupied the upper layers and the sides of rocks and cliffs. These incomparably aerial vertebrates have exploited this mobility to make themselves dominant in the highest layers of vegetation and all other niches which demand it to the full. They have also been able to enter other environments which remain closed to animals which move on the ground, such as swamps, freshwater habitats and intertidal zones. Although the marine adaptations of birds are entirely different from mammals in this habitat they are no less remarkable. Furthermore birds are the only animals which can walk, swim, dive and fly, and some do all four with equal ease.

Birds thus occupy niches which other terrestrial vertebrates have been unable to occupy, and as a result of their mobility are clearly in many ways the most highly competitive of all vertebrates, able to colonize extreme environments, especially those in which the standing crop fluctuates greatly during the annual cycle. Mammals are ill equipped to exploit fluctuating resources. Birds in contrast are the best adapted of all vertebrates whose rapid flight, requiring minimal expenditure of energy, allows them to cover huge distances and thus to defend themselves by change of habitat.

The feeding impact of birds on their habitats

Knowledge of the standing crop biomass is fundamental to the quantitative study of ecology. Use of the raw biomass is not wholly satisfactory for estimating the energy consumed by a bird population in the form of food, and the place it occupies in the energy flows within an ecosystem. We have seen that food consumption is not proportional to the weight of a bird, but with metabolic characteristics as a whole is proportionately higher the smaller the bird (Chapter 5), so that a gram of goldcrest consumes more energy than a gram of thrush and still more than a gram of bustard or penguin. The same biomass makes a very different impact on the environment, depending on whether it is made up of many small birds or a few large ones. It is therefore necessary to introduce a correction which takes account of the metabolic differences dependent on individual size. Physiologists have shown that oxygen consumption per unit weight (and consequently energy consumption) is proportional to an exponential function of the body weight (Salt 1957) the average power of which has been established as 0·75. A new value known as the metabolic weight or consuming biomass is obtained by raising the raw biomass of each species to this power. This allows different populations of birds to be compared without falsifying the results through differences in individual weight, and to be added together in calculating energy balances.

Populations of the same raw biomass may represent very different consuming biomasses through differences in the weights of individual birds. In a single environment, the mud-flats of the Bay of Aiguillon in France, 20,000 Dunlin *Calidris alpina* of an average weight of 50g form a crude biomass of 1,000,000g and a consuming biomass of 280,000g, while 1,000 Shelduck *Tadorna tadorna* of an average weight of 1,000g form the same crude biomass but a consuming biomass of only 100,000g (Spitz). The difference accurately reflects the different feeding impacts of these consumers on their environments.

577

However, the data thus obtained can be considered only as over-simple indications of true energy consumption. While they reflect metabolic levels precisely they cannot be converted into food consumptions, which depend also on the thermal factors of the environment, on the type of food taken, and on the characteristics of each species. However, the introduction of new coefficients would be useless in the present state of our knowledge and techniques, since the errors resulting from the estimation of bird densities are more important.

Thus two values can be calculated: on the one hand the crude biomass, or true weight of birds maintained by the environment; and on the other the consuming biomass, which acts upon the environment by taking energy from it in the form of food. Consuming biomasses allow us to estimate below the amount of food taken, and the proportion of the plant or animal biomass which this represents.

a Graminivorous birds The most detailed data available concern the Red-billed Dioch *Quelea quelea*, a species which has been especially well studied because of its economic importance as a ravager of crops in the African savannas (Morel 1968). This bird is one of the major vertebrate consumers of the plant production on the grassy savannas of Senegal. Its enormous and always very mobile populations concentrate in small areas to nest colonially. The nests of a colony are arranged side by side in the trees, each tree supporting 50 to 300 or sometimes up to 1,000 nests, so that the total number may easily reach 100,000 a hectare. The following figures are derived from the weight of grass needed to build a nest (15g of dry matter), the weight of seeds brought back for each feeding, the frequency of feedings, and the simultaneous consumption of the adults:

1,500kg of dry grass for building nests,

5,544kg of seeds fed to the young,

9,240kg of seeds eaten by the adults.

Comparison of the grass harvest with the grass production per hectare of the habitat shows that this harvest represents at least 27 per cent of the grass cover, but may greatly exceed the productivity per hectare. As a result the size of the colonies is limited by the availability of sufficient nesting material. While the birds may travel far in gathering food – sometimes to more than 10km – they have to find nesting materials much closer, especially since this phase of their nesting is always rapid. Unfortunately no data are yet available on the quantity of seeds produced by the savannas harvested by these birds, but it may be assumed that they are forced to exploit great areas and make a considerable impact.

In contrast there is more information on the consumption of insects and the productivity of the habitat in this resource. Though adult queleas do not eat insects, this prey makes up a large proportion of the diet of their young, as of most graminivorous birds. A colony of average density (12,400 nests per hectare) consumes 214kg dry weight of arthropods during the 18 days of brooding. Now the total invertebrate biomass per hectare of savannas is of the order of 700g, though not all the insects are edible to queleas. Combining these values, a colony occupying one hectare daily consumes all the insects living on seventeen hectares, and those on 300ha during the brooding period. An average colony, established on 50ha and feeding within a radius of 10km, and thus having some 30,000ha at its disposal, consumes the insects present on 15,000ha. While these figures only give the orders of magnitude, queleas thus consume some 50 per cent of the insects present in the habitat they exploit. This allows the impact of these populations to be calculated, on the production of savannas at various trophic levels. In order to calculate the total harvest by all the graminivorous birds of the sahelian savannas, those of many other species must be added – from ducks and waders to doves, sand-grouse and a long series of weavers and waxbills. Although they may harvest differently, they do so from the same biomass, since graminivorous birds seem to specialize very little if at all in diet, the species being differentiated much more by other niche characteristics.

Thus the case of the queleas allows us to estimate the impact of a population of graminivorous birds on its environment, and to show that a considerable part of the plant production and a larger contribution from the insects are eaten by these birds.

b   *Vegetarian forest birds* The birds of temperate forests especially consume the seeds produced by trees in enormous quantity. Although relatively few species take this diet, they sometimes harvest large amounts (Turcek 1961). In forests of spruce *Picea excelsa*, Crossbills *Loxia curvirostra* and Great Spotted Woodpeckers *Dendrocopos major* cut up and throw down between 20 and 90 per cent of the cones, which amounts to a considerable quantity since there are from 1,000 to 130,000 a hectare. A single Crossbill can throw down up to three cones in a minute, half of which are scarcely damaged, while a single woodpecker has consumed 1,093 cones from a total production of 2,700. Nutcrackers *Nucifraga caryocatactes* in the Carpathians eat on the spot or store in their hoards from 50 to 100 per cent of the seed production of Arolla Pines *Pinus cembra*. The daily takings are very large. During the winter the daily consumption of a Great Spotted Woodpecker is from 800 to 1,000 seeds of Scots Pine, the contents of fifty cones.

579

However, birds sometimes take a more modest harvest in comparison with the biomass of seeds on which they live. Thus in Norway, though tits collect large quantities of the seeds of firs and make them into reserves which they use during the winter, this represents less then 1 per cent of the available biomass (Haftorn 1960). During a study made in Russia, the pines on 160ha produced a total of 3,212,000 cones and the spruces 892,000. However only 52,000 pine cones and 2,000 spruce cones had been attacked by Great Spotted Woodpeckers, or the negligible proportions of 1·6 and 0·2 per cent.

The same is true among deciduous trees, whose fruits are also taken though apparently in smaller amounts. In favourable years oaks produce 2,000kg of acorns per hectare, and these are eaten especially by Jays *Garrulus glandarius*, which also collect them as reserves. These are only partly used during the winter, the 'forgotten' acorns germinating or being eaten by squirrels. During one month in 1965, Jays collected and transported some 300,000 acorns across thirty-seven acres of forest. Elsewhere thirty-five Jays collected 200,000 acorns in ten days. This harvest, though it is far greater than their actual consumption, represents less than 10 per cent of the acorn production.

Beeches produce from 2,000 to 10,000kg of mast per hectare, taken by Jays, Nutcrackers, Nuthatches, tits and Bramblings. In a Czechoslovakian forest the sixty wintering Bramblings consumed 2kg a day, amounting to 120kg during their stay of two months or from 5 to 10 per cent of the production per hectare. However, their impact may be much greater. The consumption of Bramblings wintering in Switzerland, a population of at least 11 million individuals, was estimated at 10 to 12 tons of beech-mast a day or 300 to 360 tons a month, which is the production of thirty to thirty-six hectares under the most favourable conditions (Guéniat).

These examples show that consumption of the fruits and seeds of trees varies very greatly, but that in some cases a considerable part of the production may be taken by the few bird species specialized for this diet.

c *Insectivorous birds* These are large consumers and take a great number of insects, especially while breeding. Great Tits carry from thirty-five to fifty insects an hour to their young, and about 10,000 are needed to raise one nestling. A Great Tit or a Lesser Whitethroat *Sylvia curruca* may eat fifty caterpillars of the butterfly *Cacoecia murinana* in 90 minutes, which is 10 to 16 per cent of their own weight. However, the impact of bird populations on their potential prey is very variable, and many mistakes have been made on this subject – notably by those who tried to assess the 'usefulness' of insectivorous birds as destroyers of insect pests.

When an insect is rare and its populations very small the impact of predatory birds is almost nil: the chances of capture are very low, since a bird searches most enthusiastically and effectively for more abundant insects.

However, when an insect is represented by medium-sized populations birds make a large impact on it, taking a toll of up to 30 to 50 per cent of the stocks in a forest biome (Tinbergen 1949). In a Dutch pine forest, the tits alone destroyed from 24 to 37 per cent of the caterpillars of *Panolis griseovariegata*, 10 per cent of the adult butterflies of *Bupalus piniarius*, 23 per cent of the caterpillars of *Ellopia prosapiaria*, 10 per cent of the larvae of the wasp *Acantholyda pinivora* and 3 per cent of those of *Neodiprion sertifer*. Other insectivorous birds in the same environment took about a further third.

Woodpeckers take only a minute proportion of the larvae of *Pissodes piniphilus* when the populations of the beetle are small, but when they exceed a certain threshold the toll reaches 95 per cent of the larvae and pupae in the canopy. In Canada the Downy and Hairy Woodpeckers *Dendrocopos pubescens* and *D. villosus* eliminated from 50 to 90 per cent of the 'apple worm' larvae of *Ernarmonia (Carpocapsa) pomonella* in the orchards of Nova Scotia. Predation by tits on the caterpillars of the related butterfly *Ernarmonia conicolana* in Great Britain increases when the density of the prey exceeds a threshold of ten larvae per fifty cones. Yet though it may reach 50 per cent of the wintering larvae, it cannot control the prey populations (Gibb 1966). In Japan Tree Sparrows *Passer montanus* and Grey Starlings *Sturnus cineraceus* take from 40 to 50 per cent of the lithosiid butterflies *Hyphantria cunea* which fly before daybreak when the males' sexual flights cease, probably in response to predation (Hagesawa & Ito 1967).

Finally, when an insect is very abundant and especially when it is swarming the impact of bird predators again becomes very small. Despite the number of insects taken, these are an extremely small proportion of the whole population. During a swarming of the Gipsy Moth *Lymantria dispar* in a Czechoslovakian oak-hornbeam forest only 1·5 per cent of the total mass was consumed, despite a high density of insectivorous birds. Wood-warblers (*Parulidae*) feed their young partly on the larvae of *Choristoneura fumiferana*, whose density may reach 20 million a hectare, but the toll they take amounts to less than 1 per cent of the prey population.

It is well known that swarms of orthoptera, especially locusts, attract great numbers of birds from starlings to storks, and these take an absolutely but not proportionally large toll. During an invasion by the grasshopper

Melanoplus differentialis in California the insects were estimated at about 40 million per km² but birds ate less than 50,000 a day in the same area (Bryant in Lack). In eastern Africa, where about a hundred species of birds take locusts as part of their diets, the most voracious take up to 3,500 a day. However this predation only reduces the stocks of the prey by from 0·25 to 6 per cent which can have no appreciable effect during the locusts' gregarious phase. It does however take effect when the populations decline, and may be the decisive factor in putting an end to this phase (Elliott). One of the rare effective actions known against a swarm of insects concerns a district of a fir forest, entirely defoliated by the tenthredinid saw-fly *Pristiphora abietina*, whose larvae were entirely eliminated by starlings (Bruns 1960). Such eradication is possible only within a limited area and under very special circumstances. Moreover, a predator's 'interest' does not lie in exterminating the prey on which it feeds, but much rather in living in equilibrium with it.

These facts raise the problem of the part which insectivorous birds play in controlling the populations of those insects which form their potential prey. Birds are certainly incapable of keeping down a swarm once it is established. At this time there is an obvious disproportion between the number of predators and their toll on the one hand, and the standing crop biomass on the other. In contrast when the insect populations are at an average level, the birds take a large part of them and can thereby prevent them from swarming, putting an effective brake on the population expansion before it can give rise to an explosive increase. These observations show that action to protect crops, and especially forests, should encourage the multiplication of birds which can control pest insects before they multiply beyond a certain threshold. Artificial increase in the number of nesting sites, which is a limiting factor for hole-nesting birds, by putting up nest boxes, can contribute along with other management techniques to the effectiveness of insectivorous birds. Thus under certain conditions birds may be one of the factors limiting insect populations, together with insect parasites. Observations indicate that birds prefer uninjured rather than parasitized caterpillars of the butterfly *Porthetria dispar*. Since parasitism and predation thus bear on different sections of the caterpillar populations, the birds' predatory action is added to that of the insects which lay in the caterpillars' bodies (Koroljkowa).

d *Carnivorous birds* Almost the same rules on trophic relationships with their prey apply to raptors (Blondel 1967, Frochot 1967). On the whole these birds take less food in unit time than other consumers, since theirs is of greater energy value, and have no regular feeding rhythm but undergo a succession of fasts and feasts. Some large raptors go several days without food and may eat only four to five times a month. However, these meals are large, a Golden Eagle swallowing up to 900g of flesh at a time. This regime is clearly adapted to the hazards of the chase on which these birds depend. The average amount of food taken per day is about 230 to 300g for the Golden Eagle, 160 to 180g for the Goshawk, 150g for the Buzzard, and 100g for the Barn Owl. However, the tolls they take are larger than these figures indicate, since raptors 'waste' a large proportion of the prey they capture, reaching up to 30 per cent by the Golden Eagle and 35 per cent by the Sparrow Hawk.

The impact of raptors can be significant when the populations on which they prey are at average densities. The Sparrow Hawk *Accipiter nisus* catches 8 per cent of the sparrows within its feeding area. During the summer it is responsible for half the *mortality* of the sparrows, and a quarter of that of the Chaffinches and Great Tits (Tinbergen 1946). The Goshawk effectively controls the populations of its prey, notably Crows and Ruffed Grouse *Bonasa umbellus*. Thirty per cent of the grouse mortality is due to this predator, the males being especially vulnerable when they drum on logs during their displays. Eagles on the other hand scarcely affect the populations of birds and mammals on which they feed.

The impact of predators is greatly reduced when their prey swarms, which is especially common among small mammals. In arctic regions the Rough-legged Buzzard, Snowy and Short-eared Owls take a large number of the lemmings which play such an essential part in these biocenoses. In four months a family of Snowy Owls, consisting of the parents and nine young, take up to 2,600 lemmings from a territory of 4km²; yet this predator removes only 20 per cent of the total population of the rodents, and scarcely affects their increase (Watson). In Scotland Short-eared Owls capture only 3 to 13 per cent of voles at the beginning of the breeding season, and proportionately even less when the rodents have multiplied (Lockie). When the small mammals proliferate the effect of predation becomes infinitesimal, so that Short-eared Owls took only 0·05 per cent of the voles proliferating in Scotland (Chitty). Here the number of predators is obviously disproportionate to that of their prey.

Thus, the same three levels appear in the impact of predators on their prey as with insectivorous birds. Between the two thresholds of prey density – known as the *lower and upper limits of vulnerability* – predation does have decided influence, although it can seldom be held solely responsible for controlling a population. Various density-dependent factors, linked more or less to the trophic limiting capacity of the habitat, reduce or even

Q

eliminate populations. Predation is only one of these factors and often not an important one because of the fundamentally different population growth rates of predators and prey, such as owls and voles, or Sparrow Hawks and small passerines.

These points have to be carefully considered in assessing the influence of predators, and especially of raptors which are too often condemned out of hand as 'pests' (Chapter 15). Situated as they are at the ends of food chains, their numbers must always be very small. They act at the species levels (attacking the most abundant for obvious reasons of 'profitability' and thus tending to equalize the stocks); of populations; and of individuals, especially through preferential elimination of the sick, deficient or abnormal. Furthermore, predation bears especially on the less experienced young animals, of which there is a surplus population which is condemned to disappear in any case. The regulatory part played by raptors is thus difficult to assess and usually misinterpreted.

e *Water Birds* There are few data on the tolls taken by water birds, and more especially on their proportion of impact on the populations of their potential prey. The amounts taken are large, as we have especially noted of waders feeding on annelids and molluscs in the intertidal zone (Chapter 20). It has been suggested that 10 per cent of the prey populations is taken, but this estimate can only be very approximate. Oystercatchers sometimes take 20 to 70 per cent of the cockles on which they feed (Davidson).

Off the snout of a glacier in Spitzbergen the Kittiwakes *Rissa tridactyla* alone took 6 million individuals of the crustacean *Thysanoessa inermis* (Hartley & Fisher). Fish-eating birds also take great quantities of prey. There are about 16 million guano birds on the Peruvian coast, of which 80 to 85 per cent are Guanay Cormorants *Phalacrocorax bouganivillei*. One of these cormorants eats about 430g of fish a day and the birds together take 2·5 million tons annually – almost entirely Anchovetas *Engraulis ringens*. Scarcely any information is available on the size of the Anchovetas populations, but the impact of the huge number of birds may be judged by comparing their toll with the 7·5 million tons fished annually by the Peruvian fleet (Schaeffer). Before man competed in exploiting the stocks of fish, birds estimated at 28 millions must have taken some 4 million tons of Anchovetas a year (Jordan).

There is also little information on fresh waters. Some studies have been made in Canada, on a river in New Brunswick which lacked a natural population of salmon (Elsar 1962). Young salmon were repeatedly introduced while the predators, Goosanders *Mergus merganser* and Red-breasted Mergansers *M. serrator*, were eliminated as far as possible. The proportion of salmon surviving from introduction to the parr stage was 4·8 per cent in the year before predators were controlled, while in succeeding years under control it was successively 9·1, 19·1 and 23·1 per cent. However, since other factors were involved, no final conclusion can be drawn from this study. In the Dombes in France the annual fish consumption per family of herons – both adults and young – was as follows (Lebreton 1964):

> Heron *Ardea cinerea* (2·5 young on average): 270kg
> Purple Heron *A. purpurea* (3 young on average)[1]: 80kg
> Night Heron *Nycticora nycticorax*[1]: 38kg
> Little Egret *Egretta garzetta*[1]: 25kg

Birds as a source of available food

Besides the part they play in biocenoses as consumers, birds also act as intermediaries in energy transfers by serving as prey to certain carnivores. Few birds are themselves predators on birds, although some raptors such as Sparrow Hawks, Goshawks and Merlins do live on them entirely. In all biocenoses the important predators are mammals. Various cats take birds, and many of these such as the serval are more or less specialized in catching them, while canids such as foxes and jackals, and viverrids such as genets and mongooses, are also predators on young or on adult birds. Oviparous reproduction, though so well adapted to the life of birds, also constitutes a serious handicap since the eggs are especially vulnerable throughout incubation. Nest robbers are found among birds such as corvids and raptors, and among mammals such as small carnivores, insectivores, primates and rodents. Reptiles, though of negligible influence in temperate and cold zones where they are rare, play a considerable part as bird predators in the tropics, especially as nest robbers taking eggs, young and even adults.

While all birds are subject to intense predation at the egg stage, as adults the terrestrial species are the principal victims. These include many plant-eating species noteworthy for their very high rates of reproduction, compensating for the more intense predation, while their low trophic level allows them larger populations than birds at higher levels. As a result the game-birds, pigeons and doves in terrestrial, and the ducks in aquatic habitats are important links in food chains, and by transforming plant into animal matter allow flourishing communities of predatory animals to establish themselves in many environments.

[1] These three species only stay for part of the year, and take distinctly broader diets than the Heron.

Quantitative evaluation of the place of birds in ecosystems

While birds thus occupy qualitatively important positions in ecosystems, their contribution should also be quantitatively studied. Unfortunately this is still more difficult, since the data are clearly inadequate, while the energy balances of ecosystems remain to be worked out. Thus the energy flow which passes through birds cannot be precisely evaluated.

Biomass is low in comparison with other animals, and as a result the part played by birds may seem small, but it varies from case to case. Large plant-eating birds certainly play an important part in some biocenoses, being responsible for a considerable fraction of the transfer of energy to higher trophic levels. Furthermore, the comparison of biomasses alone does not allow a true assessment of the place of an animal group in its ecosystem, since biomass reflects only the static aspect of the energy balance. Productivity is a much more important consideration in measuring the flow of energy, though unfortunately our information in this field is still more limited than for other aspects of quantitative ecology.

From the point of view of productivity, birds like other animals divide into two categories. Some such as the Procelleriiformes and raptors are characterized by low fecundity, long life and an extremely low rate of population renewal. Obviously their annual production of living matter must be small, and few animals can live off such producers, most of which are at the tops of the food chains in their communities.

Other birds in contrast are characterized by great fecundity, very short lives and high rates of population renewal. Graminivorous passerines and game birds are examples of this, and it is noteworthy that most of such birds are vegetarian. Insectivorous birds with this type of population dynamics are found only among species like the tits, exploiting a source which is abundant and available throughout the year. Thus the productivity of such species is considerable.

The place of birds in the living world

We should tend to underestimate the part played by birds in living communities, if we considered only what fraction of the energy transferred within ecosystems passes through them. Birds are of greater functional significance than their mass of living matter would suggest. Because of their extreme specialization, they fill unusual positions where they are irreplaceable in the circulation of energy.[1] Although they sometimes live on the same foods as other vertebrates there is usually no true competition.

Each species of bird occupies within the general scheme a habitat to which it is adapted in terms of the physical and biotic factors of the environment. One of the principal conditions is the availability of sufficient food. Specific structural and functional characters must also be satisfied, and the species must find shelter from enemies and from the adverse factors of the environment. The density and nature of the plant cover, and the availability of suitable sites for nesting, singing and display are the immediate determinants. The composition of the fauna is equally influential, since the presence of a better-adapted species characteristically excludes all other birds from a particular niche. During interspecific competition in a given terrain one species is always ecologically dominant (Hilden 1965), while habitat selection plays a part in speciation. Every species thus has its particular niche. Some show great ecological flexibility, adapting to very diverse conditions, and are correspondingly successful and widely distributed. Others in contrast have much stricter needs and as a result are confined to highly specialized niches in particular environments. Few of these niches and the species which occupy them are interchangeable.

The density of species is much higher in equatorial regions, and especially in rainforest, than in any other environment. As one moves away from the equator into environments whose climates are increasingly hostile and seasonally contrasting, the species density diminishes rapidly, whereas the population of each species tends to increase. Tropical ecosystems are characterized by very large numbers of food chains among which the energy flow is rather evenly distributed. In the simpler ecosystems of the cold regions on the other hand, there are fewer food chains and the greater part of the energy flow of the system takes place by certain favoured routes. This applies only to terrestrial habitats, since the situation in the oceans is reversed.

These variations in the density of species and the complexity of biocenoses has of course a quantitative aspect. The productivity of the various habitats is very faithfully reflected in the biomasses of the birds which occupy them. This again raises the problem of the regulation of bird populations (Chapter 15). It is indisputable that the amount of food

---

[1] It should also be remembered that birds are most important in assuring the pollination and seed dissemination of many plants, while they should also be capable of disseminating the eggs of fishes. This transport function is closely related to their ability to travel on a grand scale.

available is of the greatest importance (Lack). There is a close correspondence between the standing crop biomass and the quantity of food. The number of consumers cannot exceed a threshold set by the energy which is available in the form of food, depending on the 'efficiency' of the transfer from one trophic level to another.

However, the organization of a biological community is not subject solely to the laws of thermodynamics like a physical system. Many other factors intervene, and usually make their action felt long before the carrying capacity of the habitat is reached. Highly developed ethological characteristics, such as territorial behaviour patterns, contribute towards limiting populations, and such physiological factors are of the greatest importance among birds.

Ecosystems thus seem to function below their potential capacity, with the populations reaching equilibrium well before the trophic-carrying capacity of the habitat is reached.

To return to a comparison with mechanisms, an ecosystem functions like a motor running below its maximum performance. This gives a safety margin and a regularity which are decidedly favourable to the species in maintaining their population at optimal levels. This margin varies greatly according to the habitat, being greater within complex communities such as those of the tropics, and reduced in simple communities like those of the arctic regions or those created by man. This no doubt explains the 'jerky running' characteristic of the simpler system. These statements, which deserve to be verified by quantitative researches in depth, also apply no doubt to animals other than birds, and may well be basic to the functioning of ecosystems as a whole.

Chapter 30

Migratory Movements

ONE of the essential characteristics of birds is their ability to travel through the air, with an economy of effort unknown to ground-level vertebrates, allowing them to deal with rapid changes in environmental conditions, by seeking at a distance whatever they lack at a given place and time. As we have seen, these animals are closely dependent on the environment. Their very high level of metabolism compels them to find an abundance of foods, often highly specific, while despite their efficient thermal regulation relatively small size handicaps them in combating low temperatures. Birds are thus the class of vertebrates in which most migrants are found and seasonal movements are greatest. Ecologically, consumers which can move rapidly over great distances are the best placed to exploit fluctuating food resources, and to make use of surpluses which are available for a limited period each year. Their ability to travel according to a regular rhythm explains the almost world-wide distribution of birds, and their presence at relatively high densities in regions almost deprived of other vertebrates because of the short period during which they are habitable.

A clear distinction has to be made from the first, between migrations proper and other movements. The term 'migration' is applied only to regular journeys, in a cycle which is most often annual. A population resorts to a breeding territory, arbitrarily termed its home, and then when the young have been reared goes into a different area, called wintering range or winter quarters, where it passes a longer or shorter part of the year outside the breeding season. Thus typical migrations include two annual journeys, the post-nuptial migration to winter quarters and the prenuptial migration to nesting sites. The notion of 'home' is entirely conventional, although the impact of birds on their biocenoses may there be greatest in the breeding area, because of their increased need for food. The term *wintering* zone is also often inappropriate, since the species sometimes resorts to the opposite hemisphere, where it is then summer. However, these terms are sanctioned by long usage.

Observation of temperature avifaunas might at first suggest that sedentary birds are radically distinct from migrant ones – one group remaining during the winter while the other leaves to take shelter far away. However, the attempt to make a clear demarcation is misleading, since every intermediate stage exists. During nesting all birds are attached to a territory which they defend against intrusion by their fellows, while at other times they all show much less intense attachment to a given place. Some wander within a larger area, showing gregarious tendencies but keeping within a few kilometres. These birds are termed *sedentary*. Others disperse much more widely during the winter, and are termed *nomadic*. The most highly evolved stages of nomadism sometimes show favoured directions of travel, related to climatic conditions. This thus leads on to true *migrations* between breeding and wintering areas – which may partly overlap, or be very widely separated by territories over which the birds merely fly in the course of their journeys. One can also speak of the *altitudinal migrations* of mountain birds.

All stages between sedentary and migratory behaviour may thus be shown between the various populations of a single species. Such differences even appear within populations of *partial migrants*, some individuals of which migrate while others remain confined to a restricted area. There is similar variability in migration routes, migratory behaviour, the control of migration and its physiology, so that no strict divisions can be recognized. There is no single mode of migration. By keeping the individual in the most favourable conditions for that species, they form a group of special cases, and are themselves evidence of the astonishing evolutionary plasticity of birds.

Migrations occur in all groups of birds, except those confined to the ground by loss of their powers of flight, and are not restricted to particular regions. The most spectacular journeys begin from the cold regions of both hemispheres because of their large annual fluctuations in climate. Only the birds of tropical rainforests show scarcely any tendency to travel, because of their unvarying environment. Despite the hazards of long-range movements, resulting in severe losses which decimate migrants, these are outweighed by the advantages of leaving a particular area for part of the year. As a response of the organism to fluctuating conditions, migrations are of considerable ecological importance, and ecology is the key to their origins, controlling factors and routes.

Migrations have been exhaustively studied by very diverse techniques: simple observations on the behaviour of migrants in nature; marking by means of a light metal ring attached to the leg and allowing individual recognition; and most recently observation by radar. A few examples will show the different modes of migration, and especially the intergradation between sedentary and migrant birds.

'Partial' migrations

These, as opposed to movements over longer distances, better regulated and affecting whole populations, show no doubt how these arose in the course of evolution in response to environmental fluctuations.

a *European passerines* The passerines nesting in Europe offer many inter-mediates between residents and migrants. Tits in western Europe are on the whole sedentary, and during the winter they merely assemble in small flocks which wander about locally in search of food. However, certain populations of northern and north-western Europe are more unstable and take part in seasonal movements which have the characteristics of migration, being preferentially orientated towards more clement western regions.

The Blackbird *Turdus merula* is a parallel case. The same is true of the Robin *Erithacus rubecula*, one part of whose populations winter in the breeding areas while another, clearly migratory, travels to North Africa, Egypt and Iran. The Song Thrush *Turdus philomelos* shows still more definite migration, travelling north-east and south-west. Only the British nesting populations are partly sedentary, the remainder going to winter in the Mediterranean regions of Europe and North Africa.

The warblers are still more clearly migratory. The Blackcap *Sylvia atricapilla* winters in large numbers around the Mediterranean, while part of the European population goes to tropical Africa reaching as far as Tanzania. The Garden Warbler *S. borin* and Whitethroat *S. communis* are more demanding, since their whole populations migrate to tropical Africa and some individuals reach the Cape. The Chiffchaff *Phylloscopus collybita* winters in the south of France and the Mediterranean countries, while the Willow Warbler *P. trochilus* is also occasionally found wintering on the Mediterranean coast and slightly more often in North Africa, but mainly spreads throughout tropical Africa, from Senegal to the Cape.

Thus the migratory tendency is progressively developed and depends partly on diet. Graminivorous birds such as the finches migrate very little, since their diet allows them to find even in winter the seeds which satisfy their needs. In contrast, the greater part played by insects in the diet of a bird, the less it can survive in Europe during the winter and the more decidedly migratory it is – like the buntings (Emberizinae) which are much more insectivorous than the other finches.

b *Variation in migratory behaviour with geographical area, age and sex* We

591

have seen that the populations of a single bird species distributed across the whole of Europe can show very variable migratory behaviour from region to region. On the whole, those breeding in the north and north-east of the continent exhibit more decidedly migratory behaviour than those of western Europe, the populations of the British Isles being especially sedentary.

These differences appear in many other species, and especially in the Starling *Sturnus vulgaris*. The individuals which nest in Great Britain content themselves with wandering within the British Isles while those of France and the Netherlands also wander within a relatively limited area, after congregating in large flocks. In contrast the populations of Germany and eastern and northern Europe are clearly migratory, and while wintering spread over western Europe from Great Britain to the Mediterranean, and to North Africa where they reach high densities in Tunisia. Migrating Starlings seem to preserve the individuality of their geographical populations. In contrast to certain other birds, such as the Black-headed Gull, which mix in their winter quarters, the migrant individuals from a given area migrate in the same direction to pass the winter in a particular area, and so partly prevent mixing. The Rook *Corvgus fruilegus* also shows different migratory behaviour according to the district. Almost sedentary in western Europe, in eastern Europe it migrates following the great European plains to winter in Great Britain, the Netherlands, Belgium, France and northern Italy.

The northern limit of winter distribution for each species depends on its ecological demands. It varies from year to year in accordance with the rigours of the winter and, especially among ducks, during a single winter as a result of the waves of colder and milder weather. Thus modes of migration are not characteristics of the species, but of the various local populations of which it is made up.

These modes also often vary between individuals, since the whole population does not always react in precisely the same way: some individuals may be sedentary and others migratory. These partial migrations, which imply much individual variation in physiology, are very widespread among birds. They occur in the Song Thrush, Starling, Pied Wagtail, Blackbird, Robin, Lapwing and Woodcock among British species (Lack 1944), and in many North American birds (Nice).

One of the best-known examples is the Heron *Ardea cinerea*, whose migratory behaviour varies greatly across Europe. While the British populations are remarkably sedentary, and those of northern and eastern Europe wholly migratory, the populations of western Europe and especially France

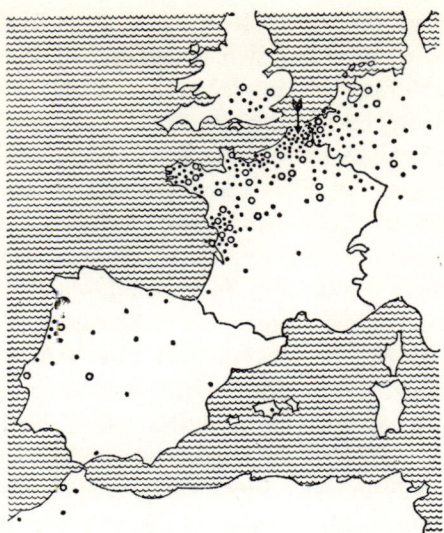

Figure 95. Winter recoveries of Herons *Ardea cinerea* ringed at Clairmarais, Pas-de-Calais (indicated by an arrow). The open circles indicate recoveries of birds more than one year old, and the dots those of birds born in that year.

show an intermediate condition. Ringing of many birds from colonies in northern France (Clairmarais near Saint Omer, Pas de Calais) has shown that certain individuals winter on the breeding grounds or nearby, while others go northwards to Belgium, the Netherlands, Germany and England. A greater proportion moves to the west and south-west and winters in western France and the Iberian peninsula, while others go to North Africa and a few to tropical Africa.

Such variations in migratory behaviour within a single species can some-times be related to genetic differences between sedentary and migratory individuals: in the Eider *Somateria mollissima* they are paralleled by differences in the composition of egg-white (Milne & Robertson). They are sometimes explicable ecologically, in terms of the various habitats fre-quented by a species within a given area, since some environments are richer in winter food than others, and so can support more individuals than can those habitats which are subject to greater fluctuations. Thus 'garden' Blackbirds tend to be more sedentary than 'woodland' ones, because artificial environments are richer in winter sources of food. Blackbirds, and other birds such as the Robin in western Europe, also show a winter shift towards the towns.

Variation in migratory behaviour is particularly marked in relation to sex

593

and age. Females tend on the whole to be more markedly migratory than males. Linnaeus called the Chaffinch *Fringilla 'coelebs'* or 'bachelor' since in Scandinavia most of the wintering birds are males, a greater proportion of the females leaving the breeding areas to migrate earlier and further. Analysis of migrating flocks shows that the proportion of females is highest during the earlier passages. They also seem to be more enterprising and form the greater part of flocks which cross stretches of sea. Female Chaffinches, Reed Buntings, Skylarks and Redstarts are more common than the males on Heligoland in the autumn. In Holland female Chaffinches appear on migration before the males and seem to go farther to winter, showing less reluctance to cross the Channel. In North America female Song Sparrows *Melospiza melodia* and Mockingbirds *Mimus polyglotos* also migrate more freely than the males. On the other hand, the males of the little Ruby-throated Hummingbird *Archilochus colubris* leave their breeding territories first (from July in the north of their range). Male hummingbirds play no part whatever in rearing the young, and are thus free to migrate while the females are still occupied. In spring both sexes sometimes leave simultaneously on the prenuptial migration, but more often the males go first so that they may delimit territories within the breeding area.

In addition young birds as a whole are more markedly migratory than adults. Most of the earliest flights of the post-nuptial migration are made by young birds which also travel farther than the adults. Most Herons and Gannets *Sula bassana* noted far from their colonies are young birds of the year. British colonies of Gannets disperse widely in winter along the Atlantic coasts as far as Senegal, but the individuals wintering in Morocco, Mauritania and Senegal are young, while the adults confine themselves to the French and Portugese coasts.

The difference in migratory behaviour between males and females and territorial behaviour and the migratory urge. Males are most often much more attached to territory than females and young, especially since in the autumn after the post-nuptial moult they often show, by song and other behaviour, a recrudescence of sexual activity. The male sex hormones inhibit the migratory urge, while in the absence of such an antagonism it can be expressed by the females and young. The results vary between individuals and populations of a single species, which explains the difference in behaviour. The division into sedentary birds, nomads and migrants may also be interpreted ecologically. The departure of part of the population when the food resources are considerably depleted reduces the impact of the birds on their habitat at the critical time, and adjusts their stocks to the winter-carrying capacity. Finally, the stronger migratory urge of the young, resulting from their weaker attachment to the breeding territories, is certainly an advantage to the species by ensuring dispersion. Pioneers which invade new territories are always young birds. This tendency also allows a more effective mixing of the various stocks, which could be harmed by over-strict geographical isolation.

Young birds sometimes indulge in special movements known as post-juvenile dispersion: their nomadic tendencies after fledging lead them in all directions from their birthplaces. Young Herons scatter over great distances, sometimes to hundreds of kilometres from the colonies where they hatched, and often in the direction opposite to what they will take later on their migration proper. Starlings ringed at the nest in Switzerland have been taken all around their birthplaces between June and September, in Belgium up to 500km from their point of departure. These movements have nothing in common with true migration, in timing or in direction. Post-juvenile dispersion must be ecologically determined by high local densities of the birds in which it occurs, all of which are gregarious or colonial. Large numbers of adults rearing young can considerably reduce the amount of food near the nesting colony. This dispersion is thus a response to partial local exhaustion of food resources.

c *Migrations of tropical birds* The migrations of tropical birds are important in understanding the origins of migration. Birds confined to rain-forest are practically sedentary, only the nectarivorous and frugivorous species moving about in response to flowering and the ripening of fruit. In contrast the birds of savannas, which are subject to large seasonal fluctuations in climate, make periodic movements which may be considered as true migrations.

Africa, the best-known continent from this point of view, encourages such movements because of its position. The graminivorous birds scarcely move about while the frugivorous, insectivorous and polyphagous ones are more decidedly migratory, in response to the wide fluctuations in the standing crop biomasses on which they live. Among these migrants are the cuckoos, rollers, nightjars, hornbills and many insectivorous passerines, as well as certain raptors.

These seasonal movements involve the birds nesting in both hemispheres, whose dates of outwards and return journeys are, as might be expected, usually reversed. The northern hemisphere species *Merops nubicus* nests from Senegal to the Red Sea between March and August and then migrates southwards. The southern species *M. nubicoides* nests in a large part of South Africa, mainly from September to March, and then leaves for the north, notably the Congo, where they remain from April to July.

595

Like the Scarlet Bee-Eaters some of these migrants seldom or never cross the equator: many migrants in northern Africa, such as various night-jars, shuttle between the sahelo-sudanian and guinean savannas. However, other migrations involve both hemispheres, with the birds going to winter on the other side of the equator. Thus the White-bellied Stork *Spheno-rhynchus abdimi* breeds only in a zone between Senegal and the Red Sea, where it remains from May to September, while after the wet season it undertakes large movements towards South Africa, crossing the East African savannas and the Congo forests. The Pennant-winged Nightjar *Cosmetornis vexillarius* migrates in the opposite direction, nesting in the southern hemisphere from September to December – a very favourable season because of the abundance of insects – and then returning northwards to winter in the savannas from Nigeria to Uganda. Thus these migrations follow increasing rainfall and the abundance of insects which hatch as a result, and once again demonstrate the influence on migratory behaviour of diet and the amount of available food.

Parallels are shown in tropical Asia by birds which occupy similar places in their biocenoses. In the Indochinese peninsula the green pigeons (*Treron* and *Sphenocercus*), parrots, broadbills, drongos and bee-eaters take part in more or less regular movements, which represent the first stages of true migration. In tropical America the same is true of the tyrant-flycatchers, whose mainly insectivorous diets make them especially sensitive to fluctuations in food resources which depend for their abundance on the rains. In the Variegated Flycatcher *Empidonomus varius* the most southerly populations (*varius*) nesting in Argentina and southern Brazil are migratory and winter in Amazonia and the surrounding districts, where the breeding race (*rufinus*) is sedentary. Thus for part of the year the density of this species is increased by the addition of a migratory to a sedentary population, which is probably not disadvantageous because of the enormous amounts of food available in Amazonia. The situation is sometimes even more complex as in the Red-eyed Vireo *Vireo olivaceus* which occurs across a large part of the New World. The migratory races of North America (*virescens*) and Mexico (*flavoviridis*) go to winter in South America from Panama to Bolivia, and on their departure are replaced there by a race (*chivi*) which is partially migratory in the most southerly part of its habitat. The wintering zone which these races share is also occupied by sedentary populations of the species. Thus the total population remains fairly stable throughout the year.

The waterbirds of the tropical regions also take part in regular migrations. The amount and accessibility of food available change periodically together with the physical conditions of the environment. Low water provides easier fishing, while many ducks and small waders roost on the uncovered sandbanks. Even some species which nest on the river banks are forced to leave by the rising waters. Thus in America the Wood Ibis *Mycteria americana* travels regularly between Amazonia and the Orinoco, so as always to be beside watercourses when their water level is at its lowest. In Africa the same is true of cormorants, anhingas, tree ducks and a few waders.

Some tropical migrants make entirely regular movements of great extent, such as the five or more species which nest in Madagascar and pass the southern winter on the African continent. The Broadbilled Roller *Eurystomus glaucurus* breeds in the rainy season from October or November onwards, and then in March migrates towards eastern Africa (the eastern Congo and Tanzania). The Lesser Cuckoo *Cuculus poliocephalus rochii* nests in Madagascar mainly from August to December, and then resorts to eastern Africa. The Madagascar Bee-eater *Merops s. superciliosus* is partly sedentary in Madagascar, where it wanders outside the breeding season from September to October, and partly migratory going to eastern Africa to hibernate from Eritrea to the Zambezi. It arrives at virtually the same time as another form of the same species (*M. s. persicus*) which leaves in order to breed between the Caspian and north-western India. Thus two distinct races, one belonging to the northern and the other to the southern hemisphere, replace one another in the same 'wintering' grounds according to the seasons.

Thus tropical birds show very variable types of migrations. All stages can be observed within a single species and even within a particular population. Some of these movements are true migrations, variable and of great extent, while some are only partial migrations. 'Wintering' areas are often poorly defined, since the birds wander widely. Thus tropical birds show remarkable opportunism in adapting their behaviour closely to environmental conditions.

Old world long-distance migrants

Partial migrations, nomadism and simple dispersal grade into much more highly developed migrations, which usually involve all the populations of a species, and take place between well defined and widely separated areas along remarkably stable routes. Here the cycle of movements attains a regularity and perfection unknown among more opportunistic migrants.

The palaearctic avifauna includes many migrants which entirely desert their breeding areas for warmer countries, such as the ducks which nest in arctic tundras where winter freezing makes their food non-existent or inaccessible. Some less-demanding species go to the milder parts of western Europe for wintering. Thus the Mallard *Anas platyrhynchos* winters in Atlantic and Mediterranean Europe and North Africa, without regularly reaching the tropics. Others in contrast seek shelter south of the tropics, following well-defined routes and often breaking their journeys by more or less fixed stages. Some of these stages are related to the post-nuptial moult, which sometimes requires special journeys. Thus Garganey *Anas querque- dula* winter in flood zones in the northern half of Africa – the lower Senegal valley, the flood-zone of the Niger, the Lake Chad depression, Bahr-el- Ghazal, and the lakes of Kenya and Uganda – gathering in huge flocks thanks to especially favourable habitats (Roux). Many shore birds migrate on the same scale. Although certain sandpipers and curlews partly winter along the coasts of Europe, much of their population follows enormous routes leading to tropical Asia, Australia and South Africa. Many gather in tropical flood zones, so that the lower Senegal valley is the main win- tering zone of European Black-tailed Godwits, many hundred thousands of which assemble there in October and November together with Ruffs and Wood Sandpipers (Roux).

Many insectivorous birds are also long-distance migrants, especially those raptors which are specialized for this diet. The Lesser Kestrel *Falco naumanni* which nests around the Mediterranean winters in much of tropical Africa, as far as the Cape. The same is true of the Red-footed Falcon *F. vespertinus*, the Hobby *F. subbuteo* and the Honey Buzzard *Pernis apivorus*, while even the Black Kites *Milvus migrans* of Europe almost all withdraw to tropical Africa.

Among the passerines, Swallows are without question the best and longest known of all migrants. Like Swifts (with which although they belong to a different order they may be associated because of their biological resem- blances), they are closely dependent on innumerable air insects. Since these insects disappear completely with the onset of winter, Swallows and Swifts are necessarily great migrants. Swallows and House Martins *Hirundo rustica* and *Delichon urbica* begin to leave their breeding area from the end of July, and the movement accelerates towards the end of September. Large flocks, which may include hundreds of birds, migrate during daylight. The swallows, arriving from mid-August in Ethiopia and towards the middle of September at Lake Chad, winter across the whole of tropical Africa as far as the Cape, travelling by many routes. This huge zone (which is also occupied by Swifts, among the earliest migrants to leave the temperate regions and

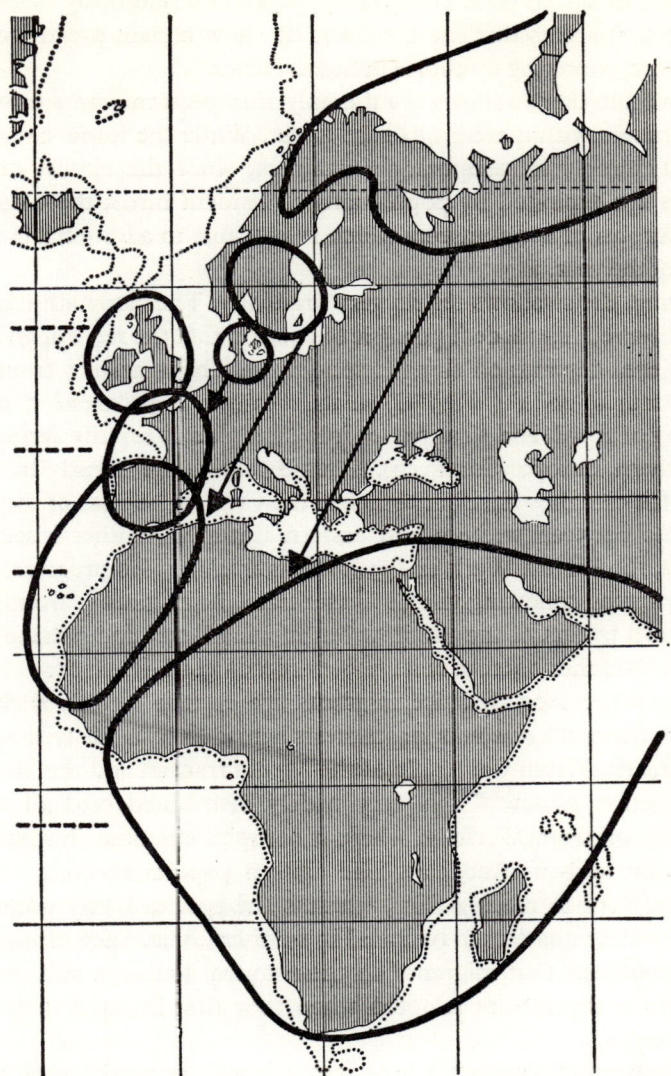

Figure 96. The breeding and wintering areas of different populations of the Ringed Plover *Charadrius hiaticula*. The two areas occupied by a population are connected by an arrow.

noted in East Africa from August) is very unequally occupied by the wintering birds, which concentrate in certain favoured regions. Although one cannot assign to each European population a rigorously defined wintering area, it seems that the Swallows do show certain preferences, with British birds wintering by choice in South Africa.

It is notable that swallows are the only European migrants to winter in large numbers within areas of dense forest. While the forest environment is not favourable to temporary consumers, since the almost unvarying resources are exploited by populations of resident birds, there does seem to be a surplus of aerial insects which is available to additional consumers for part of the year.

In spring, the swallows return very gradually. The ease with which they can be observed has made it possible to confirm that the reoccupation of the breeding area depends on climatic conditions. The migratory front follows the isotherm of 48°F ($=8\cdot9$°C) in its northwards progression from the beginning of April, and then passes it to sweep more rapidly across central and northern Europe (Southern). Thus Swallows are already in England when in the east they have only reached the Crimea, because of the climatic differences between western and eastern Europe. Another insectivorous passerine, the Willow Warbler *Phylloscopus trochilus* also progresses northwards in spring following the 48°F isotherm, its migratory front remarkably parallel to that of the Swallow but moving faster and passing the isotherm much earlier because of its less strict ecological demands.

Among other long distance migrants are certain turdids such as the Wheatear *Oenanthe oenanthe* which nests across Eurasia (as well as part of North America) from the Arctic to the Mediterranean and the Himalayas. The migration routes of this very widely distributed bird all converge remarkably on tropical Africa, where it occupies savannas (but not woodlands) as far away as Rhodesia. The eastern population cross the whole Asian continent to reach Africa, whereas the species rarely occupies the potential winter quarters to be found in southern Asia. Such examples lead to the hypothesis that migrants on their annual journeys still follow the routes which the species adopted when they first invaded their present breeding areas.

The warblers (Sylviidae), whose diet is largely or wholly insectivorous, are also long-distance migrants. Most of them cross the Mediterranean region, going to winter in tropical Africa by very various routes. The Blackcaps *Sylvia atricapilla* from most of Europe winter in large numbers in the Mediterranean region, but a minority shelters in tropical Africa, reaching Sierra Leone in the west and Tanzania in the east. The Garden

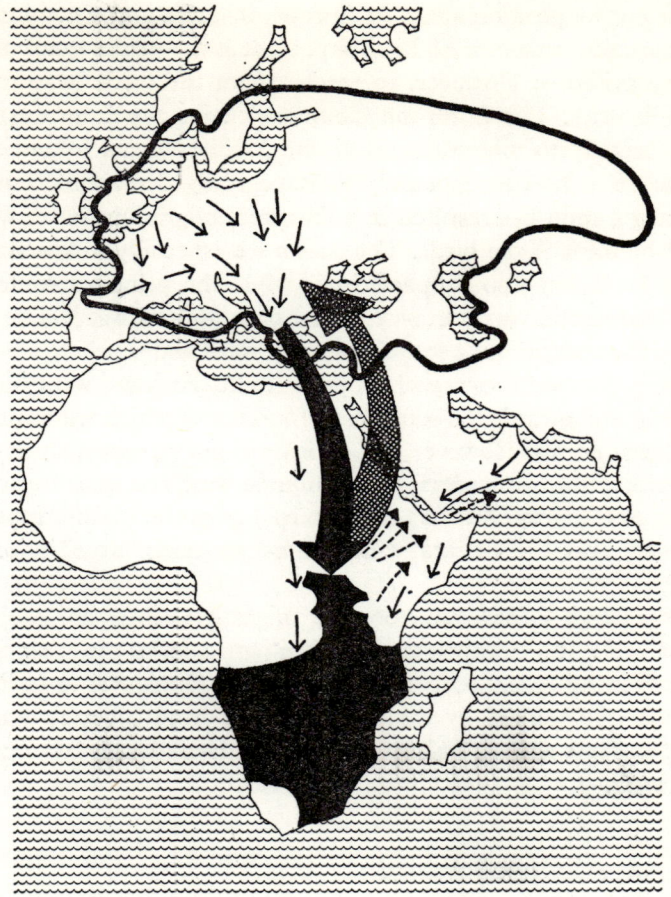

Figure 97. The migrations of the Red-backed Shrike *Lanius collurio*. The nesting area is outlined and the wintering area shown in solid black. The solid arrows show the course of the autumn migration, and the cross-hatched and broken arrows that of the spring migration.

Warbler and Whitethroat *S. borin* and *S. communis*, and the Icterine and Melodious Warblers *Hippolais icterina* and *H. polyglotta*, all cross the Sahara and reach the Cape, as do the leaf warblers except the Chiffchaff *Phylloscopus collybita*, which winters no farther away then the south of France. The flycatchers (Muscicapidae) leave their breeding areas, entirely travelling to tropical Africa and southern Asia. So do the wagtails *Motacilla*, the shrikes *Lanius*, the Golden Oriole *Oriolus oriolus*, and the Roller *Coracias garrulus*.

601

Because of its position south of Europe, its area, and a combination of favourable circumstances, Africa is important as the winter quarters for our migratory avifauna. However, to reach Africa the birds have to resolve many difficulties. The European mountain chains run from west to east, forming barriers to migration, while the Mediterranean followed by the Sahara are also hostile, especially to ecologically demanding land birds. This configuration has resulted in a few great migration routes which are followed by most of the birds. The coasts are especially followed not only by shore birds but also by passerines, while the principal routes of the western Palaearctic run south-westward or south-eastward. The greatest number of species, about 150, migrate westward and south-westward, and at least thirty-five of these reach tropical Africa. At least twenty-five other species migrate towards the south-east, fourteen of which winter in eastern and southern Africa. However, the division is not rigorous, since different populations of the same species may migrate west and east. Furthermore, a few migrants (except for the soaring birds) cross the Mediterranean at all points, and also the Sahara which most migrants cross in one stage (Moreau).

In the eastern Palaearctic, too, the migration routes are subject to a variety of ecological circumstances, and in particular to the great mountain massifs of central Asia. Some migrants cross the Himalayas to India, which is especially frequented by waders and ducks from Siberia; while those from the extreme eastern Palaearctic pass south-eastwards to tropical Asia and Malaysia.

North American migrants

In North America as in Europe, the alternation of cold and warm seasons results in wide fluctuations in the standing crop biomass. It is not surprising that a large part of the avifauna should migrate to meet its ecological needs. These needs are all the more rigorous since a greater proportion of North American than of European birds is of tropical origin. Every stage exists, their diverse wintering areas extending from the Gulf of Mexico to Patagonia.

The parallel with Europe is not perfect, since there are very important differences in the geographical configurations of the two continents. The mountain chains run mainly north and south, presenting no obstacle to migrations and even aiding them by providing guide lines. The tropical regions are wide open to migrants, while the less-demanding birds can find favourable wintering conditions in Northern America itself: California and the regions around the Gulf of Mexico enjoy sufficiently mild climates for them to survive the winters there.

The populations of certain other species concentrate towards the south-eastern United States while wintering. This is true of the American Robin *Turdus migratorius* which nests from the Canadian tree-line to southern Mexico. However, it is only a partial migrant, and its races and local populations vary greatly in migratory behaviour. Wintering birds have been recorded at latitudes as high as British Colombia and the northern United States. However, the principal wintering zone is in the south-eastern United States, where almost the whole population seems to concentrate. Some observers have recorded flocks of 50,000 individuals in Florida.

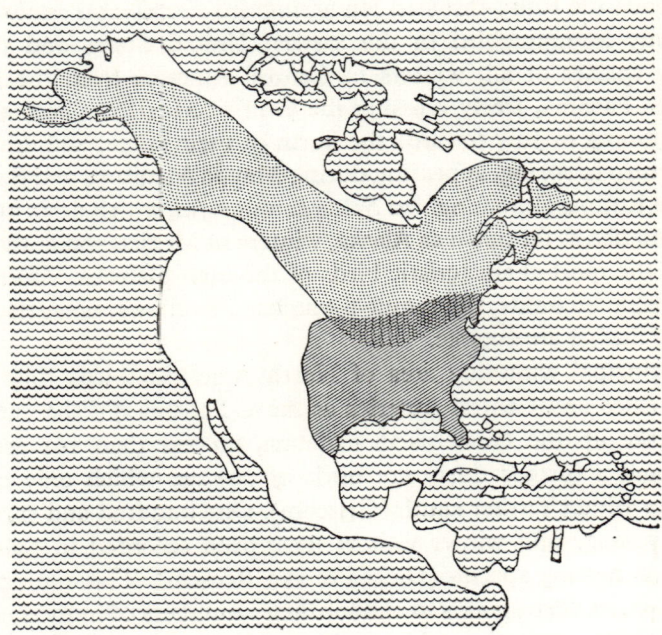

Figure 98. The migrations of the American Robin *Turdus migratorius*. Nesting area: stippled, wintering area: hatched.

Other birds reach Central America, the principal wintering area for North American migrants including the wood warblers (Parulidae). Migrations are certainly far from uniform. A few merely congregate like the Robin in the Gulf states, such as the Pine Warbler *Dendroica pinus*, which reaches high densities in its wintering zone because of the overlap between migrants and the resident birds of these territories. Most however go to Central, and some even to South America by crossing the Gulf of

Mexico or following the chains of West Indian islands. The Black-and-white Warbler *Mniotilta varia*, with a large breeding area in eastern North America, winters mainly in the Gulf states, but part of its population reaches the West Indies, Central America and even as far as Equador. Many others cross the Gulf of Mexico from the United States coast to Yucatan, while a few pass by way of the West Indies. Thus the Blackpoll Warbler *Dendroica striata* reaches its winter quarters by passing from Florida across to Cuba and then to Venezuela.

North America hummingbirds are as distinctly migratory in accordance with their diets, partly insectivorous and partly nectarivorous, which cannot be satisfied under the winter conditions of Canada and most of the United States. The little Ruby-throated Hummingbird *Archilochus colubris*, which is the only hummingbird in the eastern United States and the most northerly species of all, winters in Central America from the Mexican interior to Panama, freely crossing the Gulf of Mexico. The males leave first, beginning in July towards the north of their range, and are also the first to return in spring, appearing during May in Manitoba and Saskatchewan. The Rufous Hummingbird *Selasphorous rufus*, which nests in western North America as far north as Alaska, winters in Mexico and at that season is one of the most characteristic birds of the high plateaux. Among other long-distance migrants, the tyrant-flycatchers, swallows, tanagers, icterids and many finches are notable.

The principal wintering area of North American birds thus extends across Mexico and Central America as far as Panama. Of the 736 species which make up the Guatemalan avifauna, no less than 161 are winter visitors, and a further thirty are birds of passage which only cross the country to winter farther south (Griscom). One hundred and thirty-eight birds of passage and winter visitors have been recorded in Salvador as against 308 nesting species (Dickey & van Rossem). This multiplicity of migrant species corresponds to extraordinary numbers of individuals, since such a high concentration of wintering birds is met nowhere else in the world. This high density depends primarily upon geographical circumstances. There is a great disproportion between the breeding area of the migrants and the wintering area in which they are concentrated. The wintering area of the Chestnut-sided Warbler *Dendroica pensylvanica* is only one-sixth of its breeding area. This concentration agrees equally well with the abundance of food in Central America and its diversity of habitats, which allows the winter visitors to segregate according to their ecological preferences. The high densities of migrants affect the autochthonous nesting avifauna, limiting their populations and the duration of their nesting seasons: the nesting birds seem to wait for the migrants to return northwards before they begin to breed. The situation in the New World is in contrast to that in the Old, and especially in Africa, where the wintering migrants can spread over enormous areas, and as a result density of land birds is very high only in certain local concentrations.

In contrast relatively few North American migrants winter in South America, although as many as sixty-six migrant species are found in Ecuador, thirty-one of which are land birds including swallows, tanagers and wood-warblers. The southern countries are visited by only a small number of nearctic migrants: only five land birds from the north are recorded in Chile. However, waders and gulls are more abundant there, and the waders (Charadriidae) are the North American family with the greatest migrations. Twenty-three North American species are known as winter visitors in Chile, half of them (notably the Lesser and Greater Yellowlegs *Tringa flavipes* and *T. melanoleuca*) being regular visitors. The most interesting of all is the American Golden Plover *Pluvialis d. dominica*, which nests in the tundras of extreme northern Canada and Alaska and assembles in Labrador before setting forth on a wholly oceanic route of almost 3,800km, touching on Bermuda and the Lesser Antilles and rejoining the mainland coasts in Brazil. They continue to southern Brazil and Uruguay and then to Argentina, wintering on the pampas from September to March. On their return they follow a much more continental route to Central America, before crossing the Gulf of Mexico and ascending the Mississippi valley to regain their breeding areas. Thus their breeding and wintering areas are some 12,000km apart, and they trace an enormous ellipse across the Americas. Another race (*P. d. fulva*), nesting in western Alaska and eastern Siberia, winters in South-east Asia and Polynesia (China, Malaysia, Australia and New Zealand), together with other waders such as the Wandering Tattler *Heteroscelus incanus* and the Bristle-thighed Curlew *Numenius tahitiensis*. The specific name of the latter reflects the origin of the type specimen, which was brought back from Tahiti in 1769 by Captain Cook, whereas its breeding sites in Alaska were not certainly discovered until 1948.

A few passerines travel as far as the Golden Plover, including the Bobolink *Dolichonyx oryzivorus*. This icterid nests in Canada and a large part of northern USA and winters in eastern Bolivia, southern Brazil and northern Argentina. Following a well-defined migration route apart from most other migrants, it makes for Florida (so that the most westerly populations cross the whole North American continent), Cuba and Jamaica, from which it reaches Brazil. For wintering it seeks out swampy regions such as

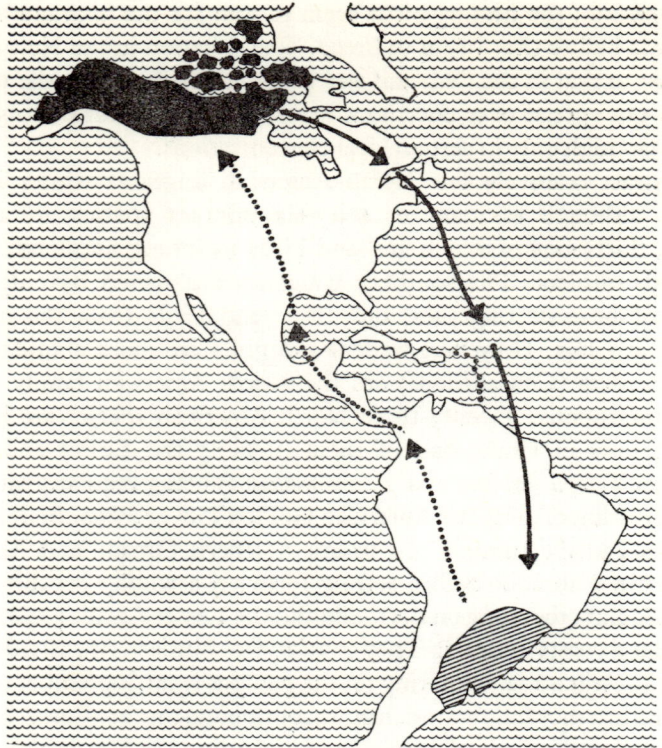

Figure 99. The migrations of the American Golden Plover *Pluvialis dominica*.
Solid arrows show the autumn migration route, dotted arrows the spring route.

those of Chaco. Thus it is more or less the only North American migrant which does not radically change habitats between its breeding and wintering quarters, since these are both situated in the temperate zones of opposite hemispheres.

Seabird migrations

Oceanic conditions at a given place fluctuate widely and this forces seabirds into periodic movements synchronized with the seasons. There is a sharp distinction between littoral and pelagic birds.

The migratory behaviour of littoral birds has the appearance of dispersion on a larger or smaller scale. Thus the European Herring Gull *Larus argentatus* shows simple winter nomadism, some individuals remaining in their breeding places, while others disperse up to 500km away. The omnivorous diet of these birds explains how they can find food throughout the

year, within a restricted area. In contrast the Lesser Black-backed Gull
L. fuscus is a migrant. Most of the individuals nesting in Great Britain
(*L. f. graellsi*) distribute themselves in winter along the coasts of France and
Spain, and others off the coasts of Senegal in West Africa, while part of the
German populations (*L. f. fuscus*) crosses central Europe and the Mediter-
ranean to the Red Sea and East Africa.

The movements of other shore birds are of the same type, often in a
preferred direction so that the birds travel towards places which are more
favourable in climate and food resources. Breeding seabirds are tied to a
particular place, determined by the disposition of suitable nesting sites.
They thus exploit a relatively small area, whose maximum radius is determined by the distance which the birds can cover during their feeding
flights while maintaining high 'efficiency'. Outside the breeding season, on
the other hand, they are free to disperse and so to exploit food resources
which are inaccessible to them while nesting, so that they can systematically
exploit the standing crop biomass.

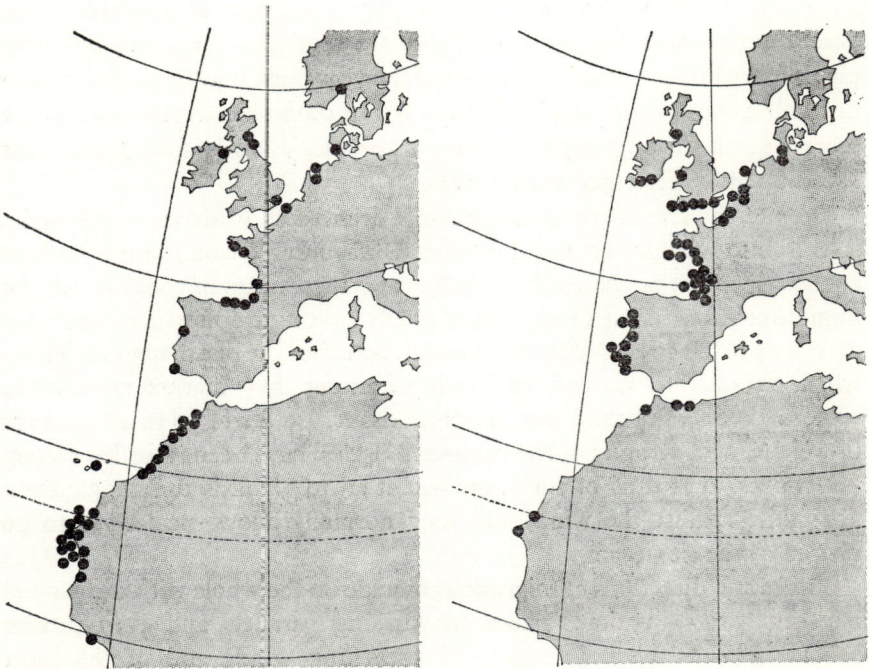

Figure 100. Winter recoveries of Gannets *Sula bassana* ringed in Great Britain.
Left birds of the year, right adults. The young often go as far as Africa, whereas
most of the adults remain on the European coasts.

It is notable that some families of littoral birds, such as the gulls, cormorants and gannets, have a few members of different habits which show a distinctly pelagic type of migration. This is especially true of the Kittiwake *Rissa tridactyla*, which frequents littoral waters and comes ashore only when breeding, while for the greater part of its annual cycle it takes to the high seas. The Siberian nesting populations then scatter across the whole North Atlantic, and individuals ringed on the coasts of Murman have been retaken a few months later in Newfoundland and on the western coasts of Greenland.

Truly pelagic birds are tied to the land only for breeding. Their migrations, usually on a grand scale, can be explained in relation to two essential factors: the amount of food available, and the general wind pattern. Departure from the breeding sites may follow from impoverishment of the fishing zones on which the birds are concentrated while nesting. Breeding presumably coincides with the optimum standing crop biomass, after which the birds leave zones where productivity is falling for those in which it is at a maximum. Climate and weather also play their parts, and whereas temperature fluctuations of the atmosphere are negligible the winds probably play a more important part. For part of the year these may be so violent and unstable as to prevent the birds from flying economically and agitate the sea so much as to hamper fishing. The pelagic birds then leave their breeding places for other marine areas.

Despite their remarkable adaptations even these birds must pay strict attention to wind speed and direction. Unfavourable conditions – such as contrary, violent or changeable winds – at critical times in zones which the migrating flocks must cross always result in catastrophes: the birds are driven on to the coasts and even inland, where they die of exhaustion. Those which have had to contend with such conditions show marked emaciation, with marked loss of fatty and muscular tissue. Losses of 30 to 40 per cent in weight, with atrophy of the pectoral muscles, have been recorded among Kittiwakes. It is thus of no advantage at all to seabirds to struggle for a long time against the wind, and they normally allow themselves to be carried by the air stream.

Times and directions of migration coincide on the whole with the general meteorological situation. There are striking parallels and even precise coincidences between the maps of prevailing winds and of the most characteristic seasonal movements of pelagic birds. At the least there is never a fundamental and systematic conflict between the general direction of migration and that of the prevailing winds at a particular time and place, except where migration routes unavoidably cross unfavourable zones. It is here that the highest losses regularly occur.

The migrants which move farthest are without any doubt Arctic Terns *Sterna paradisea*. Nesting along the northern coasts of Europe, Asia and North America, they pass the northern winter in the far south of the Atlantic and Pacific, even beyond the antarctic circle (Kullenberg and Storr), actually making the pack-ice their principal wintering zone. Thus their journeys are enormous. An Arctic Tern ringed near Murmansk on the White Sea has been retaken south of Fremantle in Australia a distance of 14,500km since it probably went around Europe and Africa before crossing the Indian Ocean. The Arctic Tern follows a definite migration route, with the post-nuptial journey of both American and European populations taking place in the eastern half of the Atlantic. The terns nesting in north America first cross the Atlantic from west to east, avoiding the warm waters of the western Atlantic, and then fly southwards in company with those from northern Europe. The prenuptial return is mainly to the west along the coasts of South and North America, though a minority of the population returns the way it came, following the coasts of Africa and Europe, to

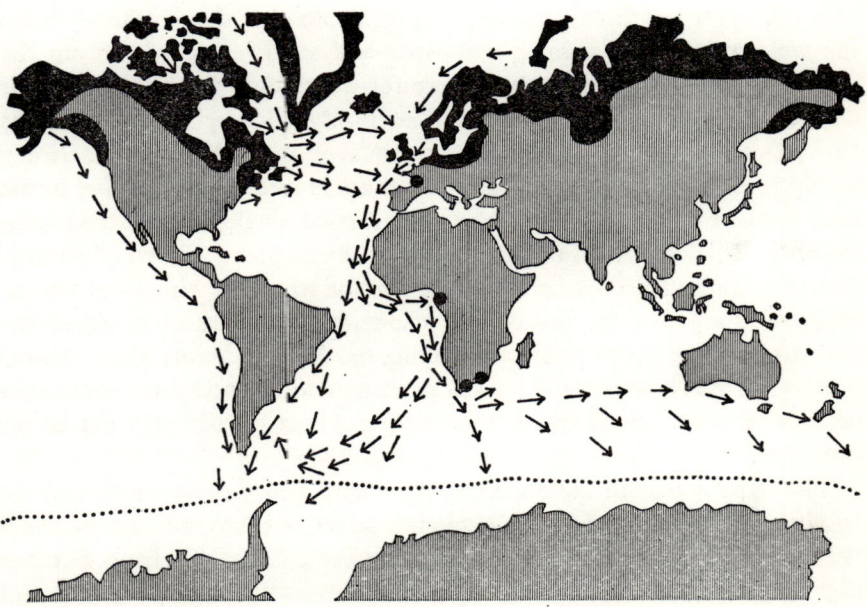

Figure 101. The migrations of the Arctic Tern *Sterna paradisea*. The breeding area is shown in black, the southern boundary of the wintering area by the dotted line, and the directions of post-nuptial migration by arrows.

spread through the eastern part of the breeding range or cross the northern Atlantic to North America. Arctic Terns nesting on the Pacific coasts of North America migrate southwards along the American coasts, to winter principally off Chile and Argentina, while the species is absent from the whole western Pacific, from Japan to New Zealand. Its presence in the antarctic sectors to the south of the Indian Ocean and Australia is probably due to drift from the Atlantic section, caused by the great westerly winds which blow at these latitudes.

Most of the Petrels are also long-distance migrants, some of whose seasonal movements have been shown to follow well-defined schedules. The populations of Wilson's Petrel *Oceanites oceanicus* nesting in the American antarctic sector (South Shetland, the Orkneys and South Georgia) have been especially well studied (Brian Roberts). In January and February all the birds are concentrated near their nesting places, from which in late March and April they begin to travel northwards, mostly following the American coasts. They move remarkably quickly across the tropics, which are poor in food and inhospitable to seabirds. By June they have completely disappeared from the southern hemisphere and then spread out over the North Atlantic, especially the part crossed by the Gulf Stream, where they pass the whole northern summer. During September they gradually leave the western Atlantic, passing eastwards and south-eastwards. From the coasts of West Africa they then make for South America to regain their antarctic breeding territories, which they mostly reach by November. Thus Wilson's Petrel covers the whole immense area of the Atlantic, describing a loop in the north of that ocean. This route is remarkably parallel to the predominant winds. The Petrels are first carried northwards by the south-westerly Trade Winds of the southern hemisphere blowing towards America. In the North Atlantic they follow the winds blowing from west to east over the top of the Azores anticyclone, and are then again led south-eastwards by the winds which blow along the African coasts. Thus there is perfect harmony between the wind patterns and the seasonal movements of these Petrels, which travel almost their whole circuit with maximum economy.

The same is true in the Pacific of the Short Tailed Shearwater *Puffinus tenuirostris*, which nests in huge colonies in the seas around south-eastern Australia and Tasmania (Serventy & Marshall). They nest from October and then depart on migration in April and May, travelling northwards and then north-westwards to follow the coasts of Japan which they reach a month after leaving Tasmania. From there they pass to their winter quarters in the North Pacific from Kamtchatka to the Aleutians, the Bering

Straits and the shores of the Arctic Ocean, where from June to August they are the most abundant Procellariiformes. On their return migration they pass eastwards and south-eastwards along the Pacific coasts of North America to California, then obliquely across the entire Pacific to north-eastern Australia whence they follow the coast towards their breeding areas.

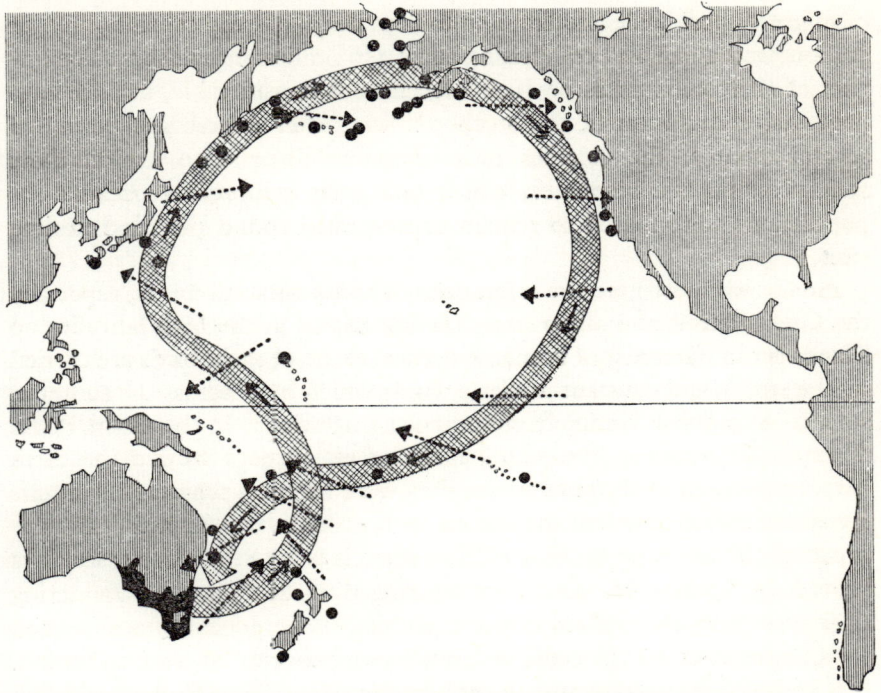

Figure 102. The migrations of the Short-tailed Shearwater *Puffinus tenuirostris*. The breeding area is shown in black, the location of recoveries by dots, the shearwaters' circuit by a broad arrow, and the prevailing winds when the migrants are passing by broken arrows.

These migrations are not only remarkable for their extent but are also admirably adapted to the general pattern of winds in the Pacific at the time when the stream of migrants pass through each sector. Superimposing the maps of migration routes and prevailing winds shows striking parallels here too, especially in the Tasman Sea where the winds explain the loop which the birds describe on their post-nuptial migration. It is not important so much that the winds should be in the direction of the migratory flight (which can even be disadvantageous, since the birds prefer not to travel with a tail wind), but much more that they should not be contrary. The only

611

point at which the general pattern of winds is unfavourable is at the north-east of Australia, where strong south-westerlies blow. Losses may be high at these latitudes, especially among the young birds.

The migrations of the Short-tailed Shearwater are probably a necessity for this species, because of its incredible densities at the meeting sites. The stocks number hundreds of millions, and observers have described streams of migrants 90m wide and from 50 to 80m deep, passing continuously for an hour and a half and comprising some 150 million individuals. The impact of these consumers on their biocenose is considerable, and leads to progressive depletion of the stocks of fish. After a breeding period of several months, the colonies must therefore disperse to exploit more widely distributed resources, which had been inaccessible because the populations were forced to remain concentrated round the fixed nesting sites.

Finally we must note the migrations of some antarctic birds, especially the Giant Petrels and albatrosses. Having nested at the high latitudes on islands set in the midst of immense oceans, many of these birds are carried by the strong and constant westerly winds which blow across the southern seas, to accomplish complete circumpolar migrations. In the Giant Petrel *Macronectes giganteus*, the young make great journeys around the earth before returning to the sites where they were hatched, whereas the adults are sedentary and content themselves with nomadic movements at the approaches to their nesting colonies. The same is true of the albatrosses, as is proved by spectacular recoveries of ringed birds. Thus a Wandering Albatross *Diomedea exulans*, ringed as a chick on Kerguelen, was recovered ten months later on the coast of Chile having covered at least 13,000km. It is known that these giant seabirds breed late in life and furthermore that their breeding cycle extends to two years. A large part of their populations accomplish circumpolar migrations of great extent, allowing themselves to be carried by the westerly 'Roaring Forties'.

Migrations are thus to seabirds a pressing necessity and a response to fluctuating environmental conditions, so necessary that even flightless birds travel in response to the seasons. This is notably true of the penguins, most of which migrate northwards during the southern winter. Gentoo Penguins *Pygoscelis papus* have been seen on passage swimming on the high seas like other birds migrating on the wing. Their migrations produce pelagic dispersion, enabling them to avoid reduction in the stocks of prey on which they feed within a restricted area while breeding, to use the resources of wider areas, and thus to exploit marine resources more rationally.

Adaptations of migratory behaviour to the environment

The study of birds' seasonal movements shows that in fact there is no one mode of migration but multiple modes, varying from species to species and even from population to population. Migratory flights involve many difficulties and considerable expenditure of energy, while each migrant tries to put itself in the most favourable condition by adapting itself to the environment and to the atmospheric and ecological conditions in force during its passage. Thus a series of adaptations appears in all modes of seasonal movements.

These adaptations appear especially in the choice of migration routes. The direction of migration is of course determined by the relative positions of the breeding and wintering areas. However, this standard direction is modified according to local conditions, so that the migrants take winding paths known as *migration routes*. While some of these are broad others are very narrow, all populations of a species from a wide area migrating over great distances along a clearly defined route. The width of migration routes varies in response to the environment, and this change in migratory behaviour precisely reflects birds' adaptations towards using all opportunities to put themselves in the best possible position.

Migrants put orographic systems and river basins to use and valleys are favoured as corridors of movement, as much for their favourable habitats as for their aerodynamic conditions such as ascending currents and the modification of prevailing winds by the relief. Stretches of lake attract aquatic birds to which the marsh vegetation gives shelter. Coasts provide direction lines which are followed especially because their interface habitats are favourable to both aquatic and terrestrial birds. Ecologically and geographically favourable zones thus provide positive factors which can determine, or at least modify, the map of migratory movements.

Besides places which *attract* birds, others *repulse* them, as seas do to land birds especially passerines. Reluctance to cross the sea is shown especially clearly by Dutch Chaffinches. Coming in autumn from Scandinavia and eastern Europe with the west-south-westerly winds, they often swerve sharply on reaching the coast of Holland and follow it south-south-westwards (though as discussed below their behaviour varies considerably with the winds). Similar situations arise elsewhere, especially at narrow tongues of land which separate coastal lagoons from the sea and which result in concentrations of migrants following the natural lines, avoiding the sea as much as possible and only reluctantly flying over it. Conversely, the land seems to play a similar part for marine migrants. Although quite capable of flying over dry land, birds such as Gannets and Eiders follow the windings of the coasts rather than take short cuts.

Such repulsions can modify the migrational geography of an entire species. The most spectacular example is that of the European White Stork, which nests in a wide band from Holland to Alsace, across Germany to Russia in the east and in the south to Spain and the Balkans. It is known that storks make use of ascending currents in their predominantly gliding rather than flapping flight; and these currents, so common over land warmed by the sun, are absent over the sea. This explains the reluctance of storks to fly over the sea, and their profoundly modified migration routes around the seas or crossing them at their least width. The European populations divide along a line from Holland to western Germany. In the autumn the populations nesting to the west of this line fly south-westwards across France and Spain, to cross the Straits of Gibraltar into North Africa and winter in tropical West Africa. The much larger populations nesting to the east of the line fly south-eastwards, to concentrate in Rumania, cross the Bosphorus and Asia Minor towards the Gulf of Suez, which they cross near Sinai, and then the Nile valley to reach East Africa. Here they are concentrated along very narrow routes. Records from the Gulf of Suez mention the passage of storks from north-east to south-west, forming a ribbon some 30m wide 1 to 3m above the sea and stretching for nearly 40km: the number of birds was estimated at a minimum of 40,000 (Schüz). Thus the migrations of European storks have been modelled by the barrier which the Mediterranean presents to birds unaccustomed to flying over sea. It is especially interesting to compare their migratory behaviour with that of cranes, whose flapping flight is well adapted for sea crossings, and which cross the Mediterranean throughout its length instead of skirting it like the storks. The migration of birds of prey in the eastern Mediterranean basin resembles that of the storks. Passages of eagles, buzzards and honey-buzzards across the Bosphorus are especially spectacular since their gliding flight prevents them too from making any prolonged excursion over the sea.

However, migrations are not always so circumscribed, since birds often take multiple routes and travel over very broad fronts. Even in the mountains observations show that birds do not necessarily follow the easiest routes. There is a certain preference for passes, but migrants do not hesitate to fly over the high chains while apparently easier passages are open to them. Similarly, birds which are habitually confined to a particular biotope readily take to habitats to which they are strangers under all other circumstances. Thus Lesser Black-backed Gulls *Larus fuscus* from the Baltic cross the whole of central Europe to winter on the Mediterranean. Many land birds fly over that sea without selecting the shortest crossings, nor those marked out by islands which would break up the flights over water. Thus birds are not wholly dependent on geographical circumstances, and can adapt themselves to very diverse conditions.

Migratory behaviour can vary very widely with atmospheric conditions, as has been shown especially clearly by observations on Dutch Chaffinches (Tinbergen 1950), which behave differently according to the wind direction and speed. Under gentle southerly winds these migrants follow the Dutch coast in loose flocks, leaving it when it diverges from their general direction of flight towards the British Isles. Under strong south-westerlies they follow it a long way, even when it takes them from their line of migration. Under gentle easterly and north-easterly winds they cross the coast directly without deflecting their course. Thus the disposition of the sites, the nature of the habitats and the atmospheric conditions provide a complex of factors which combine to determine the choice of routes.

On the whole it is the atmospheric conditions which are of prime importance. Bird migrations appear to be remarkably well synchronized with the annual climatic cycle, by which the same phenomena recur at fixed dates (subject to a certain variability) from year to year. As a result migrational and meteorological charts are undeniably related, the directions of migratory movements being closely connected to the positions of the different air masses and the distribution of high and low pressure areas which determine the prevailing winds. According to many observations made in and around the British Isles, the flocks of migrants form and pass in response to well defined meteorological situations. A country like Great Britain at a crossroad of routes thus sees the passage of migratory streams from various sources, depending on the weather at a particular time.

The meteorological situation further explains why the prenuptial migration does not always follow the same route as the post-nuptial. Many cases of *loop migration* are due to aerial currents favouring movements in one direction but not in the other. Other migrations of this type are explained by temperature factors. Thus the Black-throated Diver *Gavia arctica* from northern Russia migrates in spring by routes distinctly to the west of those which it follows in the autumn, since at the time of the prenuptial migration Russian lakes and other stretches of water are still ice-bound.

The charts of migrational movements have been modelled by the combination of geographical, ecological and meteorological factors evoked above. Although migrants may be encountered over very large areas on wide migratory fronts, they are mostly concentrated along certain highly favoured routes. Thus it has been possible to establish the lines of great migration routes on a continental scale.

The situation is especially clear in North America where (mainly from the seasonal movements of the ducks) one can distinguish four principal flyways running essentially north and south (Lincoln). Apart from those that migrate across the Atlantic or Pacific, most North American migrants follow either the Atlantic Coast Flyway, the Mississippi Flyway, the Central Flyway across the great plains which stretch east from the Rocky Mountains, or the Pacific Flyway. These flyways are not rigorously defined, and some of their component tracks intermingle, while different populations of a single species often take different routes to or from their winter quarters.

In Europe the great flyways lead south-westwards and south-eastwards because of the central positions of the alpine massif and Mediterranean, though certain migrants do cross the latter at all longitudes. Other routes lead across Asia, some towards East Africa which is the winter territory of a good part of the Siberian migrants.

Along one of these great flyways, the migrants follow a multitude of secondary tracks known as *migration veins*. The route actually followed by a particular migrant is determined by the terrain and atmospheric conditions, so that it may travel with the least expenditure of energy. This is especially clear in bad weather, when economy is essential.

Untypical migrations

The principal characteristics of migration is its perfect regularity. These movements are usually annual, since their timing depends on the seasons and the fluctuations of the standing crop biomass.

However, certain movements do not have this regularity of true migrations. A species appears in large numbers in a region, stays there a longer or shorter time, and then disappears entirely, and the interval before its reappearance is of irregular duration. Movements of this type in which the birds show no attachment to the territories from which they came and to which they do not usually return, are known as invasions or irruptions. They are linked to very great fluctuations in the populations within the breeding areas (Chapter 14). The standing crop biomass of certain environments, such as arctic, terrestrial habitats, varies so greatly that the bird population they can support also fluctuate enormously in density. When they suddenly lack food they have no alternative but to travel all together in search of it. The same is true of birds with highly specialized diets, which are dependent on a single source of food liable to large variations. This explains why invasions are especially characteristic of arctic and stenophagous birds. However they differ in controlling factors and in the way they develop.

The Scandinavian populations of the Crossbill *Loxia curvirostra*, which nest in the forests of Norway Spruce *Picea excelsa* and feed largely on its seeds, share the fluctuations of the conifers to which they are linked. When the seeds run short the Crossbills take part in nomadic movements and invade western Europe in large flocks. The same is true of the Siberian Nutcracker *Nucifraga caryocatactes macrorhynchus*, whose diet is largely based on seeds of the Arolla Pine *Pinus cembra siberica*. The very irregular rhythm of fructification in this conifer reacts on the nutcracker populations. When after a succession of favourable years the fructification fails, these birds tend to move westwards, reaching western Europe in erratic flights and not returning to their place of origin. This happened especially during the winter of 1968–9. Many other birds such as Jays, various woodpeckers, tits, bramblings and redpolls, also take part in irregular invasions. These sporadic movements are obviously controlled by diet, and appear as a means of regulating populations in habitats which show large and irregular variations in productivity.

In contrast certain invasions are much more regular, as is especially true of the Snowy Owls *Nyctea scandiaca*, characteristic of the arctic tundras. This feeds largely on lemmings, microtine rodents which show a four-yearly cycle of abundance. Each rapid fall in the lemming populations results in an invasion by Snowy Owls of more southerly zones in central Europe and the United States. These invasions are synchronized with the lemming cycle, also occurring every four years in accordance.

In contrast the movements of the Waxwing *Bombycilla garrulus* are more complex and involve several types. In the Old World these birds annually make true *migrations* to central Europe; in some years the wintering birds are much more abundant and not all return to their place of origin. These irregular *invasions* are no doubt controlled, like those of the vegetarian birds discussed above, by lack of food. In addition however (as mentioned in an earlier chapter), Waxwings regularly take part every ten years in more extensive invasions beginning earlier in the autumn and reaching the whole of western Europe.

Chapter 31

The Great Problems of Migration

THE examples in the preceding chapter demonstrate the remarkable diversity of seasonal movements among birds. Each species and even each population shows specific migratory behaviour, adapted to its environment, ecological needs and locomotor abilities. The annual cycle is closely controlled by the rhythm of the seasons, and the movements are in response to the fluctuations of the environment and especially the amount of available food.

Having examined the facts, we shall attempt to explain them and see what initiates these long voyages. Three questions immediately come to mind. The first is how migratory birds orientate themselves on their long travels. The second is how the periodic movements, which are the essential phases of a regular cycle, are controlled. The third is the 'why' of migration – its ecological reasons and the place which fluctuating populations occupy in their biocenoses. Study of this last question allows some light to be thrown on a final problem, of the origin and evolution of migrations.

The orientation of migrants

One of the most complex problems involves the migrants' orientation along the enormous routes which they cover each year. They return to nest in the same district and sometimes at the very spot, as even laymen know from the classic observations on swallows. These birds often even return to a particular place in their winter quarters.

Two parallel methods may be employed to resolve the problem of bird orientation, as well as particular experiments discussed below. Close observation of what happens in nature reveals relations with certain features of the environment. However, because of the complexity of the data, which cannot be controlled, interpretation of the natural phenomena runs into many difficulties. One is thus turned towards the experimental method, which provides more exact results. Most existing data come from experiments in displacement and *homing*. The method consists essentially of catching the birds in a particular area (usually at their nests) and transporting them varying distances. Their return is then studied. These experiments take place under entirely artificial conditions, far removed from those of migratory journeys, but the guides which enable one bird to orientate itself on migration are presumably the same as those which another uses during homing experiments. This is the justification for the use of non-migratory homing pigeons as the subjects of innumerable experiments.

These experiments have shown some birds' extraordinary abilities for orientation. Sedentary birds have only slight powers of orientation, and Great Tits *Parus major* do not return to their nests after being displaced more than about 10km. Such abilities are much better developed among migrants, as is notably shown by the comparison made in Great Britain between the rather sedentary Herring Gull, and the truly migratory Lesser Black-backed Gull. The homing abilities of the latter are decidedly superior, even when the experimental subjects are taken to areas where they have never been before. Transportation experiments have shown astounding powers of orientation in some birds. Swallows, starlings, wrynecks, shrikes and swifts have returned to their nests after being transported 1,800km away, at average speeds of 200 to 300km a day. A Manx Shearwater returned to its colony on Skokholm off the Welsh coast, after being released at Boston USA, more than 5,000km away, twelve-and-a-half days before. Laysan Albatrosses *Dicmedea immutabilis* were transported from Midway Island, where they were in danger of causing collisions with the aeroplanes using the airbase, but displaced birds returned from enormous distances. The distance record is held by an albatross released in the Philippines, which was retaken on Midway after covering 6,600km in 32 days; while the speed record belongs to one released on the coast of the state of Washington USA, which returned in 10·1 days, having covered the distance of 5,120km at an average speed of 500km per day.

Thus some birds at least possess a high-precision system of location. Analysis of many displacement experiments has shown that methods of orientation may be grouped in three types (Griffin 1955).

Type I uses only simple landmarks, for which the displaced bird seeks more or less at random until it finds itself in an area which it knows. Such a system of orientation is clearly used by certain birds such as Gannets (Griffin). Individuals caught at their nests on Bonaventure Island off the Gaspe Peninsula in Canada were transported inland to distances of up to 340km, and after release were followed in aeroplanes to observe their return behaviour. Many observations of migrants show them to use landmarks by which their routes may be diverted and their flight orientated, as is shown by the existence of flyways and migration routes (see Chapter 30).

619

Type II is characterized by the ability of the displaced bird to maintain a constant direction in flying over an unknown area, the orientation being constant whatever the direction of the point of release.

This ability to orientate and travel in a given direction has been demonstrated by a classic experiment, showing the existence of a true sense of direction ('Kompassinn') in Hooded Crows *Corvus cornix* (Ruppell 1944). First the nesting and wintering areas were precisely determined, of the populations passing on migration through Rossiten (now Rybatschi) on the Baltic coast. During the spring of the following year a large number of crows was captured on their prenuptial migration north-eastwards towards their breeding territories in the Baltic and Finland. They were taken to Flensberg 750km to the west, to Essen 1,025 km to the west-south-west, or to Frankfurt-am-Main 1,010km to the south-west, where they were released after being ringed and marked with paint so as to be recognizable in the field. The great majority of the individuals thus displaced came to nest in an area shaped more or less like their area of origin, but to the west of the latter in Denmark and Sweden. In the following autumn these crows returned to winter west of their normal winter quarters. This experiment clearly shows the existence of a true sense of direction, since the line of migration remained parallel to the true line, and the birds did not correct their course so as to regain the point which they would normally have reached.

Such maintenance of a constant direction, whatever the relative positions of the release point and target area, has also been recognized in Adelie Penguins *Pygoscelis adeliae* of the Antarctic (Emlen & Penney 1966). Birds transported to the interior of their continent took a north-north-easterly direction, which is clearly adaptive. Throughout Antarctica the sea lies towards the north, while the easterly component of their movement no doubt compensates for the drift caused by winds and currents, which tend to drive the birds westwards when they have reached the sea.

Type III is more complex, since the bird has to associate the release point and destination in order to determine the direction to be taken, which it then maintains.

This ability has been demonstrated in many displacement experiments. Even in the experiment discussed above a certain number of displaced crows, all old and therefore experienced birds, returned to the original nesting area. Similar differences in orientation behaviour have been shown by experiments on Starlings (Perdeck 1958). Individuals were trapped at The Hague in the Netherlands, from a population migrating from the

Baltic States, Denmark, northern Poland and Germany, and aiming to winter in the Netherlands, Belgium, northern France and the British Isles. After being marked they were displaced south-eastwards to be released at Basle, Zurich and Geneva. Their behaviour differed according to their age. The young birds essentially maintained their general flight direction from north-east to south-west, and went to the south of France and Spain decidedly south of their normal wintering area, having maintained the displacement imposed by the experiment. In contrast the adults returned to their habitual wintering area by north-westwards flight, very different from their normal direction. They had thus redetermined their direction of travel, and corrected for the imposed displacement. This experiment showed most of the young oriented by type II, and the adults capable of true *navigation* by type III. Thus a single species has various systems of location at its disposal and uses them either simultaneously or independently, partly as a function of age and experience.

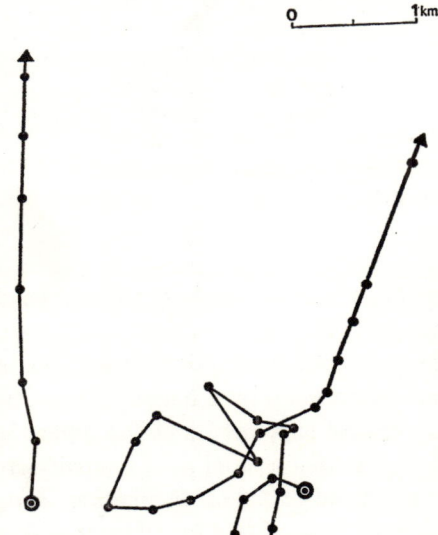

Figure 103. The points connected by each line represent the successive positions, at five-minute intervals, of an Adelie Penguin displaced and released away from its colony. A bird released in sunshine (left) went straight, whereas another released under a cloudy sky (right) first took the opposite direction, but went off in the same direction as its fellow when the sun came out.

Experiments on Adelie Penguins have shown comparable homing abilities (Emlen & Penney 1966), penguins displaced 1,900km returning to the colony where they had been caught. By following very precisely the paths they followed at the beginning of their return journeys it was shown that, despite the uniformity of the great snow-covered expanses, they took and unswervingly maintained the correct direction. The track followed by a bird returning to its colony scarcely departed from the straight line under

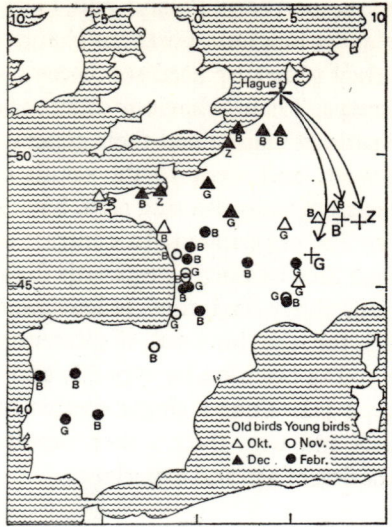

Figure 104. An experiment by A. C. Perdeck on the orientation of Starlings *Sturnus vulgaris*. Z, B & G, release points of Starlings trapped at the Hague during their autumn migration through the Netherlands. The recoveries (marked with the corresponding letter) show that the young birds continued on a displaced, southeast-wards course, whereas most of the adults regained their normal winter quarters by changing direction.

good conditions, and was longer than this minimum track by an average of only 2·5 per cent.

Birds obviously cannot choose their course, modify it in the light of natural or experimental conditions, and maintain it over long distances, by using immediate landmarks alone. A possible explanation of the orienting 'sense' lies in the earth's magnetic field. Points of equal field intensity are arranged along magnetic parallels, and those of equal declination along magnetic meridians. Thus *in theory* the two sets of data together provide a system of magnetic co-ordinates analogous to, though distinct from, our geographical co-ordinates. An animal sensitive to both characteristics of the terrestrial magnetic field would be able to navigate by it. However, this theoretical system can be of no practical use, if only because of anomalies in the terrestrial magnetic field.

Recent authors such as Yeagley have postulated a more complex system, using not only the magnetic field but also Coriolis forces, produced by the rotation of the earth, and acting on any body moving over its surface. This theory thus makes use of both magnetic and geographical co-ordinates, and demands that the bird be able to measure both the current induced in its body by its movement through the terrestrial field, and the Coriolis force produced by its movement over the turning globe. This complex theory is certainly attractive, but it does not stand up to a deeper examination, and has never achieved the smallest experimental confirmation despite the efforts devoted to this end. Not only are the forces involved extremely small, but birds would have to be capable of measuring variations rather than the forces themselves, and of distinguishing them from comparable forces which might be acting simultaneously. A bird flying at 65km/hr at a latitude of 45° N would be subject to a maximum electric force of 10^{-5} volts per cm. In moving one degree of latitude (about 112km) to the north or south, this force would change by only 1·3 per cent, or about a ten-millionth of a volt. Under the same circumstances the Coriolis force would vary by only 1·7 per cent. Further, since both the induced electromagnetic force and the Coriolis force depend on the bird's own speed, this would have to be measured with incredible precision. An error in estimation of 1 mph (1·6km/h), which must be beyond a bird's capabilities, would produce a location error of 240km. This alone would make such a system of orientation unworkable. Besides, we know nothing precise about the sensitivity of birds to magnetic fields. Most experiments have given negative results, while even the few positive ones have been challenged because the experimental conditions have sometimes been questionable. Recent experiments (Reille), based on conditioning and using the pulse rates as an indicator of reactivity, seem to show a sensitivity in the pigeon to magnetic fields of the same order of magnitude as the terrestrial field. However, no reliable deductions can be made from these trials, until they have been extended.

One has to turn to guidance systems relying on the sense of sight, which is well developed among birds and which they use in searching for reliable location points. Observation and experiments show that relationship between the percentage returns, elapsed time and the speed recorded, often mathematically fits the hypothesis of exploration better than that of true navigation (Wilkinson). The same must be true of certain migratory movements, in which the individual birds seem not to follow well-defined tracks. However, many other results, like the migratory movements which lead birds to precise points, cannot be explained in this way and imply a true navigational sense. This suggests orientation by astronomical guides, especially the sun. The hypothesis of celestial navigation has recently given rise to many studies, of which those by Gustav Kramer and his collaborators and by G. V. T. Matthews are the most notable.

Kramer noticed that captive Starlings prefer the south-western corners of their cages from the beginning of October, when they are showing the signs characteristic of captive birds under the migratory urge. The immediate surroundings of the cage seem to play no part in this specialized orientation, which the birds show even when they can see only the sky. The magnetic field cannot be involved, since experimental modifications do not affect the birds' orientation. In the following spring, the starlings kept to the opposite corners to those they had preferred in the autumn, and were thus orientated in their normal direction of flight for that time of year. The advantages of studying the behaviour of captive birds were immediately apparent. Since migrants orient themselves even in the restricted space of their cages, there is no need to observe them over long distances. Orientation can now be studied in the laboratory, and not merely in displacement experiments during which a bird cannot be followed continuously over long distances.

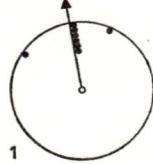

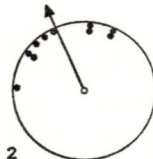

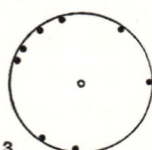

Figure 105. An experiment on the displacement of Adelie Penguins. 1. under clear sunlight the birds take the right direction. 2. under hazy sun they are in some doubt. 3. under overcast skies they are disoriented.

Preliminary observations showed that the Starlings could only be using optical guides, which they seemed to find in the sky. In order to confirm this hypothesis Kramer built a hexagonal pavilion opening by six windows, within which the cage of experimental subjects is placed. Movable shutters fitted with mirrors allow the windows to be obscured at will, and the apparent direction of incident light to be altered through predetermined angles. The bottom of the cage is formed of six sectors. During an experiment the observer notes at regular intervals (e.g. every ten seconds) the bird's position on the six sectors which form the floor of the cage. The results are then analysed statistically.

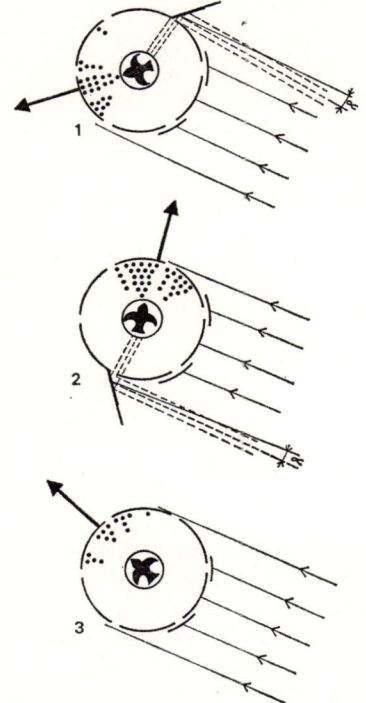

Figure 106. An experiment by Kramer on changing the orientation of captive Starlings by the use of mirrors. The shutters through which the sun's rays could penetrate directly are closed.

Kramer arranged the mirrors so that the light falling through each window was deflected by 90° to one side or the other, and showed that the birds followed the change by a parallel shift in average orientation. He next obscured all the windows through which the sun could shine directly, and trained light from around the sun on to the experimental cage by means of a mirror. Every change in the sun's apparent position was followed by a new orientation of the experimental subjects, whose headings maintained a fixed angle with the apparent azimuth. In contrast the birds were disoriented when the sky was covered by dense clouds, or the windows covered with translucent paper. This confirms what has been established in nature, that birds orient much more easily in clear than in cloudy weather. Displacement experiments on Adelie Penguins provide an especially convincing demonstration of this. These birds maintain the correct direction when the sky is clear, but when it is covered they wander at random and only regain their orientation when the sun reappears (Emlen & Penney). From all these experiments it is evident that it is indeed the sun which controls the orientation of Starlings, though it need not be itself visible: the birds can orient to a segment of sky of about 45° around the sun.

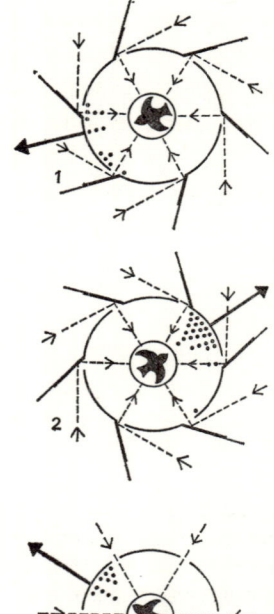

Figure 107. The experimental deflection of migratory direction in caged Starlings. 1 & 2. the sun's rays are reflected by the mirrors, and the birds alter their orientation according to the direction of the light.
3. when the shutters are open, the birds orientate towards the north-west.

Further experiments showed that birds can also take account of the time of day and the season of the year, and thus determine their direction despite the changing positions of the sun. For these experiments Kramer altered his experimental technique by training Starlings to take food from a designated receptacle, in response to cues from natural or artificial light. Although this situation is far removed from the natural conditions under which migrants orient themselves, it does allow a bird's capacity for celestial navigation to be assessed. Thus birds can compensate for the apparent movement of the sun during the day and the year. This implies the possession of an *internal clock* giving the bird the precise time it needs in order to use its astronomical observations. That such an internal clock does exist is proved by the many ways in which its effects appear. It depends upon a combination of extremely precise physiological cycles, which act in a circadian rhythm shown throughout the daily cycle. Most physiological processes which have been studied from this point of view, such as metabolism of cellular substances and indeed all functions of the organism as a whole, take place according to this well-defined rhythm.

Various experiments have been carried out, notably on Mallards, by modifying their daily rhythms through exposure for several days to experimental conditions very different from their natural ones (Matthews). One group of subjects was 'advanced' by six hours, another 'retarded' by the same interval, and a third by twelve hours so that its normal rhythm was completely reversed. When released on a sunny day, these birds behaved as predicted by theory. The control subjects which had been maintained under natural conditions left towards the north-west, the normal direction under the conditions of the experiment. Those whose internal clocks had been altered went in different directions: south-eastwards for an advancement of six hours, north-eastwards for the corresponding retardation, and south-eastwards (the opposite direction from the control group) for an alteration of twelve hours. Each group thus oriented as though the sun was in the direction appropriate to the time recorded by the birds' internal clocks.

In contrast we are by no means sure how birds use the sun to locate themselves. Kramer has suggested that the desired direction is determined by measuring the horizontal angle (azimuth) along the horizon from the vertical projection of the sun's position, in which case the celestial object serves only to maintain a fixed direction (Type II navigation) like a compass, and not to fix geographical location. Matthews on the other hand suggests that birds use the *sun arc*, estimating the angle which at a particular place the plane of the sun's apparent motion makes with the horizontal. This angle remains constant throughout the year, since it is dependent only on latitude and the angle of the ecliptic. According to this hypothesis, the bird must know precisely the various solar characteristics at its home location at the time indicated by its internal clock, which in turn is regulated by the solar rhythm. When displaced it would observe the sun and construct from observation of a small segment the whole curve at its new location. Measurement of the maximum elevation of this curve and its angle with the horizontal, and comparison with those familiar at home, would give the new relative latitude; while comparison of the sun's position in relation to the highest point with what it would be at home at the same time, would give the precise relative longitude. All this is unfortunately still hypothetical. We know almost nothing about how birds read the indications given by the sun, though we can be certain that it does serve them for reliable location.

THE ORIENTATION OF MIGRANTS BY NIGHT

Whereas storks, raptors, doves, bee-eaters, swifts, swallows, and crows migrate mainly by day, many other migrants travel only by night – especially the waders, ducks, tyrant-flycatchers, most turdids, warblers, wood-warblers and many other insectivorous passerines. Although at other times they are strictly diurnal, during migration their daily cycle undergoes a revolution and they become nocturnal as well. At twilight they become inactive after their normal diurnal activity, but become active again during the night, as is shown by captive individuals. This migratory activity is added to the normal activity and ends at dawn. Observation of transits (whether by means of a telescope trained on the moon against which the birds are silhouetted, or by radar) clearly shows the volumes and timings of nocturnal movements. During migration these birds thus have two very distinct periods of activity, the one during the day being devoted to foraging and the nocturnal one to migratory flight. The necessity of feeding during the day no doubt imposes this rhythm on insectivorous passerines, except the swallows which feed on the wing during their diurnal migration flights.

These nocturnal migrants too can orient themselves precisely. They might do so only during the day, and merely maintain their direction at night by using immediate guiding marks. That they use the moon, has been disproved. It has also been suggested they are sensitive to infra-red radiation (which is especially intense during the night) and use this for orientation, a hypothesis which has never been confirmed. However, these rudimentary methods are insufficient to account for the orientation of nocturnal migrants, which is as remarkably precise as migration by day: radar observations have shown that the movements take place in perfectly definite directions which the birds maintain over long distances.

Experiments on the homing of Mallards at night have shown that on release the birds immediately take the right direction if the sky is clear, whereas when it is covered their directions of departure are randomly distributed and they seem to be completely disorientated (Bellrose). These very revealing experiments suggest that at night too birds use astronomical guides, steering by the stars. This was conclusively proved by F. and E. Sauer in repeated experiments. The Sauers' method was similar to Kramer's. They put nocturnal migrants, especially warblers, in cages of the type used by Kramer and arranged these in the open in such a way that the birds took the right direction as soon as part of the sky was clear. Blackcaps *Sylvia atricapilla* and Garden Warblers *S. borin* especially took the south-south-westerly or south-westerly direction which is normal to them at this time of the year. Similarly in the following spring they oriented themselves towards the north-north-east, the normal direction at that season. It is clear that the birds oriented themselves by the stars, since this was the only means available to them under the conditions of the experiments, while they lost all ability to orientate as soon as the sky was covered by dense clouds.

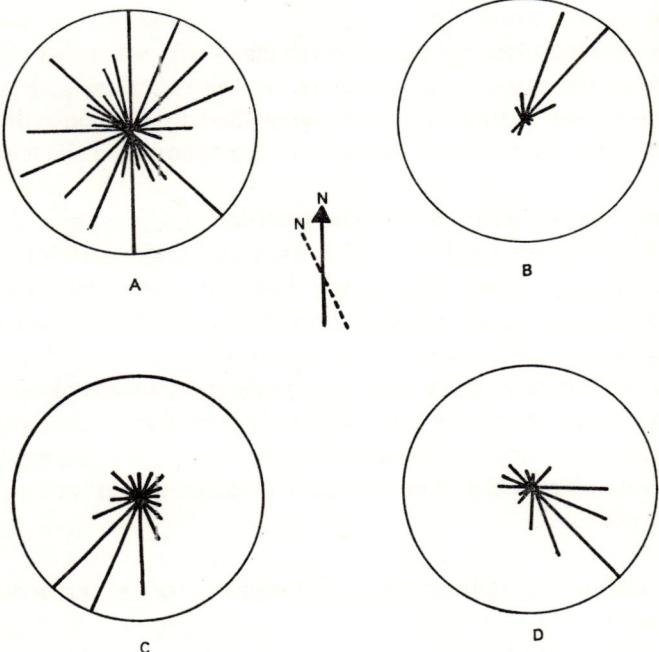

Figure 108. An experiment by the Sauers on the orientation of warblers.
A. When the planetarium is diffusely lit, the birds go in all directions.
B–C. The planetarium shows the image of the sky correct for the season.
B. The orientation of a Blackcap in spring.
C. The orientation of a Blackcap in autumn.
D. The orientation of a Whitethroat in autumn.
The birds always take the direction they would follow if free.

The Sauers then studied the migrants' reactions under the artificial sky provided by the dome of a planetarium, which has the advantage of allowing the experimental conditions to be altered at will. Under uniform illumination of the dome the experimental subjects wandered at random. When in the spring an artificial spring sky was shown to Blackcaps, they immediately took a north-north-easterly to north-easterly direction, whereas they oriented towards the south-west under an autumn sky. Lesser Whitethroats *Sylvia curruca*, on the other hand, headed south-south-east to south-east under the autumn sky, thus showing the same differences from Blackcaps as they do in nature. Experiments were continued with Lesser Whitethroats, by throwing on to the dome a sky which would be seen at the same longitude but at much lower latitudes. They reacted immediately by steering towards the south, exactly as happens during their post-nuptial migrations when birds arriving at relatively low latitudes alter course to the south.

All these experiments make it clear that birds can orient themselves at night by the stars, just as well as by the sun during the day. Not only do they use the guiding marks provided by the constellations, but they alter their behaviour in response to the stimuli and the season of the year.

Experiments designed to alter the internal clock, as already described for ducks, have been conducted with nocturnal migrants with completely different results. All the subjects, whatever the time dislocation to which they had been subjected, steered towards the north-west which is their normal spring direction. In contrast to what happens during the day, when birds use only the sun and interpret its position by means of their internal clocks, nocturnal migrants seem able to ignore that part of the celestial vault whose appearance varies according to the time and place, and to concentrate without regard to their internal clocks on that part of the sky whose appearance is fixed.

CRITICISMS OF THE EXPLANATIONS DEPENDENT ON ASTRONOMICAL GUIDANCE

The orientation of birds by the sun and stars is at first sight astonishing. Some biologists are sceptical of the explanations given by the partisans of astronomical orientation, and have analysed the data in other ways which have sometimes yielded directly contrary conclusions. A few have even questioned whether, despite all the evidence, birds do navigate by celestial guides (Wallraff).

Others have cast doubt upon the sensory precision which would be required for astronomical navigation (Adler 1963). Despite greater power of accommodation and a wider visual field birds' vision is not distinctly superior to our own. It can be calculated that, at our latitudes, an error of one degree in estimating the elevation of the sun above the horizon causes an error in localization of about 110km. The same is true of the precision of internal clocks. Their existence in birds is not in doubt, but as biological mechanisms they cannot be as precise as chronometers. At our latitudes an error of 20 minutes in estimating the time causes an error in localization of 424km westwards or eastwards. In the face of such sources of error one must ask whether the birds' sense organs are such as to give them the information necessary to account for their performance as navigators.

As a result certain authors have questioned or even denied the existence of astronomical orientation, and a few have even returned to theories calling upon the terrestrial magnetic field or unknown senses. This seems to be the only conclusion to be drawn from certain experiments on Robins *Erithacus rubecula* which, normally migrating south-westwards from Germany in the autumn, seem not to use astronomical guides (Fromme 1961). They maintain their orientation even when the sky is covered, or when placed in a totally dark room where no optical guides are available. In contrast they are completely disorientated when placed in a steel chamber though they often take the right direction when there is an opening in the wall. The 'radiations' which the Robins seem to perceive are not magnetic, since alteration of the magnetic field even by a powerful electromagnet does not change their behaviour. Nor does previous mistiming of their internal clocks have any effect. It does thus seem that yet unknown factors are involved in the orientation of these birds. It has been suggested that they perceive radiations with unknown biological effects, even cosmic rays or short waves from the stars. Comparable results have been obtained by experiments which indicate the terrestrial magnetic field as the guiding system (Merkel & Wiltscho 1965), but others have contradicted these results (Perdeck 1963).

Despite these uncertainties, the criticisms of theories involving astronomical guidance certainly do not seem well founded. The results of repeated experiments by knowledgeable biologists cannot be ignored, while astronomical orientation has been demonstrated with certainty in other animals. It is not impossible that guides whose nature is still unknown are also involved, though our ignorance of certain modes of perception is no reason to deny the evidence and facts which have been irrefutably demonstrated.

HOW MIGRANTS MAY NAVIGATE

By combining the results so far obtained, we can try to survey as a whole the means by which birds orient themselves.

It seems that migrating birds take a *primary* direction which is probably genetically determined. This is their response, when they are in the pre-migratory state (as we shall see below) at a particular time in their annual cycle, to stimuli from astronomical guiding marks. It calls for the possession of an internal calendar, which has been partly proved to exist as an annual rhythm of certain glands – especially the pituitary – under the influence of photoperiod. At the phase in their cycle when they undertake the pre-nuptial migration, the birds become especially responsive to a certain image of the sky (whereas throughout the year they respond to varying images in locating themselves during natural or forced displacements). The existence of a specific response is demonstrated by experiment: at a given time the same image of the sky causes different species to orient in different directions, while a single population responds to a given celestial pattern by orienting in opposite directions in spring and autumn. The birds must be in some sense attuned to a certain image of the sky. In addition astronomical orientation demands an *internal clock* giving information throughout the daily cycle.

These facilities would thus allow birds to perceive among astronomical guiding marks a certain image of the sky, under specific conditions to which they were receptive. Under the stimuli of the migratory urge they would depart on their seasonal journeys in the primary direction, maintaining it over long distances. Nocturnal migrants, arriving in the morning at a given place, feed throughout the day by short flights or walks which are as a whole oriented in the direction of the migratory flight, so that the birds continue to progress in the direction of their migration (if this is topographically and ecologically possible).

This primary direction of flight is repeatedly altered in response to local conditions, resulting in *secondary* directions. Immediate guiding marks play a considerable part in these readjustments and in the choice of actual migration routes. Analysis of the return routes followed during displacement experiments have shown that birds have many methods of exploration at their disposal.

Some guiding marks are topographical. Birds unquestionably use irregularities of terrain, drainage systems, mountain relief and coastlines as guides. Other markers are ecological, such as great forest blocks and stretches of water. Still others are meteorological – the direction of prevailing winds and the existence of air masses of different temperatures and humidities which are more or less stable at particular times. These are reliable guides which birds use to a much greater extent than was hitherto believed. Observations by radar (Bellrose 1967) have shown that small birds migrate in clearly defined directions but apparently without the use of astronomical guidance, at least during the prenuptial migration. When the winds are favourable their density on passage depends less on the clarity of the sky than on the wind direction, for other than aerodynamic reasons. The birds can determine the direction and strength of the winds, and since these are constant from year to year at the same time they provide a simple and effective guidance system. Birds further locate themselves by recognizing landmarks within the nesting area which they left after the preceding breeding season, and to some extent within the winter quarters, for which they certainly depend on visual memory. Displacement experiments too have shown that when a bird comes within a certain distance of its nest it begins to move with certainty.

Finally it should be remembered that migratory movements are group activities, undertaken in flocks or family parties by many birds. Within the flocks the dominant individuals are the most experienced, which have learned on earlier journeys to orient themselves The training of racing pigeons shows the part played in orientation by learning, which not only teaches what are usable guides, but also gives the urge to use them. The same must be true of wild birds on their regular migrations, although in their orientation there is a large innate component: Whitethroats *Sylvia communis*, raised from the egg in soundproof chambers at constant temperature and under artificial light, readily orientate. None the less during their first migratory flights young birds, flying with experienced adults, learn the direction to take and the guides to use (Murray). Further, in displacement experiments too much emphasis should not be placed on especially striking results. In some ways spectacular returns are 'happy accidents', picked out from what may be a high proportion of failures, since many of the subjects do not succeed in regaining their home grounds. The same must be true of migrants with only part of the population succeeding in the journey, while a sizeable fraction loses its way. For survival of the species it is necessary only that a sufficient proportion of the fittest birds survive.

The orientation of migrants, and generally of displaced birds, thus makes use of two distinct orders of guidance. The first and most rigid is astronomical and involves conditioning of the birds to a particular image of the sky. The second is much more flexible since, being immediate, it allows the birds to alter their route or maintain their direction, placing themselves to the best advantage and covering what at first sight seem excessive distances as economically as possible. These two types of guidance vary very much in importance, depending on the species, since some birds seem much more dependent than others on the immediate factors. Orienting faculties also vary greatly and each species seems to possess just those necessary to assure its success. In orientation as in other aspects of their biology, birds show remarkable flexibility and ability to adapt. There is still much to be learned but there is no doubt that different guidance systems are combined in bird navigation.

Control of the migratory impulse

In every region the cycle of migrations is synchronized with that of the seasons: in temperate countries the birds return in the spring and depart in the autumn. As a result the control of migration was formerly falsely attributed to a simple response to climatic variations. The physiological control of migration can be understood only by considering the whole annual cycle, of which breeding and moult are the other essential phases. These all take place in rigid relative order, and are clearly under the influence of the same types of factors, belonging to two very different though complementary groups. The first are internal factors, bringing into play the hormonal mechanisms which control the whole organism. The others are external, involving the physical and biotic conditions of the environment.

MIGRATIONS AND THE SEXUAL CYCLE

At first sight migrations seem to be linked to sexual activity: birds return each year to nest, and depart again when their young are able to provide for themselves. This suggests a relation of cause and effect between breeding and the migratory urge. According to Rowan, to whom the first theory based on experiments and physiological observations is due, the migratory urge would be caused in spring and autumn respectively by the development and regression of the gonads. This cycle is itself controlled by variations in photoperiod. There would be no effect from the maximal or minimal development of the gonads, whereas the intermediate stages would have the determining influence. To confirm his theories, Rowan undertook a series of experiments. By artificially varying the photoperiod he caused physiological changes in Slate-coloured Juncos *Junco hyemalis*, which nest in Canada and winter in the central and southern USA. It is known that an increase in photoperiod causes development of the testes, which can be induced to attain their spring dimensions even under the very different temperature conditions of winter. Rowan then released in mid-winter at Edmonton, Canada, juncos at different stages of sexual development. A series of traps arranged around the release areas allowed those subjects which remained in the neighbourhood to be retaken. It turned out that individuals whose gonads were at a *minimum* or a *maximum* showed no inclination to leave the release area, although they were far from their normal winter quarters and subject to a rigorous climate. In contrast, all the birds at an *intermediate* stage of development or regression disappeared from the release area (though unfortunately their directions of departure were unknown), showing that they were impelled to leave by the migratory urge. Rowan's theories have more recently been taken up again by various experimenters. However, none has been able to show that gonadial developments control the migratory urge.

MIGRATIONS AND THE ENDOCRINE CYCLE

It has been noticed that at migration time birds of migratory species held in cages show a distinctive restlessness ('Zugunruhe' of German authors), because they cannot express their migratory urge. The daily and annual cycles of such birds correspond exactly to those of their free relatives, and may be easily studied and experimentally modified through various factors. The physiology of migrants may thus be studied through the degree of agitation of experimental subjects.

Physiological study of migrants has revealed facts of the greatest interest. During the premigratory phase their metabolism undergoes profound changes, reflected especially in the deposit of fat. Body-weight, which reaches its minimum at the end of breeding, increases after the post-nuptial moult and reaches its maximum with the beginning of premigratory restlessness. This variation in weight due to the deposition of fatty reserves may be great: Whitethroats weigh on average 12 to 13g during the nesting season, 16 to 19g in autumn, and 20 to 22g in winter. This fattening has long been known to gourmets, who seek migrants at this season when their flesh is more delicate and richer in fat, to which the culinary fame of Ortolan Buntings is due. The accumulation of these reserves in the form of fat is most advantageous to birds, since these substances have the highest possible energy value for their weight, allowing the birds to carry enough 'fuel' for their long flight stages. Blackpoll Warblers *Dendroica striata*, weighing 20 to 23g at the time of their post-nuptial migration as against 11 to 12g at normal times, have enough reserves for 115 hours of flight which would allow them to fly directly from the north-eastern USA to South America. Estimates on other species have shown that the weight loss per hour is between 0·56 and 1·3 per cent of total weight on migratory flights, and 0·63 per cent in the flight of racing pigeons back to their lofts. Passerines' average expenditure of energy is 0·076 kcal/g body-weight/hr. Thus birds' accumulated reserves are ample for the expenditures involved in their longest migratory journeys (Nisbet, Drury & Baird 1963).

However, every migratory flight exhausts the migrants which arrive emaciated at the staging points. These birds show increased intake of food (hyperphagy) and especially an increased ability to replenish their reserves. These profound changes in the physiology of the bird, unique to migrants, correspond to what has been called its *migratory disposition* ('Zugdisposition'). Though sedentary birds do show a much smaller increase in weight after the breeding season, they do not have the same ability to regenerate their reserves since they do not show the true metabolic cycle of migrants.

The deposition and mobilization of fatty reserves and the appearance of migratory behaviour may be simultaneously explained by the action of various endocrine glands. The known action of the thyroid on metabolism was invoked from the first, notably by Merkel. Injection of a *large* dose of thyroid extract or thyroxin results in a noticeable loss of weight, since the thyroid hormone accelerates oxidation, mobilizes the fatty reserves, and to some extent suppresses the premigratory state: the restlessness is reduced or even entirely suppressed. A *small* dose of the same substances, on the other hand, seems to increase migratory restlessness though still producing a loss of weight. Thyroxine, acting on birds in the premigratory state, thus does obviously affect the release of migration, though it has no effect on birds which lack fatty reserves.

What is known of wild birds' annual thyroid cycle, especially in temperate climates, agrees with the seasons when action by this gland is needed. Thyroid colloid which has accumulated during the summer is resorbed in the autumn, liberating the hormone which had been fixed in the vesicles of the gland. The hormone initiates the post-nuptial moult and affects the premigratory metabolism. Colloid again accumulates during the winter, and is resorbed during the spring prenuptial period. The cycle of thyroid activity varies rather widely from species to species, especially between migrants and sedentary birds. In sedentary species and races resorbtion is shorter and less sharply marked, with the spring resorbtion sometimes entirely eliminated.

While the thyroid thus has an intrinsic cycle, it is also affected by the external temperature. Cold stimulates its functioning and so mobilizes the reserves, and can cause birds in this physiological state to depart on migration. In the spring, and when birds are sheltering in warm countries, the temperature in contrast is rising. Although the birds are generally less fat, their thyroids expand instead of tending to regress as in the autumn, and pour out into the organism sufficient hormone to mobilize the reserves, without requiring an external stimulus.

Despite the incompleteness of the information and the obvious contradictions, the thyroid clearly plays an important part in controlling migration. However, there is nothing to suggest that it actually originates this control. While it forms part of a whole system which causes and directs migration, its action cannot be more direct than that of the gonads. One thus turns towards another gland, the pituitary. The hormones from its anterior part are the best known: the thyrotropic hormone stimulates the thyroid, while the gonadropic ones act on the genital apparatus. The action of the pituitary is to some extent independent of environmental factors, with an intrinsic rhythm, and from some points of view may be considered as a timekeeper (Zeitgeber) to the organism. However, it is also dependent on the cyclic changes of the environment, especially those concerning light, as discussed above in relation to the gonads (Chapter 8).

This explains the results obtained by Rowan who produced development or regression of the gonads at will by altering the photoperiod which he believed to be fundamental to the control of migration. The connection between the sexual and migratory cycles is in fact the pituitary, which controls both simultaneously. Many studies (by Wolfson and others) have dealt with this aspect of the question, and many experiments have shown the close relation between photoperiod on the one hand and the physiological condition, development of the gonads and deposition of fats on the other. The state of fatness, and the general metabolic level which this reflects, are ultimately at least partly dependent upon the pituitary. Injection of Antuitrine G, an anterior pituitary extract consisting mainly of somatropic, produces a distinct response from the organism, including massive deposition of fats.

It must be stressed that the control of migratory urge is not dependent on any single gland, but rather on the combined action of several.

HOW THE MIGRATORY URGE MAY BE CONTROLLED

Although the gaps in our knowledge may still defeat attempts to understand the migratory urge, a hypothesis may be put forward which takes account of the results obtained so far. It is especially necessary to place migration and its causation within the birds' whole annual cycle, with the same organs and the same external factors acting in different ways at different stages. The migratory urge may be attributed to the interplay of three different orders of factors, which can either reinforce or cancel one another. This explains the innumerable variations in migratory behaviour.

1   The pituitary has its own intrinsic rhythm, wholly independent of external factors. For example, the migration of Short-tailed Shearwaters is so well regulated that laying in Australia takes place every year within a few days of the same date. The return to their breeding territories of certain long-distance migrants such as the Swifts is as regular.

Where migrant populations mix while wintering with sedentary populations of the same species, each has its own rhythm of physiological development. Of the juncos at Berkeley (California) in winter, the sedentary ones undergo earlier development of the testes than the migrants, although all are subject to the same environmental factors. Exposure of both to lengthening daily illumination resulted in rapid growth of the testes but no fattening among the sedentary birds, but in accumulation of reserves with only slow growth of the testes among the migrants (Wolfson). The same is true among birds which winter in the tropics. No fewer than five races of the Yellow Wagtail *Motacilla flava* hibernate in the same areas of the Congo, but each behaves differently although all are subject to the same environmental conditions (Curry Lindahl 1958). Their 'physiological calendars' are independent of the environment, and seem to be genetically fixed in relation to the climatic conditions of their breeding areas. Sexual activity begins in February, when part of the wintering population begins to moult and fatten while its gonads develop. These birds all belong to the nesting populations of southern Europe (*M. f. feldegg*), with a precocious prenuptial migration. In March and April the members of the central European race *M. f. flava* pass through the same phases and leave. Finally in April and May this development affects the population nesting in Scandinavia (*M. f. thunbergi*), and these leave last for their breeding area which becomes habitable much later than those of the more southerly populations. This chronology proves the existence of an internal rhythm, independent of environmental conditions and linked to the latitude of the European breeding area.

2   The pituitary is influenced, both directly and indirectly, by variations in the photoperiod. Innumerable experiments have shown that, acting through this gland, the gonads can at will be made to regress or recrudesce, and individuals from migratory populations to fatten as a result of profound metabolic changes. In nature it is the rhythm of the seasons, with its cyclical variation in the duration of daily illumination, which causes these changes and results in the fattening so characteristic of migrants before their departure, especially in the autumn post-nuptial phase.

The pituitary may be stimulated by light either directly (Benoit) or indirectly (Wolfson). In the latter process, the increasing days keep the birds awake for longer, the hypothalamus (which plays a large part in the state of wakefulness or sleep) controls the secretions of the pituitary, and the birds' daily rhythm – acting through the nervous centres – can thus influence its whole metabolism. Possibly the direct and indirect stimuli provided by light act together.

Combination of the intrinsic rhythm of the pituitary, including a 'migratory phase' at certain times of year, and the influence of photoperiod causes the development of migratory disposition in migrant birds, with visible signs including the deposition of fats and the characteristic restlessness which has been well studied in captive birds. However, Rowan had already put forward a cogent objection. Consider a migratory bird in the migratory phase caused partly by the autumn shortening of days, which departs on migration towards its winter quarters in the opposite hemisphere. It is suddenly exposed not only to longer days but, if its migration carries it to high latitudes (as for a Swallow leaving Britain for the Cape in September), to days which are increasing in length. Its gonads should therefore be stimulated to develop by the lengthening photoperiod, causing it either to try to nest in the winter quarters or to leave again immediately on a return migration. That this does not happen is due to the intrinsic pituitary cycle, which in many species shows a *refractory phase*. Thus immediately after breeding and during the post-nuptial migration these birds are not sensitive to the effects of longer photoperiod, as can be seen for example in the Oregon Junco *Junco oreganus* (in contrast to the White-crowned Sparrow *Zonotrichia leucophrys* which does not show the phenomenon). During this refractory phase the pituitary is insensitive to light, and thus sends no 'orders' to the organs under its influence. The testes maintain their sensitivity throughout the year, since the injection of gonadotropic hormone causes them to develop even during the refractory phase.

Migration involves a physiological revolution, often accompanied by profound changes in behaviour – gregariousness and nocturnal activity – which are equally necessary for the journeys.

These changes always take place differently at the post-nuptial and pre-nuptial migrations. In the autumn (at least among northern birds, which alone have been adequately studied) the energy balance is more favourable than during the breeding period and the moult. The photoperiod shortens, reacting on the whole organism until finally the pituitary enters its refractory phase.

In the spring conditions are different and the migratory state may

639

include two distinct phases (Wolfson 1959). A *preparatory* phase, which occurs at the end of summer and in the autumn, is brought about in nature by the short days and has been obtained experimentally by a daily cycle of nine hours' illumination and fifteen hours of darkness. Since at this time of year longer days inhibit this phase, it must be the duration of darkness which causes it. Thus the short autumn days prepare for migratory behaviour which does not occur until six months later. The second, *progressive*, phase then begins – in November and December for northern migrants. While it is considerably accelerated by lengthening of the photoperiod, it is not inhibited by short days. This is of the greatest importance to equatorial and transequatorial migrants, since birds wintering near the equator may react to days about twelve hours' long, whereas those which reach the opposite hemisphere are exposed to days which are not merely short but become shorter before the prenuptial migration; but these have no inhibitory action at this stage. For migrants which winter in the same hemisphere as their breeding area (in a warm temperate region for example) the days lengthen again, thus accelerating the development of the gonads, and no doubt the whole metabolism as well since the pituitary has by then come out of its refractory period.

In either case, whether at the post-nuptial or prenuptial migration, the birds are in the premigratory state (Zugdisposition).

3   The factors discussed under the two preceding headings might control migration with almost mathematical precision. The intrinsic cycle of the pituitary seems to be a well-regulated timekeeper, independent of all external influences, which varies in a regular astronomical cycle. Although species vary very much in sensitivity to illumination, whose data they in some way 'interpret', photoperiod itself is mathematically regular. If migrants were subject only to its influence, they would travel at absolutely fixed dates. This would be very disadvantageous, and such rigid timing would be in contrast to all the ecological mechanisms. Nowhere does the arrival of favourable and unfavourable seasons depend solely upon the astronomical cycle, and the birds could thus be caught unawares by an early onset of winter, when they were not yet in a state to migrate.

In fact, the combined influence of the intrinsic pituitary cycle and the variation in photoperiod merely seems to put most birds into a premigratory state. They still need immediate stimuli, without which they do not move. There are important differences between types of bird from this point of view, which indicates varying sensitivities to the different causal factors.

The amount of available food plays an obvious part, by stimulating departure when it runs short. This is especially true of Swallows, who readily wait in Europe in the autumn, as long as the weather allows insects to proliferate in a late season. Study of the migratory restlessness of captive birds shows that Blackbirds calm down when food is abundant, whereas Redwings and warblers show the same impatience even when they are well fed. Temperature also has a considerable effect on migratory restlessness, which is increased by cold and quietened by warmth. Thus a drop in temperature stimulates many birds to depart on post-nuptial migration, no doubt through its effect on the thyroid. Psychological factors resulting from gregariousness are also important: most migrations are a group activity, and there is progressive stimulation in the flocks as dominant individuals, more intensely stimulated, lead on the others. The effects of these various factors act together.

These immediate adaptive responses of birds which, being in *Zugdisposition* or premigratory state, suddenly leave on migration when they are put into *Zugstimmung* or migratory disposition, also explain their very different speeds of movement. While some migrants travel quickly most tarry on their migration routes, halting according to circumstances. For example the Golden Oriole stays a long while in the Mediterranean region during its autumn migration, when the figs ripen. The European Roller leaves its breeding territories in August or September at latest, but only arrives in South Africa from the beginning of December. It reaches Kenya fifty-five days after its arrival at Cairo 3,400km away, an average of only 62km a day, and thus passes the greater part of its annual cycle in transit. However, the prenuptial migration is generally faster, sometimes twice as fast as the autumn journey, as though the birds were pushed on by the need to nest, with the development of their gonads reinforcing the migratory urge proper.

Under the stimulus of sudden and intensely adverse factors, on the other hand, migrants immediately leave on rapid journeys often of great extent. The advent of a cold spell in Europe or North America causes birds to depart over long distances. Thus a Wheatear covered 940km from Skokholm off Wales to Landes in south-western France in forty-three hours, while a Turnstone ringed on Heligoland was retaken twenty-five hours later on the Channel coast of France, 820km away. Ducks which winter in temperate countries are also almost immediately driven from their wintering grounds by a cold spell. In North America rapid massive movements of this kind take place every year, especially along the Mississippi Valley flyway. During the 'grand passage' ducks and geese fly directly from Canada to Louisiana, covering 2,400km in one-and-a-half days or sometimes less (Bellrose & Sick). Nomadic birds, ducks and waders actually remain in this receptive state the whole winter, so that they can respond immediately to environmental fluctuations by an ecological adjustment which is vital to their survival.

This immediate response to environmental conditions explains the very variable dates of arrival of many migrants in their breeding areas. Those which German authors class as Wetter Vögel or 'weather birds' have arrival dates which vary according to the earliness of the spring. The different modalities of their journey truly depend on the thermometer and barometer. In Europe these birds include the Woodcock, snipe, Lapwing, larks and the Starling, while tropical migrants also belong to this group.

In contrast other migrants, which German authors call Instinktvögel or 'instinctual birds', show greater independence of environmental changes. Their dates of arrival and departure, not being regulated by meteorological conditions, are conspicuously regular from year to year. These birds are thus ruled by their internal cycle and the variations in photoperiod. One of the best examples is provided by Swifts. This must reflect a lower degree of adaptation, and is ecologically less advantageous. It is among these birds, which cannot react by local migrations or retromigrations, that severe losses are caused by a sudden drop in temperature soon after their arrival. Whole populations may thus be decimated by exceptionally cold spring spells. Many seabirds also migrate at fixed dates, of which the best known is the Short-tailed Shearwater *Puffinus tenuirostris* of Australia, which arrives at its breeding territories at the same time every year (Marshall & Serventy).

Thus apart from a few exceptions migrants are remarkably adaptable. Control by two orders of factors – the first fixed and dependent on internal and astronomical cycles, the other variable and dependent on immediate environmental conditions – gives them great flexibility. Here again is the same balance between factors of very different orders as is seen in the control of reproduction, a fact of great ecological importance.

All the above is still largely hypothetical, for the regulatory mechanisms of migration are still very poorly understood.

GREGARIOUS BEHAVIOUR OF MIGRANTS

Most migrants are highly social – even those species which are fiercely individualistic while breeding – so that even raptors may gather in large flocks. For example no less than 7,000 birds of prey, especially buzzards, have been counted at Hawk Mountain in Pennsylvania within the hour on a September day when migration was in full swing. Some birds such as swans, geese and storks travel in family parties while others including many waders and passerines, form uncountable flocks of one or more species, which may total thousands and even tens of thousands of individuals. Some of these groups have a social structure while others are merely concentrations of individuals. Despite this general sociability some birds, such as cuckoos, nightjars, bee-eaters, and some raptors and passerines, remain solitary during migration. Many, including shrikes, fly-catchers and some raptors and waders, defend a territory even while on the move.

The gregariousness of most migrants corresponds physiologically to the regression of the gonads, especially the testes. Since aggressive behaviour and the establishment of hierarchies are controlled by the sex hormones, they are not shown at this time when the hormones are at their lowest level. Furthermore, the eclipse plumages worn by many migrants are devoid of the conspicuous colours and patterns used as sign-stimuli in territorial and other aggressive behaviour. The plumages of juvenile or female type worn by the males allow them to associate peaceably without arousing rivalry.

The origin of migrations

The origin of migrations has been much discussed, but the explanations remain hypothetical. They must clearly be related to birds' exceptional adaptability and mobility, evolving in response to the geographical and climatic modifications which have taken place since the Tertiary.

The origin of migrations certainly cannot be understood solely from the most highly perfected examples. A much more promising approach is to concentrate on the early stages shown in migrations which are still fluid, and to examine this seasonal behaviour in its ecological context.

It is important to remember that the ranges of birds (as of other animals) are not fixed once and for all, but are ceaselessly adapted to variations in the environment by expansions and contractions. There are many examples of this in Eurasia and North America, where the recent warming allows many birds to extend their ranges towards the north. The change in climate and natural habitats, and its influence on the distribution of birds, provide a frame of reference for the study of migrations and their specialization. Migrations must be considered primarily as the birds' response to environmental fluctuations. Apart from the sole exception known to date, the American Poor-Will *Phalaenoptilus nuttalli* (Chapter 6), birds are incapable of living at a slow pace. The obligation to maintain their metabolism at a very high level forces them to dispose of a large amount of food and to eat at frequent intervals. They are thus narrowly dependent on the surrounding environment, and very sensitive to its fluctuations; but in contrast to other animals they can travel far and fast, so that their best means of survival lies in regular migrations.

The fluctuations of standing crop biomass within particular habitats vary very much in amplitude, depending on the type of food under considera- tion. Thus populations of insects, especially the aerial forms, and the amounts of some plant produce, notably fruit, vary from great abundance to almost complete absence. In contrast seeds (a dormant form of vegeta- tion) are much more stable, and a considerable stock is maintained through- out the year. Insect-eaters are the most affected by the fluctuating numbers of their prey, so that migrants include a high proportion of insectivorous and polyphagous species, while the same is true of fruit-eaters. In contrast graminivorous birds can find sufficient food during the winter, so that very few are true migrants.

Food must be not only sufficient but accessible. This is the situation of birds whose habitats disappear under thick snow in winter, and also of water birds where the depths of water varies seasonally, whose food though still present may be inaccessible when the water is too deep. The local migrations and nomadism of ducks and small waders shows their search for optimal water conditions.

Migration has probably developed gradually, as is suggested by the differing stages shown by birds at the present time. Some, like the Robin and Blackbird in western Europe, merely change habitats. Others move nomadically, principally in directions which obviously reflect a search for the most favourable conditions and optimal use of the environmental resources. Such early stages of migration are only partial, not all the popula- tion taking part in the regular movements. When part of a population leaves, the number of individuals left behind is adjusted to the carrying capacity of the area during the most unfavourable season. It is mainly the females and young birds which leave the breeding territories, since their low positions in the hierarchy put them at a disadvantage. The high rank of the adult males within their communities favours them in taking food when it is insufficient for the whole summer population.

Partial and irregular movements no doubt represent the first stages of migrations, controlled as they certainly are by shortage of specific foods brought on by the decrease in standing crop biomass. Those individuals best able to travel easily and at high speed have tended to seek elsewhere the resources which are lacking in their breeding area for part of the annual cycle. Mortality in the sedentary part of such a population would be high, so that natural selection would favour these migratory tendencies, at a rate dependent on the environmental pressures. Migrants show a notably different annual cycle from that of sedentary birds, and we must suppose that the characters involved have been progressively acquired, with selection acting to fix those which have arisen by mutation and recombination. Migration is certainly very old, and most of the birds which now migrate over great distances, such as the waders and ducks, are members of groups which already existed in the Tertiary. Although the distribution of climates must then have been very different from what it is now, highly contrasting climates already existed. There must have been a pattern of migrations, though a different one, with bird populations and species presumably already moving in accordance with environmental fluctuations.

The Quaternary vicissitudes of climate must have had profound effects on patterns of migration, especially in the northern hemisphere in Europe and North America where the great glaciations made huge areas inaccessible to birds. However, the migrations are much older than the glaciations, though their present patterns have been much influenced by them.

Palaearctic and nearctic migrants fall into two categories. Members of the first now breed in their original areas. Excluded from these for part of the year, they take shelter nearer the equator only to return with the favourable season. This category includes the waders and many anatids (geese, swans and ducks), which obviously developed in the northern cold regions. Members of the second category now winter in their original areas, migrating northwards to colonize and breed in areas where the climate is more changeable, though favourable for part of the year. Relatively few European birds are of truly tropical affinities, since even in the remote past the Sahara and Mediterranean formed effective barriers to colonization from Africa. Examples are the bee-eater, roller, hoopoe, swifts and quail. Birds of this category are more common in eastern Asia (especially Japan and to a lesser extent Siberia). The origin of migrants is generally reflected in somewhat different behaviour. The 'tropical' birds often leave earlier in the autumn and return later in the spring than the others, in response to their more stringent needs for warmth and food.

Migration thus results from the invasion of a new area in which the species is unable to remain all the year, returning to its place of origin to avoid a harsh season. Such invasions are still taking place. The breeding area of the Serin Serinus serinus has extended during the past century over the whole of continental Europe, as far as the southern coasts of the Baltic. Formerly a strictly Mediterranean species, it has remained sedentary in southern Europe, but has become migratory in the north because of the more rigorous winter climate.

Birds seem to travel on their annual migrations along the same routes as they used on their original invasions. In Europe this applies especially to those migrants which arrive by the 'oriental' route, such as the Red-backed Shrike *Lanius collurio* which arrives in spring through Syria and Asia Minor and spreads out over Europe, going both north and south round the Alps. The Wheatear *Oenanthe oenanthe*, which from Europe has invaded Greenland and north-eastern Canada (Labrador), migrates every year across the western Atlantic towards tropical Africa (like a few other birds such as the Ringed Plover *Charadrius hiaticula*, which winters along the European coasts). The Wheatear has also expanded to northern Siberia and even Alaska, from whence it migrates obliquely across the whole of Siberia to Africa, rather than directly to tropical America or South-East Asia. It thus follows what were undoubtedly its invasion routes to these outlying areas during the Quaternary. Correspondingly, the Grey-cheeked Thrush *Hylocichla minima*, a typically American species which has extended its nesting area to north-eastern Siberia (like a few other birds such as the Pectoral Sandpiper *Erolia melanotos*), returns to pass the winter in the warm regions of the New World. Thus in autumn birds cross the Bering Strait in both directions, returning to the areas where they originated. Still more strikingly, the Arctic Warbler *Phylloscopus borealis*, which nests in the arctic from Scandinavia across Siberia to Alaska, migrates in the autumn to winter quarters which extend over South-East Asia and Indonesia. Even the Lapland populations return to this area, crossing the whole of Asia in the opposite direction to the Wheatear, in accordance with their presumed origin in north-eastern Siberia. Thus this small bird covers a distance of some 12,000km. However, not all migrations follow the original invasion routes, since some species have adapted to new conditions so that nothing can be deduced from their itineraries.

Migratory behaviour has imposed on birds certain characteristic morphological traits. Migratory flights demand considerable efforts and favour the best flyers. Most avian groups which include many migrants such as bee-eaters, swallows, swifts, terns and waders, have very pointed wings. Comparison between migrants and sedentary birds within a single group is still more striking, with neighbouring species and sometimes even subspecies often differing widely in wing formula. Thus the orioles have their centre of differentiation in the Old World tropics where the birds are sedentary, and where their wings are very rounded with the primaries a little longer than the secondaries. The migratory species such as the Golden Oriole, which leave the tropics to breed but return there to winter, have much more pointed wings with long primaries.

Thus migratory behaviour must have appeared progressively during evolution, birds having developed the migratory habit under the pressure of environmental factors which have forced them to make these periodic movements. Perhaps the chances of nesting success are higher in the breeding than in the wintering areas, because of greater space for the establishment of territories, and longer days for feeding their young.

The benefit which migrants derive from leaving their breeding areas when these become uninhabitable must be greater than the disadvantages of long and dangerous journeys. Despite their choice of the most favourable migration routes and halting places they cannot be sure of finding everywhere sufficient sources of food. They have to fly over wide seas, apparently out of all proportion to their own size and power, while deserts too are formidable barriers. The birds cannot afford to make mistakes, and a high proportion of the migrants which fail to cross in one flight are lost, except where there are suitable staging points along the route. The heavy losses suffered by migrants crossing the Sahara prove the difficulty of these flights, while the same is true of mountain crossings. Storms, fogs, and the tropical tornadoes in winter quarters which are so deadly to certain migrants, all add to the dangers, so that migrants suffer higher mortality rates than sedentary birds. This is compensated by higher fecundity, with distinctly larger clutches. Thus sedentary birds and migrants differ in population structure and dynamics. The balance must be positive, between the energy required for regular long-distance flights and the food energy at their disposal within their huge total range. The migrants disperse and avoid one another in their winter quarters, thus avoiding the undue competition which would otherwise no doubt cause the carrying capacity to be exceeded, if migrant and sedentary populations actually bred side by side. The fact that migration has developed proves its advantages for a large part of the avifauna, even though these are not directly measurable.

Furthermore migrations, involving great mobility, allow populations to mix and are one of the factors involved in extending the favourable breeding area of a species.

The ecological significance of migration

Migrations must also be considered from an ecological point of view, since they allow the rational use of some natural resources which, in the absence of fluctuating populations, would remain unexploited.

While breeding, pelagic birds are restricted to the seas near islands and other land suitable for nesting. Thus they can exploit only the resources of a limited area, to a radius set by their flying speed and feeding frequency. By the end of nesting, they have more or less fished out this feeding zone, whereas other parts of the sea remain unexploited because of the lack of nesting sites.

Migrations also play an essential part in the annual cycles of complex ecosystems characterized by large fluctuations in the standing crop biomass. Within a community, only those birds at the trophic levels where large fluctuations occur take part in migrations. This applies to the breeding zone, and also to the winter quarters. Palaearctic migrants take refuge for the winter in the savannas and woodlands of Africa and tropical Asia. They avoid the rainforest, an environment of virtually constant productivity where the absence of seasonal surpluses prevents fluctuating populations from establishing themselves (except for swallows and swifts which live on super-abundant aerial insects). The migratory populations take advantage of the higher carrying capacity at the time of their stay, resulting from seasonally increased productivity, and leave when the standing crop biomass drops and a period of scarcity sets in – which may lead to the total disappearance of food suitable for a particular species in the habitat it has been occupying.

The case of migrants which pass from arctic breeding grounds to tropical winter quarters is especially interesting. Both these regions are characterized by enormous fluctuations in biomass. In the arctic zones the amount of plant and animal food passes through a very marked summer maximum but then decreases rapidly. In those tropical savannas of very contrasting climate, the fluctuations which follow the seasonal cycle is almost as marked, with a maximum at the end of the rains and a minimum at the end of the dry season. The density of a sedentary species necessarily depends on the limiting capacity of the environment at critical periods and can therefore only be low – especially in arctic regions, where as far as birds are concerned it is virtually nil. The enormous food surpluses which vary in time at a given place can be exploited only by a consumer population which itself varies in space during its annual cycle. This is the ecological justification for migratory behaviour.

The water birds – ducks and waders – provide the most striking example. They nest in large numbers, near stretches of water and in the sub-antarctic marches, when the standing crop biomass is at its maximum. When it diminishes they set forth to winter in the tropics, especially the great flood zones of the African northern-hemisphere savannas, which they reach when the rains have produced a proliferation of plants and animals. The timing of their movements is very precisely adjusted so that their cycle

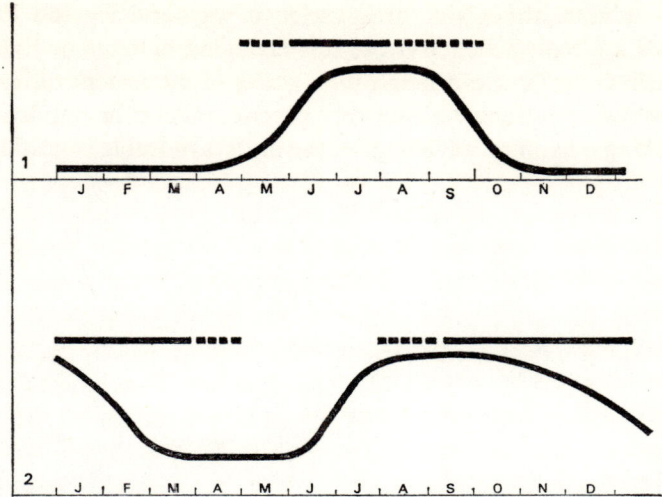

Figure 109. The migration of the wood sandpiper *Tringa glareola* and the cycle
of the quantity of food available. The seasonal variation of standing-crop biomass:
1. in the arctic. 2. in the savannas of the Sudan. The period when the migratory
birds stay in each zone is indicated by the solid lines.

agrees closely with those of the habitats which they frequent at different
seasons, reflecting their close adaptation to these habitats. Their population
sizes are determined as much by conditions in their wintering zones as by
those in their nesting territories.

It is not only the amount of food which is important, but also its availability. This is especially clear with dabbling ducks and waders. Hence the
interest of the great flood zones of tropical Africa, where the advance and
retreat of the waters cause changes in depth and resulting movements by
the wintering birds. These huge areas are exploited in sequence, so that the
birds can find the most favourable conditions throughout their wintering
period.

Migration allows large populations to subsist in regions which would
otherwise remain closed by their inhospitable conditions during part of the
year, and also allows the exploitation of the seasonal food surpluses. During
their evolution birds have no doubt gradually perfected their seasonal
movements in response to ecological pressures, among which the amount of
available food is the determining factor.

Thus bird migrations as a whole are well-regulated mechanisms, ceaselessly and flexibly adjusted to environmental conditions. A high proportion
of avian species and populations – far more than the famous long-distance

649

migrants – migrate, though in a great variety of ways and degrees. Thus only a few birds are truly sedentary, the rest changing habitats or living areas during their annual cycles though their scales of movement differ widely. These seasonal movements, some very precise, reflect the opportunism of birds in taking maximum advantage of the most favourable conditions.

Chapter 32

Man's Ravages Against Birds

Human influence on bird populations

In their natural communities, birds are in equilibrium with the environment. However, man's interference has had a very profound impact on nature. Such disruption of the biological equilibrium has profound effects on animal populations, and has caused the ruin and eradication of many.

The destruction of nature and the reduction or even the extinction of wild species are now proceeding at terrific and accelerating rates. Some species had already disappeared during primitive stages of human development, but the process increased markedly after the great explorations, when Europeans spread out over the world and especially into North America. The richness and variety of the lands which opened to the colonists contrasted with the impoverishment which had already taken place in Europe and the Mediterranean basin, and this gave rise to the myth of the inexhaustible richness of the earth – especially of the tropics – still tenaciously held by some people. Then began a plundering of natural resources, of which birds and mammals were the principal and most spectacular victims, which accelerated in the eighteenth century, and unhappily still continues despite all the efforts and achievements in creating reserves and in regulating the exploitation of natural resources.

Birds were the foremost animals to suffer human action. It is estimated that no less than 150 species have already disappeared for ever, at a rate which has increased since a precise inventory has been possible. About ten forms (species and subspecies) became extinct before 1700; a score during the eighteenth century; another score from 1800 to 1850; fifty from 1851 to 1900; and another fifty from 1901 to the present. Thus an average of one form a year has disappeared during the past century.

Animals can of course also disappear as a result of natural changes. Since evolution is continuing one may expect some species to become extinct simply because they are at the end of their natural term, showing a kind of senescence. This cause of extinction has been invoked for the Takahe or

651

Notornis *Porphyrio mantelli*, a rail apparently closely related to the Purple Gallinule and formerly distributed throughout suitable biotopes – marshy meadows and lake shores – in New Zealand. It is probable that man contributed largely to its reduction by disturbing the environment, and that climatic changes unfavourable to the species accelerated its gradual disappearance. However, the lower adaptiveness of the Takahe, shown in reduced fecundity and various other behavioural traits, is very obvious when it is compared with related species, and provides evidence of some specific senescence. This bird had even been considered as extinct for fifty years, until in 1946 it was rediscovered in a few secluded valleys of the Murchison Mountains of South Island. However, no more than 200 to 300 individuals survive, so that the species remains at the mercy of the smallest destruction and could disappear almost from one day to the next. Such cases are rare, and almost all primary factors of extinction arise from the direct or indirect action of man so that they are entirely artificial. In practice, man is guilty of almost all the extinctions of birds, as of other animals.

The factors which have resulted in the disappearance of the most vulnerable species have reduced many other birds to rarity. Apart from a few anthropophilous species which have actually benefited from the transformations produced by man, most species in the world have diminished and many even rank among what have been described as 'tomorrow's fossils'. One can only fear that this process will accelerate since detailed analysis reveals certain regularities. The majority of extinct or gravely threatened species occupied islands in Polynesia and the West Indies. These strictly endemic species have evolved in closed systems and as a result they are highly specialized, and lack defences against competitors and predators including man. This is especially true of flightless birds.

Small population sizes have also been important on islands, especially those of small size. The size of a population is one of the most important factors for its survival, since for each species there is a threshold below which a population's chances of regenerating are greatly reduced. A small population within a restricted range is very vulnerable to local changes, while a low population density over a wider range makes pairing difficult. Finally, group stimuli are important in the breeding behaviour of gregarious and even markedly social birds, and their breeding success is markedly lower in colonies of less than a threshold size.

In general, bird populations of less than 2,000 individuals are to be considered as in danger. This limit is of course arbitrary, since smaller populations established in certain areas sheltered from human interference are clearly in equilibrium with their environments. However their localization makes them vulnerable to the most restricted natural disaster, as the history of the Great Auk eloquently proves. Populations larger than 2,000 may also be threatened, especially by adverse conditions in their breeding areas. The populations of certain birds are much smaller than this limit. The Giant Pied-billed Grebe *Podilymbus gigas* on Lake Atitlan in Guatemala totals scarcely more than 100 individuals, the Japanese Ibis *Nipponia nippon* only nine, the California Condor *Gymnogyps californianus* about forty, the Monkey-eating Eagle *Pithecophaga jefferyi* of the Philippines less than a hundred, the Mauritius Kestrel *Falco punctatus* less than ten, the American Whooping Crane *Grus americana* forty-two, and the Hawaiian Crow *Corvus tropicus* between twenty-five and fifty.

The causes of reduction and extinction are varied and various factors usually combine to affect a single species, so that it is sometimes difficult to isolate one from among many. However, a principal factor can usually be held responsible for the disappearance of a particular species. Usually the most obvious cause is the direct action of human predation, including thoughtless hunting, slaughter for plumes, egg collecting and vandalism. The destruction of natural habitats is perhaps even more damaging. Many species, such as forest or water birds, are tightly bound to a definite habitat. This is also true of stenophagous birds, whose survival is dependent upon a particular type of food and which disappear when their source of supply is altered, sometimes even when the change is almost imperceptible. Reduction of other bird populations results from more subtle ecological changes. Especially in fragile or simple ecosystems, man may set off disturbances of equilibrium. This is especially true of oceanic islands on each of which a highly specialized fauna, including a high proportion of endemics, has developed as a simple and extremely fragile ecosystem, with few components and food chains.

Direct destruction

Direct destruction has been the principal factor in the disappearance of many birds. Some birds were sought for their flesh, which was eaten especially in impoverished regions where all other resources were lacking. This is without doubt what caused the disappearance of the Great Auk *Alca impennis*. This bird which reached 75cm in height, with much reduced wings incapable of flight and thus resembling the penguins of which it was more or less the arctic equivalent, was widely distributed along the European and American coasts of the North Atlantic. Easy to catch, it

653

formed part of the diet of the coastal peoples from Neolithic times, as is proved by the abundant skeletal remains found in kitchen middens. More recently and up to the middle of the nineteenth century, Great Auks were regularly hunted by sailors and fishermen who ate it or used its flesh as bait for fish. As a result the colonies disappeared one after the other. The last to survive were on islets off Newfoundland (Funk Island) and Iceland (Eldey Rock, where the two last-known specimens were taken in 1844). These were exterminated by fishermen while one in between them, on the Geirfuglasker Rocks off Iceland, was destroyed by a volcanic eruption in 1830.

Many other seabirds have suffered severe tolls from sailors and fishermen, in inhospitable countries where resources and especially fresh victuals are rare. Many Procellariiformes were captured in huge numbers as fish bait, and travellers testified that the fishing boats were covered with many hundred corpses hung from the yards ready for use (Murphy). Furthermore, young petrels remain in the nest for a long time, fattening to considerable size, and when fully grown have been systematically taken from their burrows by the inhabitants of many islands where resources were limited, while the adults were also taken and split and salted. The oil contained in the stomachs of the young petrels was collected as a lubricant or a medicine, and another oil obtained by boiling their bodies. In Tasmania 750,000 Short-tailed Shearwaters *Puffinus tenuirostris* are taken every year from the huge colonies of this species, and sold as 'mutton birds'. This exploitation still continues, under the control of the authorities, and despite its scale the populations continue to flourish (Serventy). Such practices recur in various parts of the world, as in the Hebrides with the Manx Shearwater *P. puffinus* and on Tristan da Cunha with the Great Shearwater *P. gravis*. They have not necessarily harmed the bird populations since the cull, being limited to the needs of the inhabitants who did not commercialize it, was within the tolerance of the breeding population.

This however was not always the case, especially when other causes of destruction such as the introduction of predators were added. This accounts for the severe reduction of the Bermuda Petrel *Pterodroma cahow*, the Black-capped Petrel *P. hasotata* of the Antilles, and the Hawaiian Petrel *P. phaeopygia sandwichensis*. Commercial exploitation is sometimes much more dangerous. For example, nesting populations of Cory's Shearwater *Calonectris diomedea* have been exploited for a century in the Salvage Isles off Madeira. The populations were so large that from 20,000 to 22,000 birds could be taken annually without affecting them. However, exploitation was not maintained at this reasonable rate, and now the birds are destroyed without check or limit, often for 'pleasure', through the development of motorboating and tourism. These birds are thus being progressively exterminated and a natural resource destroyed (Jouanin).

Seabirds elsewhere have also suffered from the direct action of man – especially the guano birds of the Peruvian coasts. Guanay Cormorants, Peruvian Boobies and Chilean Pelicans, with some other species, form incredibly dense colonies on the desert islets off Peru and northern Chile, where their faeces accumulate as guano. This fertilizer was exploited on a grand scale during the second half of the last century, without consideration of the producers and their breeding season, while the birds were hunted without respite and exterminated for pleasure. Serious depletion of the populations resulted at the beginning of the present century, which could have led to their total ruin if measures had not been taken for their protection. These have allowed the stocks to regenerate, though with a notably different balance of species. Man had thus all but deprived himself of an important natural resource – which he then endangered afresh by over-exploiting the marine resources, and so depriving the guano birds of the abundant food such huge populations need.

Many land birds have suffered equally thoughtless destruction by man, especially some of those hunted for food and sport such as the Eskimo Curlew *Numenius borealis*. This nested in North America in the tundras bordering the Arctic Ocean, and migrated as far as Argentina to winter. Natural causes, especially disastrous losses from the cyclones which overtook them during their long autumn migrations across the Atlantic from Labrador to South America, may also have contributed to their disappearance. However, the essential damage was done by the massacres, which took place during their spring migrations along more continental routes. Millions were sold in American markets and 7,000 were killed in a single day on Nantucket Island in 1863. The Eskimo Curlew is now very rare and indeed almost extinct.

While other factors may have contributed, excessive hunting has also caused the almost complete extinction of several other sporting birds. In 1967 only forty-seven individuals of the American Whooping Crane *Grus americana* remained, which are very difficult to protect because of their migrations between Canada and the Texan coast: rearing in captivity seems to be the best safeguard for this species. The Ne-ne or Hawaiian Goose *Branta sandvicensis* has happily been saved by this technique which allowed its reintroduction to its original habitats in 1960; and in 1966 the world population amounted to about 500 individuals of which 285 were living free in the Hawaiian islands. Other birds threatened by excessive hunting are the Hawaiian Duck *Anas platyrhynchos wyvilliana*, several races of the Canada Goose *Branta canadensis*, the Cereopsis Goose *Cereopsis novaehollandiae* of Australia, and many Asiatic pheasants. Other game species have also suffered considerably from this cause combined with the transformation of their habitats.

Other birds have been over-hunted for their feathers. Egrets were sought for their famous plumes, and their colonies devasted during the breeding season since these are nuptial ornaments. The birds of paradise are still hunted for their gorgeous plumes, despite the laws intended to protect them in New Guinea. A feather trade developed towards the end of the last century to serve feminine fashions, which brought to Europe (mainly to Paris and London) tens of thousands of birdskins. These came mainly from Malaysia, Japan (though the Japanese collected throughout the Pacific), Senegal, Trinidad and South America (Bahia and Bogota). Consignments contained birds of drab plumage mixed with hummingbirds, glossy starlings, parrots, birds of paradise and egrets. This trade, though despite its volume it seemed to have no serious effects on South American birds, was disastrous for those of the Pacific. Perhaps the saddest example and the most spectacular was that of the Short-tailed Albatross *Diomedea albatrus*, which wandered over much of the Pacific and nested only on a few islands of the Bonin archipelago. There it was hunted to excess by Japanese collectors, who killed the birds at the nest by striking them with sticks. Between 1887 and 1903 more than 500,000 albatrosses were massacred for their feathers, and in 1962 only forty-seven birds could be counted on Torishima, the sole island on which the species still nested.

Other species are sought as cage birds, which have long been prized. Certain brightly-coloured Australian parrots, notably the Orange-bellied, Turquoise and Golden-winged Parrots *Neophema chrysogaster*, *N. pulchella* and *Psephotus chrysopterygius* are gravely threatened by excessive nest robbing. Small birds such as bengalese and waxbills must suffer heavy losses, and also despite the abundance of many species their numbers must be locally reduced. The capture of rare birds for exhibition in zoos also endangers some of them, such as the Monkey-eating Eagle *Pithecophaga jefferyi*. This species endemic to the Philippines is valuable both alive and as prepared skins, so that it is eagerly hunted and its total population is now less than a hundred.

A final excuse, to add to those which man all too readily seizes to kill or catch birds, is that of combating species which can damage crops. This was responsible for the total extinction of the Carolina Parakeet *Conuropsis carolinensis*, which was abundant in the south-eastern United States. It was then considered to be a crop pest and was therefore systematically exterminated, while many were caught as cage birds. This parrot has entirely disappeared, the last individual dying in captivity in 1914. Supposed harmfulness has similarly been used to justify the systematic destruction of raptors, against which a real war has been declared, and which have as a result diminished considerably – throughout the world but especially in eastern Europe. Since such predators are at the ends of their food chains they are naturally scarce, and many have been pushed below their critical thresholds of population density. It has been established, without avail, that this human attitude to raptors is completely unjustified.

The most dramatic disappearance of all was that of the Passenger Pigeon *Ectopistes migratorius* from the deciduous forests of the great North American Atlantic plains. This was undoubtedly the most abundant land bird in the whole world. Until 1871 an area of 2,200km² in Wisconsin alone supported not less than 136 million birds, nesting colonially so densely that the branches sometimes bent and even broke under the weight of nests. During migration the flocks, on the way to their winter quarters in Gulf States, passed for days at a time, obscuring the sky. During the nineteenth century these pigeons were massacred as systematically as the bison. Their flesh was greatly esteemed, justifying all the hunting parties which were also often organized for the sake of the spectacle. Trees were felled in order to collect the young, as fat as could be wished in their nests. As a result the species declined rapidly, the last colonial nestings being noted in 1885, and the last isolated nests in 1894 in areas near the Great Lakes, though a few individuals were seen after this. The last survivor died in 1914 in the Cincinnati Zoo. Thus man exterminated an incredibly flourishing species. The destruction of the great deciduous forests which were its preferred habitat certainly contributed to its reduction; but it was hunting – or rather systematic massacre – which must be held primarily responsible.

Destruction by pesticides

The abuse of pesticides and especially insecticides during the last few decades has been as harmful to birds as to pests. Many of these toxic substances, which are often massively distributed, have quite low lethal doses. Insectivorous birds which feed on prey charged with insecticides can absorb from them sufficiently large doses to kill themselves and their broods. In intensively treated areas of the United States, where the abuse of insecticides has been especially deplorable, losses up to 80 per cent or even 97 per cent of all the birds have been recorded. American Robins and many other birds are especially sensitive to the paralysis of the central nervous system produced by these products. Especially high mortalities of raptors from the ingestion of pesticides have been recorded.

Furthermore, insecticides are readily concentrated in the sexual organs of birds, so that sublethal doses have equally disastrous results. Partial or total sterility of the breeding birds, resulting from disturbances in the secretion of sex hormones, reduces egg production, while the proportion of fertile eggs and the survival of young also fall off. In the USA the Bald Eagle *Haliaetus leucocephalus* is hit by partial sterility, and the Osprey's brooding success is markedly low. Observations on the Golden Eagle *Aquila chrysaetos* in Scotland show that the proportion of breeding birds has fallen, from 72 per cent of the total in 1937-60 to 29 per cent in 1961-3. The eagles also showed abnormal behaviour, breaking and eating their own eggs, as has also been recorded of Peregrine Falcons. The Sparrow Hawk is in the same situation: in southern and eastern England where pesticides are used on a large scale, four eggs were recorded during 1955-63 in an area where fifty were laid in 1940-54 (Cramp). Raptors, already greatly reduced by more or less systematic destruction, are thus the birds most threatened by the use of synthetic insecticides.

These substances act to poison birds directly and also after a delay, especially by being concentrated in passing through the food-chains. The toxic product may be absorbed by an organism which, having high specific resistance, can concentrate it without serious ill effects, and then pass into the body of a susceptible predator. The best-known case is that of American Robins in the United States, mentioned above. DDT has been massively spread to protect elms from a disease transmitted by insects. Falling to earth it is ingested by earth-worms, which are resistant to DDT but concentrate it in their tissues. The Robins which eat the worms in abundance take in large amounts of the poison, from which a high proportion die. Some ornithologists have ventured to assert that this species, still so abundant in North America, actually risks extinction like the Passenger Pigeon.

Concentrations of toxins also take place in aquatic biocenoses, along the food chains leading from the plankton to fishes and water birds, as the classic example of the grebes on Clear Lake, California eloquently proves. DDD applied at a rate of 0·042 parts per million reappears at 5 ppm in the plankton, at from 40 to 300 ppm in the plant-eating fish, and at 2,500 ppm in the grebes. These last succumb, so that the breeding stock of grebes on this lake has fallen from 1,000 to thirty.

These effects even make themselves felt in the sea, and DDT has been found in the fat of Antarctic penguins although these are still more remote from the centres of application. This is also true of the Bermuda Petrel *Pterodroma cahow*, already long gravely threatened. Until the beginning of the seventeenth century this bird nested in enormous colonies in Bermuda, but measureless slaughter reduced its populations until it was believed extinct. However in 1951 a few individuals were rediscovered, breeding on islets where they have since been carefully guarded, and where twenty-four pairs bred in 1966. However, since 1958 the breeding has been unsuccessful, and the species has declined by some 3·35 per cent a year, towards imminent extinction. Analysis shows that its eggs contain high doses of DDT and other organochloric insecticides, which block embryonic development. The near extinction of this species, which had already once been only just saved, proves the gravity of pollution, by which the planet is poisoned and from which even the seas are not immune (Wurster & Wingate). Many pelagic plankton-feeding birds are potentially threatened in the same way.

The destruction of habitats

Man may have even more profound and almost irremediable effects by destroying habitats. It is easy to understand that birds decline and disappear with the environment which is indispensable to them. Throughout the world two types of environment are especially threatened, and in the most diverse regions have been massively reduced: forests and aquatic habitats.

Forests everywhere have been systematically exploited for timber and cleared for cultivation. Bird extinction is more or less proportional to deforestation, as has been demonstrated especially in the West Indies (Greenway 1958). Hispaniola where there are still 2·3 hectares of forest per inhabitant still has all its species, whereas between two and four species have disappeared from islands such as Jamaica, the Lesser Antilles and Puerto Rico where the forested area is less than 0·40 acres per inhabitant. The disappearance of many West Indian birds is thus at least partly due to the destruction of the forest which sheltered them, as well as to the introduction of predatory animals. This is well shown by three species which have almost disappeared from the heavily deforested island of Martinique, though they were formerly widespread there. The House Wren *Troglodytes aedon* (represented by endemic races) is now rare, as on Guadeloupe, whereas it is still common on St Lucia, St Vincent and Granada where the forests are of sufficient area. The White-breasted Thrasher *Rhamphocinclus brachyurus* was even considered extinct, but survives in a single wood of a few hectares in the Caravella Peninsula, where its tiny population is at the mercy of any change in this vestigial habitat. The same is true of the Trembler *Cinclocerthia ruficauda* which certainly survives only as a very small population at Trois Ilets (Pinchon).

The effect of forest destruction in reducing bird populations is equally clear in North America, whose Atlantic seaboard was covered with very extensive broadleaved forests. Of at least 174 million hectares which were timbered at the time of European colonization, only 7·7 million hectares now survive, and these certainly in a much-modified condition. This deforestation, together with intensive hunting, has been responsible for the extinction of the wild Turkey *Meleagris gallopavo sylvestris*, which long ago disappeared from the area. Other species are now narrowly localized, such as Kirtland's Warbler *Dendroica kirtlandi* which breeds only in pine forests of scarcely 400km² in Michigan, where its total population cannot exceed 1,000 individuals. Although it may not be directly threatened at present, the survival of such a local species may be vulnerable to the smallest change in the natural equilibrium of its confined environment. The destruction of forests is equally responsible for the virtual extinction of the Ivory-billed Woodpecker *Campephilus p. principalis*, formerly distributed through the riverine and swampy forests of the south-eastern USA. Loss of this habitat, and with it of the particular tree species bored by this woodpecker for its specific prey, led to the gradual reduction of this magnificent species, which has not been seen recently. The endemic Cuban race *C. p. bairdi* is also practically extinct for the same reasons.

In other parts of the world, deforestation has led to the same train of ruin. On Mauritius and Réunion it, together with hunting and the introduction of exotic animals, has resulted in the extinction of most of their autochthonous species. In Asia many pheasants confined to dense forests, of various types, have regressed with them, while intensive hunting has hastened their decline. Thus the White Eared Pheasant *Crossoptilon crossoptilon* of southern China. Tibet and Assam, the Brown Eared Pheasant *C. mantchuricum* of north-eastern China, Swinhoe's Pheasant *Lophura swinhoei* of Formosa, Edwards' Pheasant *L. edwardsi* of Palawan in the Philippines, the Palawan Peacock Pheasant *Polyplectron emphanum*, Elliot's Pheasant *Syrmaticus ellioti* and Cabot's Tragopan *Tragopan caboti* both of south-eastern China, are all now very rare and some of them virtually extinct. The same is true of the Australian Scrub-Birds *Atrichornis*, confined to dense scrubs and unable to survive their clearance. One of the two species, the Noisy Scrub-Bird *A. clamosus*, is represented by only about fifty individuals and for sixty years was actually thought to be extinct.

On the whole birds of open habitats have suffered less from man's activities than those of forests, since we have transformed biotopes primarily by attacking closed habitats. However, a few open-country species have suffered from the severe disturbance of their habitats by cultivation. North American grouse are classic examples of this, and especially the Greater Prairie Chicken *Tympanuchus cupido*. The nominate race *T. c. cupido*, which once occupied the Atlantic seaboard of the north-eastern USA, has disappeared, while the race *T. c. attwateri* from Texas southwards to Louisiana has considerably diminished. Its surviving populations estimated at about 750 in 1965, are no more than 1 per cent of what they were a century ago, while its range has contracted to 7 per cent. The other races are equally reduced and fragmented. Transformation of habitats by cultivation has been the principal factor in reducing populations and areas, especially by depriving the grouse of certain plants which were their principal food.

Waterbirds have also suffered severely. For the past century the

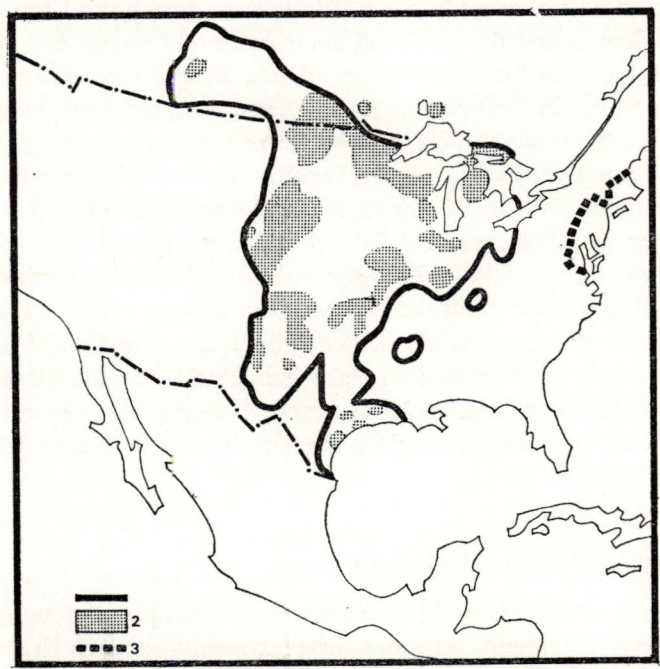

Figure 110. The contracting range of the Greater Prairie Chicken *Tympanuchus cupido*. 1. the distribution at the beginning of European colonization. 2. the present distribution (the races *T. c. pinnatus, attwateri* and *pallidicinctus*). 3. the former boundary of the eastern race *T. c. cupido*, now extinct.

661

destruction of aquatic habitats has been accelerating catastrophically, especially in the temperate zones of the northern hemisphere. The bird populations dependent on these environments have declined in parallel. This is especially true of migratory species, which need not only adequate breeding territories but also winter quarters, and migration stages along their extensive routes. While the first have been only slightly affected, especially in the immense arctic tundras, man has profoundly altered the second and third. Besides his direct effects, pollution has considerably modified their ecology by reducing productivity. The effects of destroying these aquatic habitats have been aggravated by hunting, often uncontrolled and almost always too intensive.

All this has considerably reduced the northern aquatic avifaunas, consisting of ducks, geese, swans and small wading birds. According to censuses undertaken by the International Wildfowl Research Bureau the world populations of Pink-footed and Barnacle Geese amount to 40,000 and 30,000 individuals respectively, while the European populations of the Brent Goose do not exceed 20,000. While these may seem large, they are far lower than before the profound environmental changes resulting from human action. Furthermore the population densities of many wild fowl species breeding in Scandinavia do not seem to agree with the extent of favourable environments. Thus ducks, geese and waders evidently do not achieve the production potentials of which they are ecologically capable. The same is true in North America, where the breeding territories are used to only a tenth of their capacity (Gabrielson, Curry-Lindahl).

Nor have the wading birds been spared. While many have greatly diminished, the situation of the Japanese Ibis *Nipponia nippon* is uniquely bad. This was common eighty years ago in Japan, China and Manchuria, but there are now no more than eight individuals on Sado Island, and a further one has been seen on Noto Island. Hunting has been responsible for this near-extermination, while the survivors are threatened by deforestation and drainage.

Upset of biological balance

Most of the cases considered above are relatively simple, although destruction of the environment has more complex results through the profound ecological upset it causes. More subtle disturbances of the food chains may sometimes have quite as serious repercussions on certain species as the most spectacular destruction. An especially regrettable example is provided by the Lammergeyer *Gypaetus barbatus* in Europe. This effectively occupies the apex of complex food chains, above the large carnivores on whose kills it normally lives, breaking open the bones to feed on the marrow. These predators having disappeared from a large part of Europe, the Lammergeyer fed like other vultures on the carcases of domestic animals left in the pastures. Now however, changes in pastoral methods deprive it even of this food, and its reduction is hastened by poisoned baits set out to destroy wolves and other terrestrial predators, and by the stupid prejudices of man who slaughters this harmless bird. The Californian Condor *Gymnogyps californianus* – a living fossil whose remains are known from the Pleistocene of the whole western USA, Texas, Florida and north-eastern Mexico – has also fallen victim to the decline in stock-raising in California, together with senseless hunting.

The ecosystem can also be upset by the artificial introduction, whether accidental or deliberate, of foreign elements. Innumerable bird species have been acclimatized: as ornaments by colonists nostalgic for the fauna of their country of origin, or as game, or to combat crop pests. Attempted introductions are often unsuccessful, the introduced population being unable to resist physical and biotic factors of the environment, which form an effective barrier and prevent it from expanding. In contrast, when there is an empty ecological niche or when the introduced species has superior vital potential (greater adaptability, less strict ecological demands, higher fecundity and breeding success) it begins to multiply in the area to which it has been artificially transported, where it may undergo a real population explosion and often becomes a pest.

Such proliferation almost always takes place at the expense of autochthonous species. There are two possible situations. In the first case it subjects the autochthonous populations to predation for which they are not prepared. Thus the human action amounts to introducing a foreign, very high trophic level to the ecosystem, immediately changing its equilibrium. This happens especially to wingless and otherwise flightless birds. It was the situation of the Dodo *Raphus cucullatus* of Mauritius, the Réunion Solitaire *R. solitarius* and the Rodriguez Solitaire *Pezophaps solitarius* who disappeared soon after European colonization, the Dodo around 1680 and the two others during the eighteenth century. Without doubt direct human predation was partly responsible for the disappearance of these over-evolved and maladapted birds, which were taken on board ships as fresh provisions; but dogs, cats and pigs, all introduced and feral, were the main cause. This same is true of the flightless rails in the Pacific, which have evolved as a series of endemic forms each confined to a particular island. The introduction of mammalian carnivores, especially cats and dogs, overcame these remarkable birds. Many are already extinct and the others have become so rare that their early disappearance is to be feared, the more since many of their forested and marshy habitats are being transformed.

One cat was apparently responsible for the disappearance of the Stephen Island Wren *Xenicus lyalli*. This small passerine of the New Zealand endemic family Acanthisittidae was confined to Stephen Island, where its population must always have been minute, confined as it was to a wood of about 2km². This explains how this nocturnal bird, flying little and nesting on the ground, could be exterminated in 1894 by the cat belonging to the keeper of the lighthouse which had been established on the island.

Indian mongooses, introduced to the West Indies to control snakes (especially the dangerous Fer-de-lance) have caused the ruin of all the ground-nesting birds in this area, which has already been so profoundly disturbed by human interference. They are responsible for the disappearance of several species of rainforest passerine. Rats and even mice are also responsible for intense predation on terrestrial birds and their young and eggs. Procellariiformes nesting on the ground or in burrows are especially vulnerable to these rodents, accidentally transported by man across the whole world. Their effects have been especially intense in the antarctic islands. especially South Georgia, Amsterdam, St Paul, Kerguelen, Crozet and Macquarie. On Amsterdam Island only four species of bird still flourish, while innumerable bones testify to the many species of petrel (*Bulweria, Puffinus, Pachyptila, Pelagodroma, Pelecanoides*) exterminated by the rats (Jouanin & Paulian). In the northern hemisphere, the puffins and other auks have sometimes suffered to the same extent. A colony of Puffins *Fratercula arctica* estimated at 250,000 individuals was established on the little Scottish island of Ailsa Craig. Rats landed accidentally after a shipwreck in 1889, and by 1947 only about thirty of these birds remained (Fisher & Lockley).

Even the introduction of fish may be harmful to birds. Young Atitlan Grebes *Podilymbus gigas*, an endemic species restricted to this Guatemalan lake, were taken by Perch *Micropterus salmoides* after the introduction of those voracious fish in 1956. This, together with competition for food, human hunting and the changes produced by cutting the reeds, has considerably reduced the populations, which in 1968 were estimated at no more than 116 individuals (Bowes).

Introduced animals may also prove to be dangerous competitors. In New Zealand, competition from the European Starling has led to the reduction of autochthonous pipits, kingfishers and the Tui *Prosthemadera novaezelandiae*. This active and enterprising bird monopolizes their food and occupies their nesting sites, dominating them by its vitality, as the mynah from south-east Asia does in the various territories from South Africa to Oceania to which it has been introduced. No fewer than twenty-three species have been introduced to Mauritius, mainly from Asia, among them the Red-whiskered Bulbul *Pycnonotus jocosus*. The descendants of six individuals introduced in 1892 now occupy almost the whole island and have ousted the autochthonous birds, especially the Mascarene Olive White-eye *Zosterops olivacea chloronotos* which it clearly dominates. The endemic Seychelles Owl *Otus insularis* has become very rare, and may even be extinct, largely as a result of competition with the introduced Barn Owl.

Finally, certain species have disappeared as pure stocks by hybridizing with closely related introduced forms. Thus in New Zealand the native Grey Duck *Anas s. superciliosa* has hybridized with the introduced Mallard *A. platyrhynchos*, as has the Mexican Duck *A. diazi*.

The introduction of pathogens foreign to the ecosystem may also be harmful to birds. This has perhaps contributed to the far advanced disappearance of many Drepanididae from Hawaii. These endemic passerines have always been hunted unrestrainedly by the natives, who made large cloaks and headdresses from their feathers. However, it was not this but the activities of colonists from America which seem to have caused their disappearance. Clearance and deforestation removed the environments on which they were dependent and where they found their specific foods (insects and the nectar of native trees). The Drepanididae also succumbed to avian diseases transmitted by introduced mosquitoes. Many species have disappeared and others are on the verge of extinction, having persisted only in the high zones which retain their plant cover and have not been occupied by the mosquitoes. Drepanidids caught in these zones and taken to the lowlands have soon died of diseases.

Conflict between autochthonous and introduced species, expressed either in predation or in competition for food or nesting sites, or even resistance to introduced diseases, is usually linked to the transformation of habitats. Thus on Mauritius the surviving native birds are confined to the remnants of forest, where the intruders acclimatized by man have not penetrated. To understand the mechanism of such competitions, it is necessary to consider the association of a given species with its environment. The conservation of habitats is of great importance for the protection of threatened species, since nature forms an undissociable whole.

Chapter 33

Birds and Man

WE have seen how disastrous human activities have been to birds, as to wildlife as a whole. Except for a few species which have benefited from man and the changes he has brought to natural habitats, birds are disturbingly involved in the general retreat of wild nature before the increasingly severe impact of man and his techniques. In contrast birds have contributed much to humanity since the earliest times. More than other animals and at least as much as mammals, they have caught man's artistic and emotional imagination by their powers of flight, their beauty, and their conspicuous place in the most varied landscapes.

A few representations of birds have survived from prehistory, such as the owls of the Grotto of Three Brothers, Ariege, which dates from the Magdalenian, and they have inspired artists of all civilizations up to our own. They also play a most important part in legends and myths, as the traditions of the most diverse countries eloquently show. At this epoch of the conquest of the moon, it is appropriate merely to recall that according to an Eskimo legend birds brought men to this planet, an idea repeated in whimsical European prints of the seventeenth and eighteenth centuries. Birds still retain their evocative power as symbols in many ways, such as on seals, national emblems and postage stamps. However, their practical importance is completely different.

Domestication

Man had succeeded in domesticating certain birds at least by the beginning of the Neolithic in some countries, though they appeared only later in western Europe to which they were probably imported. As with mammals, domestication probably arose from the practice of keeping captured animals as a kind of living larder until they were needed for food. Some became tamer in captivity, at rates depending on circumstances and on the temperament and needs of each species. Some must have proved untameable, since we know that attempts were made to keep many more species than those now domesticated.

All birds which man has succeeded in domesticating have certain common traits. The first is a tendency to gregariousness. While flocks of gamebirds and wild geese show a rigid social organization, whereas ducks live in looser associations, all these together with pigeons are markedly sociable. Wild and solitary birds can be domesticated only with great difficulty, first needing to be ethologically conditioned, and are therefore of no practical importance. All domesticated birds are also either vegetarian – graminivorous like gamebirds and pigeons or herbivorous like geese – or omnivorous like ducks.

Domestic fowl originated in south-east Asia, where their ancestor the Red Jungle-fowl *Gallus gallus* is still found from Kashmir to southern China, Malaysia, Sumatra and Java. Though the three other species of the genus are very closely related they do not seem to have played any part in the origin of the domestic fowl. Among existing races the most 'primitive' show exactly the coloration of the wild species, from which man has obtained strikingly different breeds by long-continued selection. The wild stock can still easily be kept in captivity, where it grows larger with a bigger crest, though most individuals remain intractable.

The fowl was already known to the Indus Valley civilization around 2,500 BC, among whose figurines it appears, though it does not seem to have been reared for the table. From north-western India it was taken eastwards, appearing in China towards the fourteenth century BC, and westwards, reaching Egypt in the fifteenth and fourteenth centuries BC. It later disappeared from Egypt, but during the first millennium BC rapidly expanded again across Persia, Mesopotamia and Asia Minor, becoming common in the Mediterranean basin towards the sixth century BC. Besides its symbolic and religious role as an offering to the gods, its flesh and eggs were considered as food resources. It reached central Europe very early, around the sixth century BC at Hallstatt.

There were already many distinct breeds in Roman times, and others were obtained by selection during the Middle Ages, when the domestic fowl became increasingly important. There are now very many breeds, with over 200 described in the American Standard of Perfection. Most of these however are interesting only as ornamental birds or curiosities which show the extraordinary plasticity of the species. Rearing for the table has become an industry on a large scale, and has led to selection on the one hand for heavy races, and on the other for those which produce flesh quickly because of their rapid growth and high efficiency in the transformation of vegetable foods into animal proteins. Rearing techniques have been

667

industrialized and are strictly controlled from incubation to the product, the eggs or the table fowl.

Selection has considerably increased the laying period and the egg production, from the well defined annual breeding season of the wild birds during which each lays six to eight eggs. The average for good layers was 112 eggs a year as recently as 1925 but is now 220, while the record is 361 eggs in 365 days. Selection and artificial lighting are responsible for the increase and the regularity of production. Flesh production has also been considerably improved. 1kg of chicken, which still needed 4kg of corn to produce in 1948, now needs only 2·5kg, while eight to ten weeks are sufficient to produce a marketable bird. The fowl and its eggs thus continue to play an important part in human feeding. In the USA in 1965 they met 3·9 per cent of the annual need for calories, 12·6 per cent for proteins, 9·6 per cent for phosphorus and 9·2 for iron (Shaklee in Stefferud 1966).

The Guineafowl *Numida meleagris* originated in tropical Africa and arrived in Greece before the fifth century BC. After the classical period it disappeared from Europe and was reintroduced from West Africa by the sixteenth century Portuguese navigators. The Pheasant *Phasianus colchicus* and the Peafowl *Pavo cristatus* were also introduced: the first probably from the Caucasus where it abounds, the second from India by way of Persia and Mesopotamia, only becoming common after the conquests of Alexander the Great. The Pheasant remained a game-bird, being reared for the purpose only quite recently, though it was established in France and England as early as the tenth century. The white collars of our present stocks are due to the introduction of races from China and central Asia.

The Turkey *Meleagris gallopavo* is the only North American bird to be domesticated, probably in the south-west, long before the European discovery of the New World. Turkeys were first introduced into Europe in 1523, and twenty years later raising them was already a flourishing business. Broad-breasted breeds, have been produced by selection since the white flesh of the pectoral muscles is especially esteemed. Although the Romans raised Quail *Coturnix coturnix* they did not eat them, believing them to be poisonous. Quail are now of some importance, since the eastern races brought from Japan sixty years ago for their flesh and eggs readily breed in captivity.

Thus the gallinaceous birds seem preadapted for domestication by their gregarious temperament, the social structure of their flocks, their poor powers of flight, and their diets. However, other birds have been very successfully domesticated, though they have remained more independent and usually more like the wild type in appearance.

This is true of the Rock Dove or Pigeon *Columba livia*, found wild across Europe and North Africa as far as India, which no doubt resorted to the earliest cultivated fields and has perhaps been domesticated since the Neolithic. In classical times it was very widely raised in Mesopotamia and around the Mediterranean, where it played a part in mythology and divination, and was used for food and to carry messages (see below). The Pigeon too seems preadapted for domestication. Despite the independence given by its powers of flight it is faithful to a fixed place; it nests in conditions easy to reproduce in captivity, on crags rather than in trees like related species; and it feeds on grain. While it retains its importance as food, it is also reared as an ornamental bird to which innumerable fanciers are devoted. Very many races have been derived from the wild type, some of which differ from it greatly. The Barbary Dove *Streptopelia 'risoria'* derives from the tropical African Collared Dove *S. roseogrisea*. Already domesticated in Roman times, it has given rise to many races which are kept as ornamental birds.

The Greylag Goose *Anser anser*, found wild from Europe to China, is known to have been domesticated in the Neolithic, probably first in southeastern Europe. The Greeks commonly reared it while the Romans also knew how to fatten it and produced white races, by selecting albinistic individuals such as occur in wild populations. Besides the Greylag, the White-fronted Goose and a few other species were held captive in Mesopotamia and Egypt from the earliest times and domesticated from the New Empire, perhaps under Nordic influences. In China another species, the Swan Goose *Anser cygnoides*, was domesticated long ago. It is reared in many countries, since it resists heat better than the Greylag and puts on more flesh while young. The Mute Swan *Cygnus olor* has been held in captivity and semi-domesticated since antiquity, originally as a source of food and later as an ornamental bird. In England they are all the property of the Crown, except those which carry on their bills the carefully recorded marks allocated to the few individuals and societies royally licensed to own swans.

The Mallard *Anas platyrhynchos* was certainly first domesticated in Mesopotamia, but China is another ancient centre from which originated one of the most important breeds, the Peking Duck. The Greeks, Romans and Egyptians raised ducks, but without truly domesticating them. In contrast the inhabitants of ancient Germany may have reared ducks from the Hallstatt period. The so-called 'Muscovy' Duck *Cairina moschata* originated in South America, where it was domesticated long before the Spanish conquest. It was brought to Europe from Peru and Colombia in the sixteenth century and quickly spread in captivity.

Attempts were made to domesticate other birds, notably the Egyptian Goose *Alopochen aegyptiacus*, which was considered sacred in Egypt, but its quarrelsome temperament is incompatible with life in captivity. Pelicans were kept for their eggs which the ancient Egyptians prized highly. The cormorants *Phalacrocorax carbo* and *P. capillatus* have been reared for fishing since the fifth century in Japan and China, their eggs being brooded by hens. Finally the Ostrich *Struthio camelus* has been reared since classical times, as it still is in South Africa.

Domestic birds have never played the predominant part of their mammalian counterparts, except in the European Hallstatt culture in which the goose and no doubt the duck were important. Domestic mammals are a fundamental link in man's exploitation of grasslands, as indispensable transformers of energy. Domestic birds are not vital as a source of food, though they are of some importance if only for their easily kept and carried eggs. They have certainly been kept partly for the pleasure they give, as is shown by the present fashion for aviculture without utilitarian ends.

Homing pigeons

Of all domestic birds, pigeons are the most independent. Certain breeds have remained the same as the Rock Dove from which they originated in form, coloration and performance. Their ability to navigate within a certain radius remains well developed. They were employed by the ancient Egyptians, Greeks and Romans who used them to carry messages announcing the results of chariot races. During the siege of Paris in 1870-1, 150,000 official messages and a million private letters were exchanged by this means. During the Second World War the armies used homing pigeons, the US Army alone having up to 50,000, and the resistance movements in the occupied countries repeatedly communicated with England using pigeons which had been parachuted to them by the allied forces.

Homing pigeons undergo very rigorous training. Birds selected for their heredity and for various physical indications are trained over increasing distances. The sport of pigeon racing is enjoyed in various countries, especially in Belgium, France and England where there are some two million racing pigeons. The meetings are often very spectacular, with courses of up to 1,000 miles. When these involve sea crossings the losses are always severe. The time taken to return is measured exactly, and often corresponds to an average speed of sixty-six, or even over one hundred km per hour when conditions are favourable. The birds fly for up to sixteen hours a day. Racing pigeons have often been used in the study of orientation and homing (Chapter 31).

Other birds have been similarly used, though on a very limited scale. The Romans caught swallows at their nests, and after dyeing them in the colours of the victorious chariot released them to announce the results of races. The Polynesians used tame frigate birds to communicate between islands. However, these techniques were rudimentary in comparison with racing pigeons, whose selection and training is a highly developed aspect of domestication.

Falconry

The use of raptors to catch animals, especially other birds, dates back to the earliest times. It probably arose in the east around 1200 BC, and China, India, Persia and Arabia remain the areas where the sport continues to be most regularly practised. Falconry then passed to the Mediterranean basin where, though unknown to the Greeks, it was practised by the Romans using the Goshawk and Sparrow Hawk. The Arabs maintained the tradition, which returned to Europe long before the Crusades (to Great Britain around 860), though it enjoyed a vogue at that time. Although this later diminished considerably, falconry still flourishes in the east, and even in Europe and North America continues to be practised by devotees.

Falconry has developed naturally from the observation of hunting raptors. Its difficulty lies in training notoriously wild and independent birds, while preserving their flying and hunting abilities to the full. Two chief methods of hunting are distinguished. In falconry or high flying, falcons (Gyr Falcons, Peregrines, Lanner and Saker Falcons) are used to hunt from high in the air. In contrast hawking or low flying uses Goshawks and Sparrow Hawks which take their prey by surprise after a short flight. Eagles are used in a distinct technique rather like hawking. Buzzards (especially the North American Red-tailed Hawk *Buteo jamaicensis*) have also been used though without great success.

Hawks and falcons are always caught wild, since no-one has yet succeeded in rearing them in good condition. They are taken either from the nest ('eyasses') or later when they have already flown, in their first or second plumages ('passage hawks' and 'haggards' respectively). Their training, carried to perfection as early as the Middle Ages, is long and difficult, a delicate art in which the falconers' patience and knowledge are severely tested. Training depends essentially on the bird's need for food.

The prey hunted varies widely: birds of open habitats such as pigeons, partridge, grouse and crows for the high-flying falcons; more diverse prey even including rabbits for the low-flying hawks. Eagles have been used by the Mongols to hunt wolves and foxes.

The practice of falconry thus demands deep knowledge of the biology of the hunting birds, together with meticulous training since they are naturally wild, rebellious and intolerant of man's presence. Falconry also requires perfect co-ordination between master and bird, and often also between them and a setter used to find potential prey. The bird must thus be trained to keep an eye on both the dog and on prey which it puts up, so as to position itself to strike the prey. These requirements make the sport a consummate and subtle art, one of man's highest achievements in the use of birds.

Cage and aviary birds

Man has kept birds in captivity for ornament or song since the earliest times – perhaps almost as long as he has domesticated them at all, though no doubt his first objects were utilitarian. Parrots were kept in the east and were carried to Europe around 300 BC by Alexander the Great's soldiers. The parrot concerned came from India and still bears the great conqueror's name as the Alexandrine Parakeet *Psittacula eupatria* (= *P. alexandri*). The Crusades and later the sixteenth century voyages of exploration revealed many exotic species. *Serinus canaria* from the Canary Islands, where no doubt it had been domesticated by the Guanches before the Spanish colonization, was brought by Jean de Bèthencourt in 1404 to the Castillian court and to Queen Isabella of Bavaria, the consort of Charles VI of France around 1406. The fashion for cage birds developed, reaching its zenith in the nineteenth century. Aviculture is by no means confined to Europeans since Asians have long been past-masters of this art, which is practised at some level by all peoples.

Attempts have been made to rear almost every bird in captivity, but most cage birds raised by non-specialist amateurs are graminivorous because they are easy to feed, since seeds in great variety are available and easy to store. In contrast nectarivorous and especially insectivorous birds are much more difficult to keep in good health in captivity, and still more difficult to breed. Graminivorous birds do not present these difficultes, which explains the great popularity of brightly-coloured finches and weavers. Cardinals, bishops, buntings, weavers, fire-finches, waxbills, bengalese, senegalese, diamond sparrows and of course canaries are among the most widespread of all cage birds. Frugivorous and omnivorous birds are a little more difficult to rear, or at least to keep in captivity, so that many are kept as cage birds, notably tanagers and bulbuls among the passerines, and above all the parrots.

Many of these birds have retained their natural characteristics. Aviculturalists have succeeded in profoundly altering others by mutation, selection and hybridization, to obtain varieties which differ greatly from the original stocks and are thus true domestic birds. This is especially true of the Canary of which there are now many breeds each with its own standards, selected for plumage, coloration or song; and also of the Budgerigar, some stocks of which have retained the green coloration of the wild Australian species, while others show colours from blue to yellow and pure white.

Aviculture has also added greatly to the knowledge of bird behaviour. The nuptial displays of pheasants and especially of ducks were first described and interpreted from observations on captive birds, before being verified in nature, and the very advanced state of avian ethology is largely due to aviculture.

Catching birds for the cagebird market certainly affected the equilibrium of some species and a few species have been endangered by excessive trapping. The collection, transportation and sale of small passerines, such as waxbills, still often involve scandalously high mortalities, and such practices contrary to nature conservation need to be controlled. In contrast, certain threatened species have been saved only by being reared in captivity. This is true of many pheasants and parrots, such as Australian parrots of the genus *Neophema* which have become rare and very local in the wild, and especially of the Hawaiian Goose or Ne-ne *Branta sandvicensis*, which was so severely threatened in its home islands that its extinction was feared. Reared in captivity, especially by the Wildfowl Trust at Slimbridge under Peter Scott, the stock rapidly increased to the point where the species is now secure and could even be reintroduced to its natural habitat.

Hunting

Palaeolithic men hunted birds and collected their eggs, and these practices are still vital to the survival of tribes which remain at a primitive stage. Hunting increased in later ages and was perfected and specialized in various ways. It was originally utilitarian, but as stock rearing developed, hunting became a sport. This is now its most important aspect, despite the special foods it incidentally provides.

Some conservationists deplore the persistence of hunting, which they consider as a barbaric and bloody anachronism. Hunting is undeniably an art carried to a high degree of perfection, based on deep knowledge of the ways of life of the game species. When properly understood and based on exact knowledge of population size, structures and dynamics, hunting is a rational exploitation of a natural resource. Under such conditions man may be considered as a predator, substituting his takings for other causes of mortality and participating in the biocenose.

There are serious problems in controlling hunting where there are large populations and many hunters. There the restrictions suggested by biologists are often considered as persecution, although they are intended to preserve the hunted stocks at their highest level.

Bird watching

These activities which first appeared in the Anglo-Saxon countries have rapidly spread across the world. Without any scientific pretensions, many amateurs greatly enjoy identifying and observing birds in nature. Instead of game bags they keep lists, recording the species they meet and recognize and drawing up life, annual or even daily totals. The 'Big Day' in the United States is a competition in which enthusiasts, alone or in groups, try to spot within a given area the greatest number of bird species inside twenty-four hours. In this way 160 species were noted in one day in an area of Massachusetts (reported by Peterson in *Birds over America*). The 'Christmas Census' is another form of this ornithological sport, from which a true ornithological tourism has developed. Some people stay in areas famous for their avifaunas simply to identify birds in their natural surroundings, and special tours to distant places are regularly organized for this purpose.

The contribution of amateurs to ornithology as a biological science cannot be sufficiently stressed, and has been decisive in ecology and the study of behaviour, and also of migrations since ringing has been carried out largely by amateurs.

Birds as a source of raw materials

Besides providing man with food products of high quality, and satisfactions of various kinds, birds have served him as a source of various raw materials.

Their feathers have been used from the earliest times for practical purposes, because of their softness and insulating properties. Wild birds were first exploited and then domesticated ones, some countries producing a large volume of feathers for export. However, the collection of feathers from wild birds, especially eiders, still persists. Like all ducks, the female eider plucks feathers and down from her foreparts and arranges them in the hollow of her nest to protect and insulate the eggs, and they are gathered by man to make clothing and quilts.

Feathers have also served as ornaments, the most decorative and vividly coloured being chosen for this purpose. The Papuans take toll of the birds of paradise, whose highly specialized plumes they use in their head-dresses and facial decorations. These plumes were formerly also exported to Malaysia and India, where they were used to make ornaments of prestige. The Hawaiians used the red and yellow feathers of many drepanidids to make the cloaks and head-dresses of their chiefs, stitching the feathers one by one to a cloth to produce remarkably effective multicoloured patterns. Amerindians also made great use of feathers, and birds played a further considerable part among the North American Indians as is shown by their dances, inspired partly by the elaborate displays of Prairie Hens. Feathers have also been used in western fashions.

Birds provide a famous nitrogenous fertilizer, *guano*, which accumulates where huge bird populations gather to nest (Hutchinson 1950). Certain fairly strict conditions must be met for this to be satisfactorily produced. First, the birds must be very abundant so that they deposit enough faeces, which implies markedly gregarious behaviour. Such a concentration of birds requires the surrounding area to be rich in food, and only the sea can supply the quantities to feed such dense populations. As a result all the guano-producing birds are marine and assemble in the most productive areas of the oceans. It is also necessary that they drop their excrement on land, and not scatter it while on the move which would greatly reduce the amount of guano deposited. The areas where they live must have little rain, which would both wash away the deposits and accelerate chemical changes (especially from nitrates to ammonia). Finally, the colonies must be on level ground, for the guano to accumulate rather than fall into the sea.

All these conditions are met at various places, notably in Peru and northern Chile thanks to the unique oceanographic conditions we have already noted. Three species of birds, amounting to some 16 million individuals in Peru alone, are responsible for the production of guano there: the Guanay Cormorant (forming 85 per cent of the total), the Peruvian Booby and the Chilean Pelican. These birds' faeces have accumulated for thousands of years, to form deposits which around the middle of the last century reached a thickness of some forty metres on certain favoured islands. Within a century these layers have been entirely stripped: between 1848 and 1875 alone, 20 million tons were sent from Peru to North

America and Europe. At present, thanks to protection of the colonies and rational exploitation, and despite the intensive fishing which competes with the birds for food, Peru annually produces about 250,000 tons of guano. Other centres of production are known and exploited around the world: notably in Lower California, along the coast of south-western Africa, and on various islands of the Pacific and Indian Oceans and the Caribbean.

It has been calculated that the birds need 10 tons of fish to produce 1 ton of guano. This manure owes its properties to its high content of nitrogen (from 8 to 20 per cent) and phosphates (from 6 to 17 per cent), which makes it a remarkable fertilizer, still of great importance despite the synthetic products prepared by the chemical industry.

Birds in science

Ornithology has kept its character of *scientia amabilis*, but is also a branch of zoology and a sector of general biology.

Birds are preferred material for many researches. Their systematics is very advanced, especially in comparison with the descriptive stage to which the specialists in many other groups are still condemned for a long time to come. Though a few new species are still being discovered, the inventory is virtually complete and the present classification accurately catalogues all the known forms.

Researches on sexuality and endocrinology have made much use of birds, some of which show much more marked sexual dimorphism than mammals. Their plumage reacts very obviously to hormonal conditions, faithfully interpreting the physiological state of the subject and allowing measurement of the effects. The influence of hormones from the pituitary, thyroid and sexual glands on the annual cycle of moult, seasonal changes in plumage, breeding and various other types of behaviour can be detected much more easily than in other vertebrates. Thus the effects of light and of variations in photoperiod were first demonstrated on birds, before being confirmed on other animals.

Ecology also owes much to the study of birds. Their adaptations in general, and of various types of bird in particular, provide a wide field for investigation. The study of bird populations – their structure, dynamics and food relations with the environment – has yielded conclusions extending considerably beyond ornithology, notably on the regulation of populations.

Birds are chosen material also for the study of migrations, which are of greater scope among them than in any other group of animals. The conclusions have then been applied to many other animal groups, thus stimulating their study.

Many ethological discoveries have also been made on birds, which were the first animals to be studied behaviourally, apart from insects whose ethological mechanisms are notably different. The fundamental concept of territory was developed for birds and then extended to other animals by comparative studies. Ornithologists were equally prominent in the recent rejuvenation of the 'New Systematics', freeing classification from its undue dependence on purely morphological and often superficial criteria. Having largely completed the analytical phase they have been able to reach a synthesis in defining new concepts such as the notion of subspecies and a species concept more in accordance with reality and in disentangling the broad lines of a 'natural' classification based on phylogeny. At the same time they studied new and very promising physico-chemical characteristics, and introduced the systematic implications of behaviour. This progress has allowed the third phase of systematics, that of interpretation, to be attained. Birds have long been used for studies on evolution.

These examples might be multiplied many times, to show ornithology as a particularly fruitful discipline making very substantial contributions to general zoology. It allows an approach to many of the great problems, and is one of the most active sectors of contemporary biological research.

Chapter 34

The Place of Birds in a World Transformed by Man

THE place which birds occupy in relation to man and his activities has remained essentially the same as in previous ages. They may interfere with our interests on the one hand, and assure us both material profit and intellectual, aesthetic, cultural and emotional satisfactions on the other. However, birds are among the first victims of man's struggle with nature, in which he irrationally destroys for his immediate ends, while often ignoring his true long-term interests.

The major problem in conservation is to allow the 'peaceful coexistence' of man and nature, and it is in this general context that we should consider our attitude towards the avian world. Not only is it morally wrong for man to destroy nature, but it is not in his own best interests. The biosphere is an interrelated whole, of which birds are an important part, and its total output and renewable resources can only be maintained at optimal levels if man respects its balance and recognizes inviolable laws.

We have the duty to conserve at the highest possible levels even those species which at first sight have no practical use. Management is necessary to assure the future of all wild birds.

The questions to be considered here are the beneficial or harmful parts played by birds from a strictly human point of view, and their place in the artificial environments created and maintained by man. We must thus mention briefly how birds have adapted to the transformations which man has wrought on the surface of the planet, and what we should do to preserve what remains of the world's avifauna.

Birds and towns

Man has completely transformed certain areas by building towns: an entirely artificial environment, although some parks and public gardens may have the appearance of a natural habitat. However, towns have been

678

occupied by a fraction of the avifauna, which has succeeded in adapting to and sometimes even benefited from urbanization.

Towns provide certain birds with nesting facilities. Thus the originally cliff-dwelling Swallow has found very favourable nesting sites on or even within buildings. The House Sparrow, which in natural biotope nests in holes in trees (as it still occasionally does in towns) now nests in the crevices of buildings and old walls, as does the Starling, whose hole nesting is perfectly adapted to buildings. The Jackdaw *Corvus monedula* has adapted to church towers where it forms small colonies. Domestic Pigeons have found in certain architectural features the equivalents of the cliff ledges on which their wild representatives nest. In certain seaside towns the same is true of Kittiwakes, which use window-sills instead of the narrow ledges on rocky cliffs where they normally nest. Tits occasionally build in letter boxes, as the equivalent of their habitual nesting sites in tree holes. Other arboreal birds find enough trees in parks to continue nesting as they do in their natural habitats. This has allowed the Chaffinch and Wood Pigeon to penetrate urban surroundings (and also the Stock Dove, though this is rare in towns).

Urban environments also provide birds with considerable quantities of food, especially in gardens and parks. The accumulation of domestic rubbish together with sewage farms attract many gulls and other omnivorous birds. Graminivorous birds find ample food wherever grain is handled. There are insects even in the great cities, if only in the lower air where swallows and martins hunt. Thus many birds need not leave the towns in order to gather food. Thus in London the Domestic Pigeon and Wood Pigeon feed within the city while the Stock Dove leaves to forage at its edges (Goodwin). The Starlings which gather in millions to winter in some towns do not find enough food there and scatter daily to the surrounding countryside, making feeding flights of more than 15km.

Towns also offer birds a much more favourable microclimate than the nearby areas, especially in winter. The winter standing crop biomass is higher, especially since many cultivated plants bear highly appreciated berries or seeds. As a result 'town' Blackbirds are less migratory then their 'woodland' relatives in the same area, while towns attract many wintering birds which benefit both from the shelter and from increased food supplies. This is true of the Starlings which gather in extraordinarily dense flocks in London and other towns.

As a result of the milder microclimate urban birds develop earlier. In the spring the gonads of Starlings, Domestic and Wood Pigeons wintering in towns enlarge before those in the country. Artificial illumination allows the birds' activity to continue longer each day.

Finally, towns protect birds from many predators, which adapt poorly to urbanization and are unwelcome to humans. However, there are predatory birds even in towns. Tawny and other owls frequent parks and wide tree-lined avenues, feeding especially on House Sparrows (up to 93 per cent of their diets) with a few rodents. More surprising is the presence of Peregrines in certain large towns of Europe, the USA and Canada, where they occur on passage and even nest on cathedral towers or other tall buildings. Kestrels too are quite common in towns (Olivier).

The limited number of species which have become anthropophilic thus forms a real urban community. No fewer than thirty-five species regularly nest in London, though in very various numbers. The House Sparrow and Domestic Pigeon form 90 per cent of the total populations, followed at a great distance by the Wood Pigeon and then the Blackbird and Starling, though many more Starlings winter there. The garden species are the Coal and Blue Tits, Song Thrush, Robin, Greenfinch and more rarely a few other passerines, while bodies of water attract several species of ducks and gulls. Between 23,000 and 45,000 pairs nest in Inner London where the average density varies between 4·4 and 8·6 birds per hectare, and may rise to 21·7 in large gardens (Cramp & Tomlins 1966).

The best adapted of these birds are in equilibrium with the urban environment which they never leave, maintaining their populations entirely by local breeding. The population structure and dynamics of other species are anomalous. Thus in Dortmund and Kiel Blackbirds show an abnormal sex-ratio of 1·7 males to every female, and while their mortality rate is lower than that of country Blackbirds their greatly reduced fecundity does not balance the losses, so that the populations can maintain themselves only by immigration from the countryside. Thus city life is favourable to the individual Blackbird, but not to the species (Erz 1964).

Some of these genuine townsmen have even become pests by reproducing excessively. Thus the House Sparrow and Starling populations in New York are estimated at 50 millions, and these birds scarcely leave the towns in the countries to which they have been artificially introduced. This is especially true of the Domestic Pigeon. Positive reactions for ornithoses due to viruses have been obtained from two-thirds of the pigeons tested in Paris. While it would be regrettable if pigeons disappeared from the urban scene, their populations must be strictly limited. The methods used are still very inadequate, including nest destruction, baiting with narcotics, the use of threads to protect monuments and important architectural details, and jelly to discourage landing.

Birds may interfere dangerously with human activities by causing accidents to aircraft, mainly on landing and take-off. The birds involved in these strikes are mainly gulls, plovers, pigeons, rooks and starlings. Some such as Rooks and Starlings, frequent aerodromes and even runways because they resemble their chosen habitat, while others find suitable biotopes in the immediate neighbourhood, especially when the site is beside the sea or marshes. The danger is still greater to jet aircraft, since birds can not only damage their airframes but may also be drawn into the engines and stop them. In 1960 at Boston, Massachusetts, an airliner taking off struck a flock of Starlings which stopped its engines, and all its passengers were killed in the crash. One in 1962 at Edinburgh succeeded in touching down safely after striking about 125 gulls while landing. Many such strikes have been recorded throughout the world, most of which have fortunately involved only material, though often severe, losses. They may even happen in mid-flight as when an aircraft crashed in Maryland after striking a flock of swans, but the great altitudes at which modern aircraft fly reduce the risks except for military aircraft, which often have to fly very low. The annual cost of bird strikes on RAF aircraft in Great Britain reached a million pounds, which illustrates the frequency of such accidents.

Many techniques have been proposed to ward off this serious potential danger. Apart from technical modifications to the aircraft, and reduction in bird populations which are out of the question at the scale which would be effective, these consist essentially in reducing the local populations by making the habitat around the aerodrome unsuitable for them. As far as possible airfields should be sited away from natural concentrations of birds, which remain attached to a particular spot even when it has been transformed by man. On Midway Island in the Pacific, where the runways cross the nesting grounds of an albatross so rare that its elimination cannot be considered, from 300 to 400 bird strikes a year, most of them fortunately not severe, occur at this very busy air base (Robbins). Bird scaring by sound signals has been successfully used, by passing along the runways before each air movement, broadcasting recorded alarm calls which frighten off the birds – especially gulls, plovers and starlings which gather in large numbers outside the breeding season. Alarm calls have the advantage that the birds do not become habituated to them, as they do to artificial noises such as explosions. Radar allows birds in the approach zone to be detected, which is especially useful when migrations or feeding flights result in large concentrations. The air safety services of various countries have set up a

681

system for spotting birds by radar and broadcasting the 'ornithological situation' together with the meteorological conditions.

Contrariwise, human structures may be lethal traps for birds which crash into them, especially during migratory flights at night. Lighthouses are especially dangerous since they attract and dazzle the birds, leading them to collide with the structure. However, illuminating lighthouse buildings has had happy results by enabling birds to avoid these obstacles. Twenty thousand migrants, largely wood-warblers, killed themselves in a single night near the Mississippi in Wisconsin, by striking a television tower 300m high. During their nocturnal hunting, Eagle Owls like many other birds are killed by electric cables.

Birds and pathology

Birds may affect the health and hygiene of man and his domestic animals by acting as reservoirs and vectors of pathogens. It has long been known that parrots may transmit psittacosis, so that their importation to certain countries is forbidden. Corresponding virus diseases known as *ornithoses*, as well as many others, are transmitted by a great variety of birds, which have even been accused of carrying foot-and-mouth disease, salmonelloses, Q fever, rabies and anthrax.

The ability of birds to travel great distance enables them to carry pathogens far and fast under ideal conditions. Much attention has recently been paid to the important and sometimes crucial part played by birds in the epidemiology of arthropod-borne virus diseases (arboviroses). Thus they may serve as latent sources of infection, and most importantly may find new centres during their movements, especially migrations. Birds are also parasitized by ticks, themselves disease vectors. It is thought that the viruses responsible for human encephalitis of various types, for West Nile fever, and for western equine encephalitis are all periodically introduced by migrant birds from the areas where these diseases are epidemic or endemic. Thus the birds form a circulating reservoir of the virus, to which biting arthropods have access at certain times in their annual cycles. During the breeding season young birds which are not yet immune provide an ideal environment for viruses to multiply.

Many birds can thus act as reservoirs, and serological tests have shown a long series of viruses in a very great number of species from all over the world. Migrants are especially important because they transport the pathogens and of these the aquatic and some forest birds are especially often infected, because biting insects are common in their habitats. Swallows are also important because of the length of their migrations and their association with man. Virological research on a grand scale has been undertaken throughout the world. It has shown some positive results everywhere, and revealed the appearance of centres of infection along the great migratory routes. This important aspect of epidemiology is at present being intensively studied, because some of the diseases involved are so serious for man and his domestic animals.

Birds and agriculture

Man has more or less profoundly transformed a series of habitats throughout the world in order to increase productivity, and above all to channel it to his exclusive advantage by increasing that part of the production which he can directly use. In so doing he has profoundly altered the biological equilibrium, and created artificial communities which are often very different from natural biocenoses. These changes have benefited certain birds, especially the graminivorous species of open habitats, but have harmed forest birds and insectivores. The amount of grain available has freed some consumers from food as a limiting factor, and allowed them to proliferate so much that these few birds have become pests.

Among the birds established among crops we are accustomed to distinguishing *useful* and *harmful* species (qualifications clearly related solely to human interest), between which are the neutral ones without any obvious effect. No bird is altogether harmful, since some compensate for the damage they do to man by undeniable benefits, while others such as the raptors have only minor effects because of their low densities. Nevertheless, from a strictly human standpoint certain birds do exert on balance a negative and others a positive effect.

Predominant among the first group are those which plunder crops and orchards, and which have indirectly benefited from man. Far from being incidental misfortunes, 'pests' must be regarded as the inevitable result of agricultural practices and of the changes made by man in natural habitats. These species transfer from their natural food plants to cultivated ones which are more productive and often preferred. Destruction by birds under these circumstances may be great.

The second group especially includes insectivorous species which help to control insects harmful to crops, the abundance of particular species varying greatly according to the environment. They are more effective in controlling the pests the more nearly the environment resembles a natural habitat.

Fields are undeniably the most artificial agricultural environment,

683

especially when they extend over great areas without the interruption of different habitats such as hedges, thickets and woods. They support few individuals and species of birds, since most cannot adapt to this environment especially if it supports a monoculture. Here the effect of birds on pests is negligible.

In this environment certain birds may themselves become pests. A number of graminivorous species have increased in proportion to the food made available to them, as long as they find sufficient nesting sites. In North America the Red-winged Blackbird damages maize crops, while various other icterids such as the Bobolink despoil growing rice. In Europe and especially England the Wood Pigeon, which has spectacularly increased, ravages cereal seedlings. House Sparrows may be real plagues in fields of cereals, as Spanish Sparrows *Passer hispaniolensis* are in Morocco. The Rook is destructive by selectively grubbing up seedlings. In eastern Asia the Java Sparrow *Padda oryzivora* ravages the paddies. In tropical Africa the Red-billed Dioch *Quelea quelea* is as damaging as might be expected from its highly gregarious populations numbering millions, especially in the paddies where it has recently become a real scourge. Its success depends of course partly on its graminivorous diet, but also on its powers of flight, resistance to high temperatures and lack of water, gregariousness, and adaptability (Morel). Other birds also ravage the African paddies, notably whistling and other ducks, Egyptian Geese and wintering godwits.

The struggle against birds which despoil fields is far from being as effective as it should. The methods used are destructive and often very large scale. Queleas have been attacked with flame-throwers or poisoned with toxic chemicals such as Parathion, but after millions of individuals have been exterminated the species remains flourishing and the survivors' prospects are even improved. The populations cannot be controlled by such techniques, except temporarily and on a local scale. Various scaring methods have been tried, such as broadcasting alarm calls near the concentrations of birds, which are thereby put to flight. Good results have been obtained in the struggle against Rooks and Starlings, and applied discerningly at the critical times these methods are certainly effective. However, success depends ultimately on being able to control the populations of destructive birds, by preventing them from multiplying excessively.

Orchards are a considerably less artificial environment than fields, which maintain a considerable diversity of birds, including a high proportion of insectivorous species.

In these environments birds may play an important part in controlling insect pests. It is of course impossible to class them categorically as useful or harmful, since the same birds hunt both pest insects and those which predate them. Thus for example sparrows and tits catch along with other insects the larvae of syrphid flies and chrysopid neuropterans, which prey on aphids. This shows how impossible it is to separate birds according to their usefulness to man. Certain birds may certainly do damage. Some, such as the omnivorous Blackbird and Starling, feed on fruit. The latter is especially harmful in the Mediterranean olive groves of Tunisia which it frequents in winter, eating many olives and wasting many more by letting them fall to the ground. Sparrows damage a variety of plants. Bullfinches may cause serious damage by nipping off buds – especially the flower buds of the fruit trees which they prefer. They have thus become a nuisance in commercial orchards, where large numbers of them can markedly reduce the future crop. Chemical repellants have not proved effective, and though protective viscose threads give better results, they are laborious and often difficult to use. Control of the populations by trapping is still the most effective method.

Forests are still an almost natural habitat, despite human interference and modern methods of sylviculture. However, entirely artificial plantations of a single species of tree are obviously much further from a natural environment than are forests which have merely been managed.

In forest environments birds occupy a favoured place, which should be preserved for them since they effectively attack the myriads of insects which gnaw the bark, wood and leaves. In our forests tits, nuthatches, treecreepers, fly-catchers and several species of warbler form an army of consumers, which can play an effective part in controlling the populations of destructive insects and keeping them down to reasonable levels. Great Tits feed their broods especially on caterpillars and other damaging larvae. Woodpeckers specialize in the search for wood-boring insect larvae – scolitids, cerambicids, and buprestids – and also catch on the bark many other insects harmful to trees. They also catch ichneumons which parasitize these pests, and sometimes do damage by boring their nest holes in sound trees, and thus opening the way to parasites and fungi. However, they mostly prefer the softer wood of trees which have already been attacked, or of dead trees: a survey of one hundred woodpecker nests in Russia showed that only three had been bored in sound trees. On balance, the activities of woodpeckers are thus clearly beneficial to the forest.

Thus in forests insectivorous birds are undeniably useful, though they are unable to control an insect plague once it is under way. They thus form part of the braking system which exists in all balanced environments, preventing the various elements in the community from fluctuating too wildly.

They can be further encouraged as in the placing of nest-boxes in German forest lots. (Bruns 1960). Within ten years the population of Bordered White Moths *Bupalus piniarius* in an oak-pine forest were from four to fifty times less in the lots where the density of birds had been increased than in the control lots. Protecting birds and maintaining their populations at optimal levels should thus be included in the rational management of forests. Further, the amounts of seeds which birds take from forest trees, though very variable, never endanger the future of the forest. Some birds, however, may harm plantations to some extent. Jays freely grub up the seedlings and Wood Pigeons, Chaffinches and Bramblings eat the sown seed. Nevertheless it is easy to protect the nurseries from their attacks.

We must briefly return to another group of birds, the raptors. Their numbers, necessarily small because of their high place in the food chains, have been further reduced by thoughtless hunting and the effects of pesticides. Many are still prejudiced against them: quite unreasonably as we have already seen, since predation is seldom a factor limiting prey populations. Also, from a strictly human point of view, many get rid of harmful animals. Most of the small species are insectivorous while buzzards eat rodents, especially field mice which are injurious to crops. A few other species may appear more decidedly 'harmful', such as the Goshawk and Sparrow Hawk which are guilty of taking 'useful' small birds. However, their influence on the populations is minor. Apart from a few individuals which specialize in hunting game or domestic poultry, and which it is legitimate to eliminate, raptors deserve our protection as the occupants of a well-defined place in complex biological systems.

Birds and hunting

Hunting (here used to include all techniques of taking wild birds) is a traditional human activity, and a rational way of exploiting the natural capital represented by populations of game-birds. However, it may be very deadly as soon as it becomes too intense, since it then draws upon the breeding stock rather than merely on the annual increase.

It is essential to know, and to increase as much as possible, the quota which may be killed each year without damage to the species. This implies techniques of managing the wild fauna which have long been applied empirically, but have only recently been based on objective ecological studies.

The next question is how this quota may increase, by enlarging the total population and reducing the natural mortality of the annual increase. Thus it is necessary to ensure that the hunting pressure is not too severe, and field studies leading to habitat management are of prime importance in hunting administration. Certain agricultural practices are known to be harmful. The ways in which pesticides are used and the dates on which they are applied are important. So are the density and composition of the plant cover, and aquatic environments may be managed so as to increase the 'production' of ducks.

Hunting itself must be controlled with due regard to the time of year. The breeding season must of course be respected, and hunting begun again only when the young birds are fully able to fend for themselves; but restraint at other times of year is scarcely less necessary for the rational exploitation of game. Thus biologists condemn spring hunting still carried on in some countries at the expense of migrating wildfowl. Autumn hunting bears on populations which have already reached their maximum, and which will in any case be reduced by the hazards of migration and the rigors of winter. It therefore involves mere substitution of one cause of mortality for another, with man acting as a natural predator among all the other adverse environmental factors. In the spring on the other hand, hunting pressure acts on populations which are at their minima as a result of the winter losses; and at this time human onslaughts are added to the natural causes of destruction. Thus it is not from sentimentality but rather from concern for the dynamic equilibria of bird populations that at this time of year biologists wish to protect the 'bird about to breed'.

Apart from a very few well-known species, the knowledge on which hunting management should be based is still very rudimentary. The situation is still more complex since some hunted species are migrants, whose ranges include widely separated areas of very different ecological conditions. Despite the difficulties, ecological research applied to hunting is essential if we are to conserve these birds and exploit them optimally. However, one must beware 'overhunting' analogous to 'overfishing'; for here too over-exploitation quickly results in depletion of the breeding stocks, and the disappearance of birds hastened by the transformation of habitats which makes their survival more and more difficult in the world of today.

The place of birds in the modern world

Man's relations with nature are now passing through a critical phase. Humanity is proud of its highly perfected techniques and believes itself independent of any natural environment. Western civilization has taken possession of the entire globe and is destroying all the remnants of wild habitats throughout the world, which within a few decades could be entirely partitioned between industry and urbanization on the one hand and mechanized agriculture on the other. We have already emphasized the disturbing aspects of the resulting regression of the wild fauna and flora, and especially of the birds. In fact, despite all our progress in technology and the faith which most of our contemporaries profess in a mechanized civilization, humanity continues to depend closely upon renewable resources and above all on the primary production.

Man must of course modify part of the earth's surface – and especially that part which is directly usable – in order to increase productivity. We can never again do without fields, meadows and managed forests. But it is not in our interest to transform the whole surface of the earth. On the contrary we should preserve the diversity of habitats, even if only for reasons of material profit, and thus safeguard all forms of life, all plant and animal species, and all the communities which they have built up in the course of evolution. This involves managing the earth rationally, with due regard to the potentialities of each area and to the marginal zones, so that on the one hand some lots should be maintained in absolutely their original state, while on the other some should be wholly transformed, according to an integrated plan worked out in accordance with the laws of ecology and not solely from technical necessity. Birds form part of these natural communities which we must preserve or manage, and cannot be dissociated from the vast machine of which they are but a few cogs.

Management has many aspects. It is first essential to conserve a complete and integrated sample of all habitats throughout the world. This is the only measure which can preserve a great many species, which are narrowly dependent on well-defined biotopes. This is particularly true of tropical rainforest, most of whose birds disappear with the habitat, and also of the intertidal zone, an extremely important environment for waders all of which are highly specialized. The ecological demands of these birds are such that the least alteration of the environment, with the resulting upset of the ecosystem, quickly causes the populations to fall. Projects for transforming coastal mudflats into polders or stretches of permanent water would entail the ruin of this habitat, the loss of considerable production, and the elimination of bird populations. The same is true of most species confined to fresh waters and to swamps.

In areas devoted to agriculture, transformed into artificial fields and meadows, there is no advantage to be gained from inflexibly modifying everything. Until recently, especially in Europe, farmers had applied a formula of exploitation which largely respected natural equilibria and had created a harmonious countryside blending fields, meadows and woodlands. This is now threatened by consolidation and the suppression of the interstitial environments represented by thickets, hedges and copses, replacing the patchwork by huge uniform areas devoted to monocultures and mechanization. This development is still more disturbing in the tropics where the extreme conditions make the soils and habitats much more fragile than in temperate zones.

Total suppression of the intermediate habitats is an agricultural mistake. Management is favourable to birds and more particularly to game, which is a source of sizeable profits in agricultural areas. While neither useful nor harmful in themselves, birds contribute to a balance which it is in our interest to preserve.

In addition to all this, certain species require very special attention since their extreme reduction in numbers, very limited distribution, and vulnerability to adverse environmental factors severely threaten their very existence. Only maintenance in captivity can save the last survivors of the most endangered species. If they multiply there it may later be possible to re-introduce them to their natural habitats, if in the meanwhile it has been possible to suppress the causes of their near-extinction. While this technique has already succeeded in several cases, it is only a last resort, since our true objective must be to ensure the survival of animals in their natural habitats.

Thus it is not only within our power to safeguard all the birds which still survive, but it is clearly understood to be in our interest since they form part of a huge natural combination on which we are closely dependent. Man's relations to birds are many and varied, and while their part in natural biocenoses is very important it is quite as much so in artificial environments. They take part in the complex biological systems of the biosphere, whose balance must as far as possible be conserved. Rational management is the key to long-term prosperity for the greater good of mankind.

Birds still have much to teach us scientifically. Their study is in its infancy, and will illuminate many essential aspects of general biology. They must continue to form part of the scenery, whether natural or artificial, for the satisfaction of our aesthetic, cultural and emotional needs. These are amply sufficient reasons for protecting animals whose very special adaptations have created one of the most remarkable types in the whole animal kingdom.

Bibliography

Chapter 19

Ashmole, N. P. (1963) 'The regulation of numbers of tropical oceanic birds'. *Ibis*, **103b**: 458–73

Ashmole, N. P., and Ashmole, M. J. (1967) 'Comparative feeding ecology of sea-birds of a tropical oceanic island'. *Bull. Peabody Mus. nat. Hist.*, **24**: 1–131

Fischer, J. (1952) *The Fulmar*. Collins, London

Fisher, J., and Lockley, R. M. (1954) *Sea-birds*. Collins, London

Lack, D. (1967) 'Interrelationship in breeding adaptations as shown by marine birds'. *Proc. XIV Int. Orn. Congr.* (Oxford), 3–42

Murphy, R. C. (1936) *Oceanic Birds of South America*. Macmillan and Amer. Mus. Nat. Hist., New York. Two volumes

Schmidt-Nielsen, K. (1959) 'Salt glands'. *Scientific Amer.*, **200** (**1**): 109–16

Wynne-Edwards, V. C. (1962) *Animal Dispersion in Relation to Social Behaviour*. Oliver and Boyd, Edinburgh and London

Chapter 20

Alm, B., Myhrberg, H., Nyholm, E., and Svensson, S. (1966) 'Densities of birds in alpine heaths'. *Vär Fägelv.*, **25**: 193–201

Belopolskii. (1961) *Ecology of sea colony birds of the Barents sea*. Jerusalem: Israel Program for Scientific Translations

Bengtson, S. A. (1963) 'On the influence of snow upon the nesting success in Iceland, 1961'. *Vär Fägelv.*, **22**: 97–122

Carrick, R., Holdgate, M., and Prévost, J. (Eds) (1964) *Biologie antarctique*. Hermann, Paris

Holdgate, M. W. (1967) 'The Antarctic ecosystem'. *Philos. Trans. Royal Society of London*, B, no. **777**: 363–83

Karplus, M. (1952) 'Bird activity in the continuous daylight of arctic summer'. *Ecology*, **33**: 129–34

Løvenskjold, H. L. (1963) 'Avifauna Svalbardensis'. *Norsk Polarinstitut Skr.*, no. 129

Mougin, J. L. (1968) 'Etude écologique de quatre espèces de Pétrels antarctiques'. *Oiseau, R. F. O.*, **38** (suppl.): 1–52

Prévost, J. (1961) *Ecologie du Manchot empereur 'Aptenodytes forsteri'*. Gray, Hermann, Paris

— (1963) 'Densités de peuplement et biomasses des Vertébrés terrestres de l'archipel de Pointe-Géologie, terre Adélie'. *Terre et Vie*, **17**: 35–49

— (1964) 'Remarques écologiques sur quelques Procellariens antarctiques'. *Oiseau, R. F. O.*, **34** (suppl.): 91–112

Prévost, J., and Sapin-Jaloustre, J. (1964) 'A propos des 'remieres mesures de topographie thermique chez les Sphéniscidés de la terre Adélie'. *Oiseau, R. F. O.*, (suppl.): 52–90

— (1965) 'Ecologie des Manchots antarctiques'. In: *Biogeography and Ecology in Antarctica. Mon. Biol.*, 15. Junk, La Haye: 551–648

Ryder, J. P. (1967) 'The breeding biology of Ross' Goose in the Perry River region, North-west Territories'. *Canad. Wildl. Service Rep. Ser.*, no. 3, Ottawa

Sapin-Jaloustre, J. (1960) *Ecologie du Manchot Adélie.* Hermann, Paris

West, G. C., and Meng, M. S. (1966) 'Nutrition of Willow Ptarmigan in northern Alaska'. *Auk*, **83**: 603–15

Chapter 21

Ehlert, W. (1964) 'Zur Ökologie und Biologie der Ernährung einiger Limikolen-Arten'. *J. Orn.*, **105**: 1–53

Recher, H. F. (1966) 'Some aspects of the ecology of migrant shorebirds'. *Ecology*, **47**: 393–407

Spitz, F. (1964) 'Répartition écologique des Anatidés et Limicoles de la zone maritime du sud de la Vendée'. *Terre et Vie*, **III**: 452–88

Verwey, J. (1966) 'The Waddenzee and its Riches'. In: *A Plea for the Wadden Sea.* Contact Comm. Nat. Landscape Press, Amsterdam: 1–24

Chapter 22

Delacour, J. (1954–64) *The Waterfowl of the World.* Country Life, London. Four volumes

Dussart, B. (1966) *Limmologie. L'Etude des Eaux continentales.* Gauthier-Villars, Paris

Goodge, W. R. (1959) 'Locomotion and other behaviour of the Dipper'. *Condor*, **61**: 4–17

Johnsgard, P. A. (1966) 'The biology and relationships of the Torrent Duck'. *17th Ann. Rep. Wildfowl Trust*: 66–74

Lebreton, P. (1964) 'Introduction écologique à l'étude de l'avifaune de la Dombes'. *Terre et Vie*, **III**: 20–53

Madsen, F. J. (1957) 'On the food habits of some fish-eating birds in Denmark'. *Danish Review of Game Biology*, **3** (**2**): 19–83

Chapter 23

Colquhoun, M. K., and Morley, A. (1949) 'Vertical zonation in Woodland bird communities'. *J. Anim. Ecol.*, **12**: 75–81

Ferry, C. (1960) 'Recherches sur l'écologie des Oiseaux forestiers en Bourgogne. I. L'avifaune nidificatrice d'un taillis-sous futaie'. *Alauda*, **28**: 93–123

Ferry, C., and Frochot, B. (1968) 'Recherches sur l'ecologie des Oiseaux forestiers en Bourgogne. II. Trois années de dênombrement des Oiseaux nicheurs sur un quadrat de seize hectares en forét de Citeaux'. *Alauda*, **36**: 63–82

Glutz von Blotzheim, U. N. (1962) *Die Brutvögel der Schweiz.* Verlag Aargauer Tagblatt, Aarau

Turcek, F. J. (1956) 'On the bird population of the Spruce forest community in Slovakia'. *Ibis*, **98**: 24–33

Turcek, F. J. (1961) *Ökologische Beziehungen der Vögel und Gehölze*. Slov. Akad. Wiss., Bratislava.

Yapp, W. B. (1962). *Birds and Woods*. Oxford Univ. Press, London

Chapter 24

Cade, T. J., and MacLean, G. L. (1967) 'Transport of water by adult sand grouse to their young'. *Condor*, **69**: 323–43

Cott, H. B. (1946) 'The Edibility of Birds'. *Proc. Zool. Soc. London*, **116**: 371–524

Dementiev, G. P. (1958) 'La faune désertique du Turkestan'. *Terre et Vie*, **105**: 3–44

Frith, H. J. (1959) 'The ecology of wild ducks in inland New South Wales'. *CSIRO Wildlife Research*

— (1962) 'Movements of the grey teal *Anas gibberifrons* Müller Anatidae'. *CSIRO Wildlife Research*, **7**: 50–70

George, U. (1969) 'Über das Tränken der Jungen und andere Lebensaüsserungen des Senegal-Flughuhns *Pterocles senegallus*, in Marokko'. *J. Orn.*, **110**: 181–91

Immelmann, K. (1963) 'Drought adaptations in Australian desert birds'. *Proc. 13th Int. Orn. Congr.*: 649–57

Keast, A. (Ed). (1959) 'Biogeography and Ecology in Australia'. *Mon. Biol.*, 8

Miller, A. H. (1963) 'Desert adaptations in birds'. *Proc. 13th Int. Orn. Congr.*: 666–74

Schmidt-Nielsen, K. (1964) *Desert animals*. Clarendon Press, Oxford

Valverde, J. A. (1957) *Aves del Sahara español*. Inst. estudios africanos, Madrid

Chapter 25

Dorst, J. (1962) 'Considérations sur l'hivernage des Canards et Limicoles paléarctiques en Afrique tropicale'. *Terre et Vie*: 183–92

Moreau, R. E. (1950) 'The breeding seasons of African birds. I. Land birds'. *Ibis*, **92**: 223–67

Morel, G. (1968) 'Contribution a la synécologie des Oiseaux du Sahel sénégalais'. *Mém. ORSTOM*, **29**: 1–179

Morel, M. Y. (1969) 'Contribution à l'étude dynamique de la population de *Lagonosticta senegala* L. (Estrildidés) à Richard-Toll (Sénégal). Interrelations avec le parasite *Hypochera chalybeata* (Müller) (Viduinés)'. Thèse Fac. sci., Rennes

Chapter 26

Brosset, A. (1966) 'Recherches sur la composition qualitative et quantitative des populations de Vertébrés dans la forêt primaire du Gabon'. *Biologia gabonica*, **2**: 163–77

— (1968) 'Location écologique des Oiseaux migrateurs dans la forêt équatoriale du Gabon'. *Biologia gabonica*, **4**: 211–26

— (1969) 'La vie sociale des Oiseaux dans une forêt équatoriale du Gabon'. *Biologia gabonica*, **5**: 29–69

Davis, T. A. W. (1953) 'An outline of the ecology and breeding season of birds of the lowland forest region of British Guiana'. *Ibis*. **95**: 450–67

King, J. R., and Farner, D. S. (1964) 'Terrestrial animals in humid heat: birds'. In: *Handbook of Physiology*. Amer. Physiol. Soc., Washington, **4** (38): 603–24

Snow, D. W. (1962) 'A field study of the Black and White Manakin *Manacus manacus* in Trinidad'. *Zoologica*, **47**: 65–104

Snow, D. W. and Snow, B. K. (1964) 'Breeding seasons and annual cycles of Trinidad land-birds'. *Zoologica*, **49**: 1–39

Chapter 27

Coe, M. J. (1967) *The Ecology of the alpine Zone of Mount Kenya*. Junk, La Haye (*Mon. Biol.*, 17)

Corti, U. A. (1955) 'Die Vogelwelt der Alpen'. *Acta XI Congr. Intern. Orn.* (Bâle, 1954): 59–71

Couturier, M. (1964) *Le Gibier des Montagnes françaises*. Arthaud, Grenoble

Dorst, J. (1955) 'Recherches écologiques sur les Oiseaux des hauts plateau péruviens'. *Trav. Inst. fr. Etudes andines*, **5**: 83–140

— (1962) 'Nouvelles recherches biologiques sur les Trochilidés des Hautes-Andes péruviennes (*Oreotrochilus estella*)'. *Oiseau, R. F. O.*, **32**: 95–126

Norris, R. A., and Williamson, F. S. L. (1955) 'Variation in relative heart size of certain passerines with increase in altitude'. *Wilson Bull.*, **67**: 78–83

Chapter 28

Amadon, D. (1950) 'The Hawaian Honeycreepers (Aves, Drepaniidae)'. *Bull. Amer. Mus. nat. Hist.*, **95**: 151–262

Bowman, R. I. (1961) 'Morphological differentiation and adaptation in the Galapagos Finches'. *Univ. Calif. Publ. Zool.*, **58**: 1–302

— (1963) 'Evolutionary patterns in Darwin's Finches'. *Occ. Pap. Calif. Acad. Sci.*, **44**: 107–140

Curio, E., and Kramer, P. (1964) 'Vom Manrovenfinken (*Cactospiza heliobates* Snodgrass und Heller)'. *Zeitschr. Tierpsych.*, **21**: 223–34

Eibl-Ebesfeldt, I., and Sielmann, H. (1962) 'Beobachtungen am Spechtfinken *Cactospizapallida*'. *J. Orn.*, **103**: 92–107

Lack, D. (1947) *Darwin's Finches*. Cambridge Univ. Press, Cambridge

Mayr, E. (1939) 'The origin and the history of the bird fauna of Polynesia'. *Proc. Sixth Pacific Sci. Congr.*, **4**: 197–216

Paulian, R. (1961) *La Zoogéographie de Madagascar et des Iles voisines*. Publ. Inst. rech. sci. Tananarive, Tsimbazaza

Snow, B. K. (1966) 'Observations on the behaviour and ecology of the Flightless Cormorant *Nannopterum harrisi*'. *Ibis*, **108**: 265–80

Chapter 29

Blondel, J. (1967) 'Réflexions sur les rapports entre prédateurs et proies chez les Rapaces. I. Les effets de la prédation sur les populations de proies'. *Terre et Vie*, **114**: 5–32

— (169) *Synécologie des Passereaux résidents et migrateurs dans le Midi méditerranéen français*. Centre rég. docum. pédagogique, Marseille

Bourlière, F., and Lamotte, M. (1962) 'Les concepts fondamentaux de la synécologie quantitative'. *Terre et Vie*, **109**: 329–50

Bruns, H. (1960) 'The economic importance of birds in forests'. *Birds Study*, **7**: 193–208

Elson, P. F. (1962) 'Predator–prey relationships between fish-eating birds and Atlantic Salmon'. Fisheries Research Board of Canada, Ottawa. Bull. No. 133

Frochot, B. (1967) 'Réflexions sur les rapports entre prédateurs et proies chez les Rapaces. II. L'influence des proies sur les Rapaces'. *Terre et Vie*, **114**: 33–62

Gibb, J. A. (1966) 'Tit predation and the abundance of *Ernarmonia conicolana* (Heyl.) on Weeting heath'. Norfolk, 1962–3, *J. anim. Ecol.*, **35**: 43–53

Haftorn, S. (1960) 'The proportion of spruce seeds removed by the tits in a Norwegian spruce forest in 1954–5'. *Kong. Norske Vidensk, Selsk. Forhand.*, **32** (1959): 211–25

Hasegawa, H., and Ito, Y. (1967) (Biology of *Hyphantria cunea* (Lepidoptera Arctiidae) in Japan. I. Notes on adult biology with reference to the predation by birds'. *Appl. Ent. Zool.*, **2**: 100–10

Hilden, O. (1965) Habitat selection in birds'. *Ann. Zool. Fenn.*, **2**: 53–75

Kilham, L. (1965) 'Differences in feeding behaviour of male and female Hairy Woodpeckers'. *Wilson Bull.*, **77**: 134–45

Lack, D. (1954) *The Natural Regulation of Animal numbers*. Clarendon Press, Oxford

Lamotte, M., and Bourlière, F. (1967) *Problèmes de Productivité biologique*. Masson, Paris

Lebreton, P. (1964) 'Introduction écologique à l'étude de l'avifaune de la Dombes'. *Terre et Vie*, **111**: 20–53

Morel, G. (1968) 'L'impact écologique de *Quelea quelea* L. sur les savanes sahéliennes. Raisons du pullulement de ce Plocéidé'. *Terre et Vie*, **114**: 69–98

— (1968) *Contribution à la Synécologie des Oiseaux du Sahel sénégalais*. Paris (ORSTOM) (Mém. ORSTOM, no. 29)

Salt, G. W. (1957) 'An analysis of avifaunas in the Teton Mountains and Jackson Hole, Wyoming'. *Condor*, **59**: 373–93

Spitz, F. (1964) 'Répartition écologique des Anatidés et Limicoles de la zone maritime du sud de la Vendée'. *Terre et Vie*, **111**: 452–88

Tinbergen, L. (1946) 'De Sperwer als Roofvijand van Zangvogels'. *Ardea*, **34**: 1–213

— (1949) 'Bosvogels en insekten. *Ned. Bosch Tijdschr.*, **21**: 91–105

Turcek, F. J. (1953) 'Ecological analysis of the bird and mammalian population of a primeval forest on the Ponana mountain (Slovakia).' *Bull. intern. Acad. tchèque Sci.*, **53**: 81–105

— (1956) 'On the bird population of the spruce forest community in Slovakia'. *Ibis*, **98**: 24–33

— (1961) *Ökologische Beziehungen der Vögel und Gehölze*. Bratislava (Verlag Slowak. Akad. Wiss.)

Chapters 30 and 31

Adler, H. E. (1963) 'Psychophysical limits of celestial navigation hypotheses'. *Ergeb. Biol.*, **26**: 235–52

Bellrose, F. (1967) 'Radar in orientation research'. *Proc. XIV Int. Orn. Congr.* (Oxford, 1966): 281–309

Curry-Lindahl, K. (1958) 'Internal timer and spring migration in an equatorial migrant, the Yellow Wagtail (*Motacilla flava*). *Arkiv Zool.*, **11**, No. 33: 541-57

Dorst, J. (1962) *The Migrations of Birds*. Heinemann, London

Emlen, S. T., and Penney, R. L. (1966) ' The navigation of penguins'. *Scient. Amer.*, **215**: 104–13

Farner, D. S. (1955) 'The annual stimulus for migration: experimental and physiologic aspects'. In: *Recent Studies in Avian Biology*. Univ Illinois Press: *Urbana*. 198–237

— (1960) 'Metabolic adaptions in migration'. *Proc. XII Int. Orn. Congr.*: 197–208

Fromme, H. F. (1961) 'Untersuchungen über das Orientierungsvermögen nächtlich ziehender Kleinvögel (*Erithacus rubecula, Sylvia communis*)'. *Z. Tierpsych.*, **18**: 204–20

Griffin, D. R. (1955) 'Bird navigation'. In: *Recent Studies in Avian Biology*. Univ Illinois Press: *Urbana*, 154–97

— (1965) *Bird Migration*. Heinemann, London

Kramer, G. (1957) 'Experiments on bird orientation and their interpretation'. *Ibis*, **99**: 196–227

Lack, D. (1944) 'The problem of partial migration'. *Brit. Birds*, **37**: 122–30

Matthews, G. V. T. (1955) *Bird navigation*. Cambridge Univ. Press, Cambridge

Merkel, F. W. (1938) 'Zur Physiologie der Zugunruhe bei Vögeln'. *Ber. ver. Schles. Orn.*, 23. Sonderheft: 1–72

Merkel, F. W., and Wiltschko, W. (1965) 'Nachtliche Zugunruhe und Zugorientierung bei Kleinvögeln—Orientierung zugunruhiger Rotkehlchen im statischen Magnetfeld.' *Zool. Anz.* Suppl. **28**: 356–61

Nisbet, I. C. T., Drury, W. H., and Baird, J. (1963) 'Weight-loss during migration. I. Deposition and consumption of fat by the Blackpoll Warbler, *Dendroica striata*. II. Review of other estimates.' Bird Banding **34**: 107–138

Perdeck, A. C. (1958) 'Two types of orientation in migrating Starlings, *Sturnus vulgaris* L. and Chaffinches, *Fringilla coelebs* L., as revealed by displacement experiments.' *Ardea*, **46**: 1–37

— (1963) 'Does navigation without celestial clues exist in Robins ?' *Ardea*, **51**: 91–104

Rüppell, W. (1944) 'Versuche über das Heimfinden ziehender Nebelkrähen nach Verfrachtung.' *J. Orn.* **92**: 106–132

Tinbergen, L. (1950) 'Der geheime Finkenzug.' *Orn. Beob.*, **47**: 164–70

Wolfson, A. (1959) 'Ecologic and Physiologic factors in the regulation of spring migration and reproductive cycles in birds.' In: *Comparative Endocrinology*. J. Wiley, New York: 38–70

Chapter 32

Dorst, J. (1970) *Before Nature dies*. Collins, London

Greenway, J. C. (1958) *Extinct and vanishing Birds of the World*. Amer. Com. Intern. Wildlife Prot. Sp., Publ. no. 13, New York

Vincent, J. (1966) *Red Data Book*. vol. II. *Aves*, IUCN, Morges

Chapter 33

Hutchinson, G. E. (1950) 'The biogeochemistry of Vertebrate excretion.' *Bull. Amer. Mus. nat. Hist.*, **96**: 1–554

Mayr, E. (1963) 'The Role of Ornithological Research in Biology'. *Proc. XIII Int. Orn. Congr.* (Ithaca): 27–38

Stefferud, A. (Ed.) (1966) *Birds in Our Lives*. US Dept of Interior, Fish and Wildlife Service, Washington

Zeuner, F. E. (1963) *A History of Domesticated Animals*. Hutchinson, London

Chapter 34

Bruns, H. (1960) 'The economic importance of birds in forests.' *Birds Study*, **7**: 193-208

Busnel, R. G., and Giban, J. (Ed.) (1965) *Le Problème des Oiseaux sur les Aerodromes*. INRA, Paris (C. R. Colloque, Nice, 1963)

Cramp, S., and Tomlins, A. D. (1966) 'The birds of Inner London, 1951–65'. *Br. Birds*, **59**: 209–33

Erz, W. (1964) 'Populationsökologische Untersuchungen an der Avifauna zweier nordwestdeutscher Grossstädte.' *Zeitschr., Wiss. Zool.*, **170**: 1–111

Giban, J. (Ed.) (1962) 'Colloque sur les moyens de protection contre les espèces d'Oiseaux commettant des dégats en agriculture.' *Ann. Epiphyties*, 13

Murton, R. K., and Wright, E. N. (1968) *The Problems of Birds as Pests*. Acad. Press, London (*Symp. Inst. Biology*, no. 17.)

Index